Biomass Waste: Processing, Valorization, Industrial Applications

Abdellaoui Hind
Editor

Biomass Waste: Processing, Valorization, Industrial Applications

Springer

Editor
Abdellaoui Hind
Polytech Mohammed VI University
(UM6P)
Ben Guerir, Morocco

ISBN 978-981-95-0990-4 ISBN 978-981-95-0991-1 (eBook)
https://doi.org/10.1007/978-981-95-0991-1

This Springer imprint is published by the registered company Springer Nature Singapore Pte Ltd.
The registered company address is: 152 Beach Road, #21-01/04 Gateway East, Singapore 189721, Singapore

Contents

Chapter 1
Introduction to Biomass Role in Industrial Sustainability

Prem Baboo

Abstract Biomass is the oldest form of renewable energy, having been used for cooking since the beginning of human history. It consists of organic materials derived from plants, animals, microorganisms, and algae, including food and agricultural waste, municipal waste, animal waste, forestry residues, and industrial waste from wood, paper, and food industries. Biomass is also known as biogas or biofuel (Kalak in Energies 16:1783, 2023; Saleem in Heliyon 8, 2022). When biomass harvesting does not exceed forest growth, it can be considered carbon neutral. The biomass industry emphasizes sustainability by managing, planting, using, conserving, and restoring forests and woodlands for human and environmental benefits (Saleem in Heliyon 8, 2022). Biomass plays a crucial role as a renewable and sustainable energy source. It can be burned directly for heat or converted into liquid and gaseous fuels through various processes. Today, biomass is used to power electric generators and other machinery such as boilers as seen in Fig. 1.1. Biomass is organic, composed of materials from living organisms, primarily plants and wood. Recycling and reusing biomass can be highly profitable for various value-added applications (Saleem in Heliyon 8, 2022; Biomass Properties and Handling Bio Renewables Education Laboratory Biomass Properties Summer Academy Instructor's Guide JB/CB 2011; Ciolkosz et al. in Renew Altern Energy Fact Sheet 5:1–2, 2010).

Keywords Biomass wastes · Recycling · Functional materials · Value-added applications · Industrial applications. biofuel · Generation biofuel · Gasification · Pyrolysis

P. Baboo (✉)
Fellow of Institution of Engineers, Kolkata, India
e-mail: pbaboo@hotmail.com

Former DGM (Production & Process), Dangote Fertilizers, Lekki, Nigeria

National Fertilizers Ltd. (A Government of India), Panipat, India

A. Hind (ed.), *Biomass Waste: Processing, Valorization, Industrial Applications*,
https://doi.org/10.1007/978-981-95-0991-1_1

1.1 Charecteristic of Biomass

Biomass requires analysis for key properties such as proximate composition (moisture, volatiles, fixed carbon, and ash content), bulk density, angle of repose, higher heating value, and volumetric energy density. It can be converted into biofuels through biochemical, thermochemical, co-firing, pyrolysis, and gasification processes. Before combustion, biomass often undergoes a process called torrefaction, where it is heated to 200–300 °C (Fig. 1.2). This reduces its moisture absorption capacity, decreases weight by 10%, and converts 90% of its energy into a usable form. Compared to carbon-based fuels like coal, biomass has distinct characteristics: lower carbon content, higher oxygen and silica levels, more potassium, greater moisture content, lower heating value, reduced density, and less friability (Ciolkosz et al. 2010). It also contains trace inorganic elements such as calcium, potassium, phosphorus, magnesium, silicon, aluminum, iron, and sodium—depending on the feedstock (Fig. 1.1).

Chemically, biomass is rich in cellulose but contains lower amounts of nitrogen and sulfur. Its moisture and ash content make it suitable for biofuel production. Factors such as temperature, moisture level, density, chemical composition, and particle size significantly influence biomass performance. These characteristics can vary widely depending on the type of biomass (Ciolkosz et al. 2010).

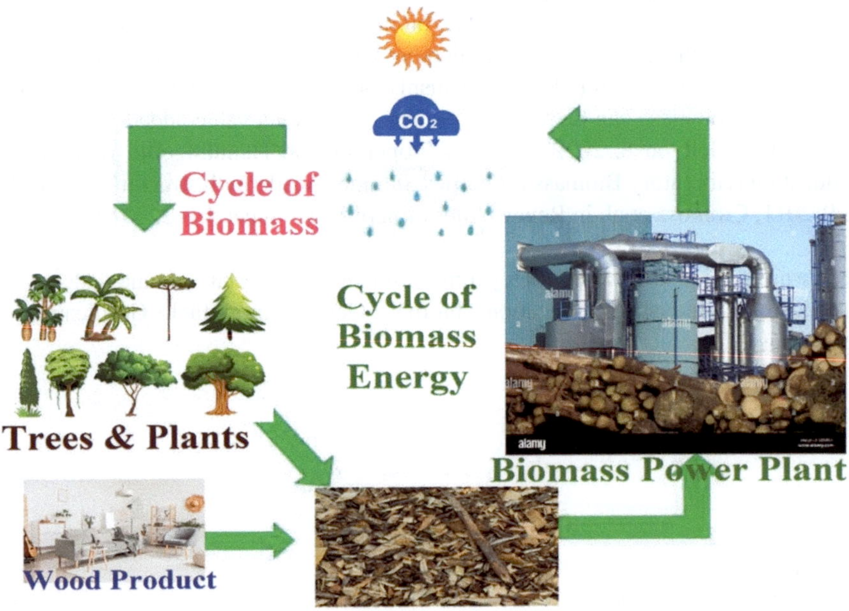

Fig. 1.1 Cycle of biomass energy (Ciolkosz et al. 2010)

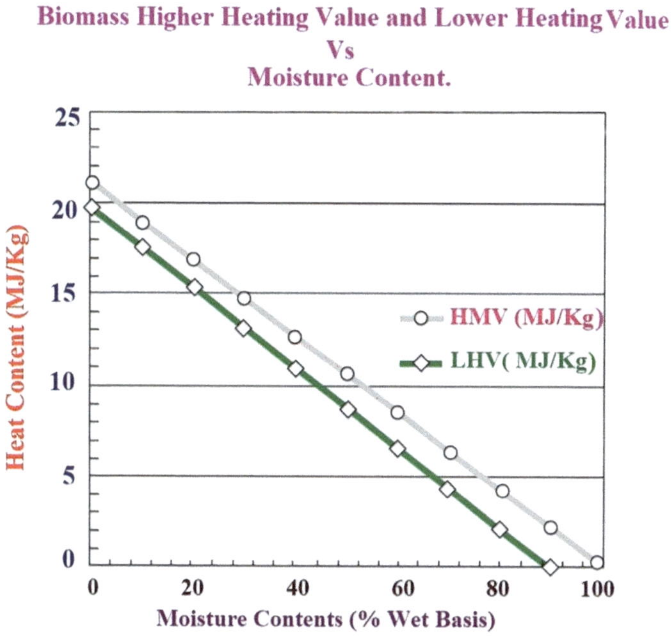

Fig. 1.2 Biomass higher heating value and lower heating value versus moisture contents (Ciolkosz et al. 2010)

1.2 Classification Types and Source of Biomass

Biomass is a heterogeneous energy source available in various forms, including agricultural waste, forestry residues, organic matter from river basins, human-generated organic waste, sewage sludge, and municipal solid waste. The supply of biomass is virtually unlimited, as trees and crops can be continuously grown, and waste will always be generated (Global Bioenergy Statistics 2022). Biomass encompasses a wide range of materials such as wood and agricultural products, solid waste, landfill gas, biogas, and alcohol fuels like ethanol and biodiesel as shown in the Fig. 1.3.

1.3 Biomass Resources and Uses

This chapter will discuss different types of biomass resources, their structural components, characteristic features, and properties. The challenges associated with biomass utilization will also be addressed. Using biomass and biofuels made from biomass has both positive and negative environmental impacts. Like any coin, each side has its advantages and disadvantages. The combustion of fossil fuels and biomass releases carbon dioxide (CO_2), a greenhouse gas. However, the plants used for biomass

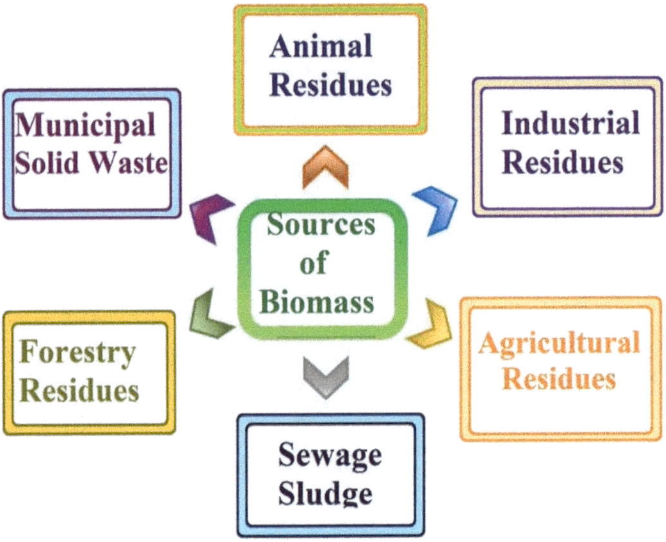

Fig. 1.3 Sources of biomass (Global Bioenergy Statistics 2022)

capture almost as much CO_2 through photosynthesis as is released when burned, making biomass a carbon–neutral energy source. In some cases, the emissions and pollutants from biomass can be worse than those from fossil fuels (Ciolkosz et al. 2010; Global Bioenergy Statistics 2022).

Biomass is a carbon-neutral energy source when managed sustainably. Plants and trees absorb CO_2 for growth, which is released when used for energy, forming a closed cycle. Sustainable forestry practices are crucial for protecting forests from conversion to agricultural, urban, or industrial use. Biomass can be used for industrial growth rather than as a direct energy source (Global Bioenergy Statistics 2022). Biofuels derived from it include pellets, wood shavings, firewood, fruit stones, and nut shells. Wood chips come from shredding forest and agricultural waste (Fig. 1.4).

Biomass waste can be effectively utilized as an energy source, aligning with current trends in renewable energy demand. Many countries already leverage biomass as a viable alternative energy solution. The primary use of biomass involves producing biofuels through mechanical, thermochemical, and biochemical processes. This provides a sustainable fuel source that can partially replace depleted fossil fuels, reduce solid waste generation, and minimize greenhouse gas emissions (Zhang 2023).

As traditional energy sources dwindle, biomass offers an abundant alternative. Unlike fossil fuels, biomass is carbon-neutral when managed properly. Waste biomass, which often spoils or rots, can be repurposed in industries. Old trees and plants can be replaced with new ones, providing a continuous energy source. When biomass is used efficiently in industries with good management, it poses no harm, similar to the effective use of biodiesel. Biomass can be sourced from various sectors, including direct land use for biofuels, industrial by-products like sugar factories, or

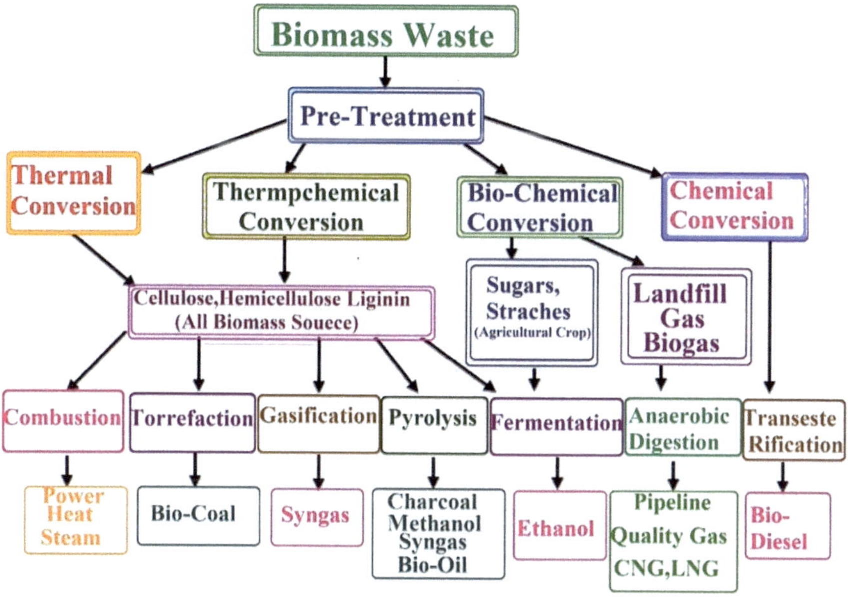

Fig. 1.4 Biomass conversion into bio products (Global Bioenergy Statistics 2022)

recovered from other human activities. This includes organic products from forests, wood waste from construction, furniture, and packaging materials. Using biomass as a liquid or solid fuel offers numerous environmental and socioeconomic benefits, provided that sourcing is done responsibly, considering life cycle carbon emissions and impacts on natural resources and living conditions (Zhang 2023).

1.4 Biomass Present and Future

Globally, there is growing awareness about converting biomass into valuable products. Waste-to-energy technology is a promising concept, transforming organic compounds into useful products like biogas and converting municipal waste into sustainable resources. Biomass is a renewable energy source that can be used repeatedly, making it a more sustainable option.

Unlike fossil fuels, biomass is carbon-neutral, as it absorbs as much CO_2 during growth as it releases when burned, making it a more environmentally friendly choice. With many fossil fuels nearing depletion, the future looks promising for biomass. Some trees and plants have rapid growth cycles, allowing for continuous renewal. Projections suggest that by 2050, biomass could meet about half of the world's energy demand and 60% of the electricity market could be supplied by renewable sources (Rebea et al. 2023) (Fig. 1.5).

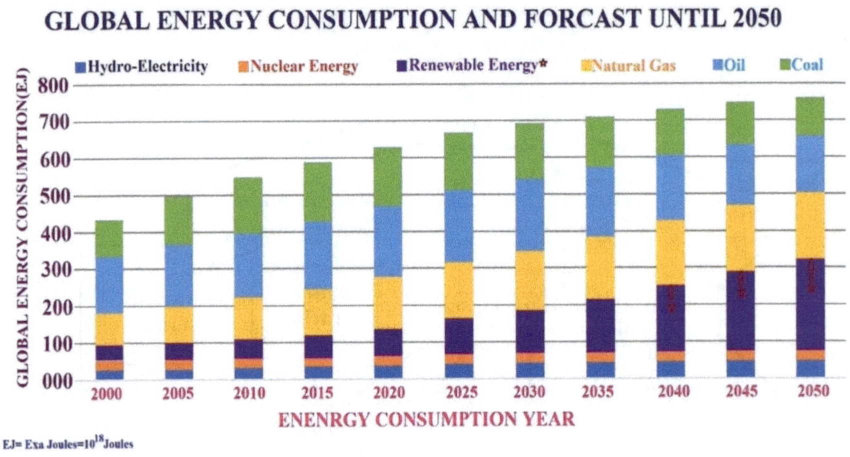

Fig. 1.5 Global energy consumption (2000–2019) and a forecast until 2050 (Rebea et al. 2023)

In recent years, there has been a growing emphasis on utilizing biomass to reduce dependency on crude oil and lower CO_2 emissions. Biomass is increasingly recognized as a sustainable energy option worldwide. Waste materials, such as discarded furniture, can be repurposed as fuel, while boilers designed for coal can be adapted to use biomass with minimal modifications. Biomass energy is derived from various sources, including trees, perennial grasses, agricultural residues, forestry waste, and landfill gases (Rebea et al. 2023). Wood pellets, a popular form of biomass fuel, are produced from agricultural biomass, grasses, wood waste, and forestry residues. These pellets can generate energy and heat or even be converted into liquid fuels (Fig. 1.6).

Sustainably managed biomass is considered carbon-neutral because it releases only the CO_2 absorbed during the growth of plants. In contrast, fossil fuels release large amounts of greenhouse gases that trap heat in the atmosphere and contribute to global warming. The environmental benefits of biomass include reducing air and water pollution and mitigating climate change impacts caused by fossil fuels.

By 2050, projections suggest that biomass could meet about half of the global energy demand (approximately 400 EJ/year) and supply 60% of electricity through renewable sources. Proper management of biomass resources ensures minimal environmental harm while offering a viable alternative to traditional energy sources (Rebea et al. 2023).

The pulp and paper industry is expected to thrive for years, with biomass playing a key role in its energy needs. Biomass emits minimal CO_2 when managed sustainably, making woody biomass a promising renewable energy source for future generations. Feedstocks for biofuel production can be categorized into four generations (Al Afif et al. 2023):

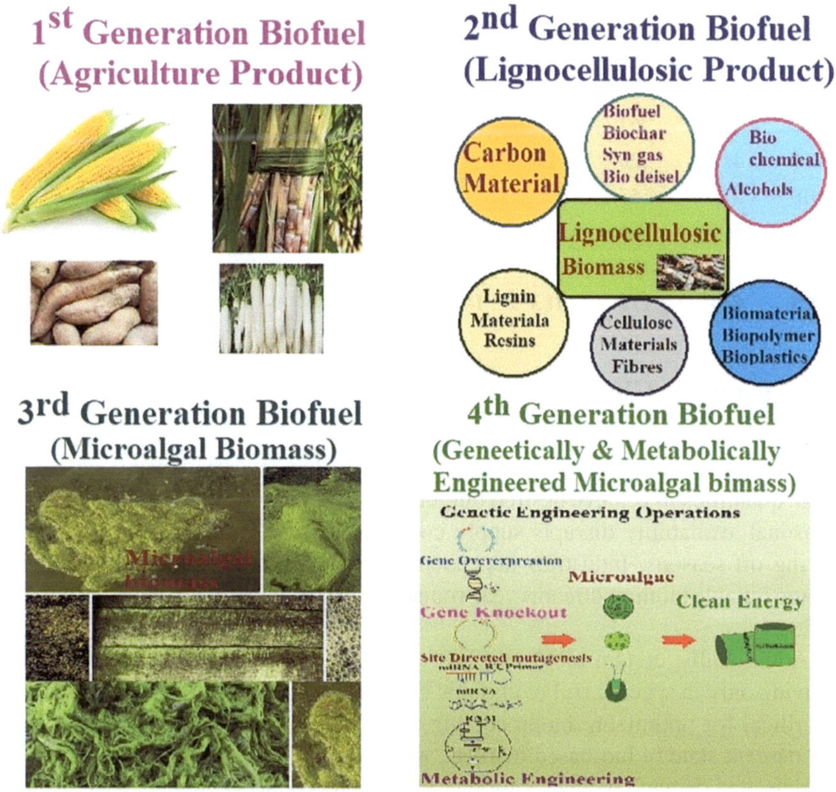

Fig. 1.6 Generation biomass form (Rebea et al. 2023)

- First Generation: Crop-based biofuels derived from wheat, barley, and sugar. However, these have faced criticism due to the "food versus fuel" debate and negative net energy gains.
- Second Generation: Non-food crop biofuels, such as lignocellulosic biomass, which proved more efficient for biodiesel production.
- Third Generation: Algae-based biofuels, offering higher yields and reduced environmental impact.
- Fourth Generation: Advanced biofuels developed using micro-algal biotechnology and metabolic engineering to maximize output.

Globally, renewable energy technologies—including solar, wind, hydro, biomass, and geothermal—are gaining traction. By 2025, these sources are projected to account for 17% of primary energy supply, marking a 1.1% increase from the previous year (Al Afif et al. 2023) (Table 1.1).

Table 1.1 Globally total primary energy supply (all value in EJ (Exa Joules 10^{18} Jouls) (Al Afif et al. 2023)

Year	Total	Coal	Oil	Gas	Nuclear	Renewable	Renewables%
2000	420	96.9	154	86.5	28.3	53.7	13
2005	421	125	168	98.6	30.2	58.3	12
2010	537	153	173	114	30.1	66.3	12
2015	569	161	182	122	28.1	75.1	13
2020	585	157	172	138	29.1	88.0	15
2025	**640**	**175**	**183**	**149**	**30.1**	**94.0**	**17**

1.5 Biomass Challenges

Biomass faces several operational hurdles. Its low energy density necessitates large quantities to achieve desired outputs, complicating storage and transportation. Seasonal availability disrupts supply consistency, requiring alternative feedstocks during off-seasons. Industries dependent on uninterrupted production struggle to secure reliable biomass streams year-round (Global Bioenergy Statistics 2022; Zhang 2023).

Additionally, certain biomass types exhibit high heterogeneity and complexity, thriving only in specific soil or climatic conditions. Accurate property measurement is critical for optimizing bioprocessing systems. Further complicating adoption is the nascent state of bio-based markets, with underdeveloped supply chains, pricing models, and demand pathways for energy and bioproducts (Zhang 2023).

1.6 Biomass Versus Fossil Fuel

The bulk density of most biomass is low, requiring more storage space. However, some liquid biomass types have better density. Biofuels differ from fossil fuels because they are produced from renewable organic materials and living organisms. The energy stored during photosynthesis is converted into chemical energy in forms like ethanol, wood, cow dung, poultry litter, and forest residues. In contrast, fossil fuels such as coal, natural gas, crude oil, and petroleum products are nonrenewable energy sources (Global Bioenergy Statistics 2022) (Table 1.2).

1.7 Biomass Role in Industrial Sustainability

Biomass is versatile in industrial applications, used in stoves, boilers, and furnaces to generate heat and steam. It can be converted into sustainable fuels like cellulosic ethanol, HVO, and SAF, reducing greenhouse gas emissions and dependence on

Table 1.2 Biomass densities (Data taken from Forest Research under the Open Government of India) (Global Bioenergy Statistics 2022)

Fuel	Net CV by mass		Bulk density	Energy density by volume	
	GJ/tonne	kWh/kg	kg/m^3	MJ/m^3	kWh/m^3
Wood chips (30% MC)	12.5	3.5	250	3,100	870
Log wood (stacked—air dry: 20% MC)	14.7	4.1	350–500	5,200–7,400	1,400–2,000
Wood (solid—oven dry)	19	5.3	400–600	7,600–11,400	2,100–3,200
Wood pellets (10% MC)	17	4.8	650	11,000	3,100
Miscanthus (bale—25% MC)	13	3.6	140–180	1,800–2,300	500–650
House coal	27–31	7.5–8.6	850	23,000–26,000	6,400–7,300
Anthracite	33	9.2	1,100	36,300	10,100
Heating oil	42.5	11.8	845	36,000	10,000
Natural gas (NTP)	38.1	10.6	0.9	35.2	9.8
LPG	46.3	12.9	510	23,600	6,600

fossil fuels. Woody biomass is processed into wood pellets or chips for power plants, biomass boilers, and pellet stoves (Rebea et al. 2023).

Biomass is considered carbon-neutral because plants absorb CO_2 during growth, offsetting emissions when burned. Sustainable practices ensure continuous renewal, helping to lower emissions and mitigate climate change. Biomass is heated through thermal gasification to produce syngas for power generation, heating, or as a feedstock for chemicals and biofuels. Thermochemical processes include gasification, combustion, and pyrolysis, which can occur at low or high temperatures. Pyrolysis is particularly promising due to its lower temperature requirements compared to gasification (Rebea et al. 2023).

Biomass offers numerous advantages as a renewable energy source, replaceable through organic processes like plant growth. It is used in various industrial applications, including power generation, process heat, and biogas production through anaerobic digesters. Biomass can also be converted into chemicals for plastics and other products typically made from petroleum. Additionally, biomass can be transformed into biochar, a highly porous and stable carbonaceous residue obtained through pyrolysis. Biochar has valuable properties, including high porosity, surface area, stability, carbon content, cation exchange capacity, and water holding capacity, making it suitable for diverse applications (Rebea et al. 2023).

1.8 Fertilizers Production from Biomass and Green Hydrogen

As the world intensifies efforts to combat climate change, renewable energy installations are increasing globally. The declining costs of renewable energy are driving advancements in green chemistry applications, such as green ammonia and hydrogen for fertilizers like green urea. Technical innovations and stricter environmental regulations are also encouraging the development of CO_2 utilization from biomass combustion (Al Afif et al. 2023).

In this evolving landscape, modular solutions for small- and large-scale urea plants have been developed, integrating seamlessly with green technologies and CO_2 derived from biomass. These solutions aim to enhance the flexibility of traditional urea production processes while incorporating sustainable elements like green hydrogen. Existing plants can be retrofitted with low-emission configurations to align with environmental goals. Small-scale urea plants, due to their reduced production capacity, are particularly suitable for utilizing waste streams such as surplus CO_2 or ammonia, as well as renewable energy sources. This makes them an ideal component of sustainable production systems. Post-combustion carbon capture technologies play a crucial role in enabling green urea production by efficiently capturing CO_2 from biomass burning. Unlike conventional methods, enzymatic carbon capture technologies offer a cleaner alternative without producing toxic by-products. These systems are environmentally friendly, safe for human health, and cost-effective due to their ability to use residual heat for regeneration (Al Afif et al. 2023).

1.9 Green Urea Philosophy

Hydrogen is produced through various methods mentioned earlier, while nitrogen is obtained from an air separation process (ASU). The hydrogen and nitrogen are then combined to form green ammonia. Carbon dioxide, produced from biomass, is used in the urea plant. The ammonia is reacted with CO_2 to form a green urea product (Al Afif et al. 2023), as shown in Fig. 1.7.

$$CO_2 + 2NH_3 \leftrightarrows NH_2COONH_4 \leftrightarrows NH_2CONH_2 + H_2O$$

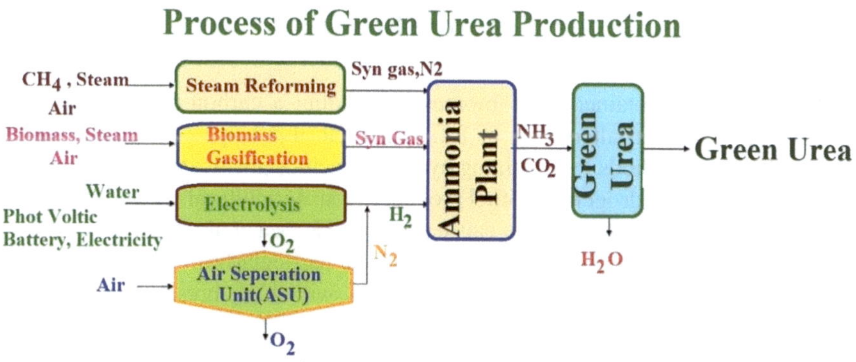

Fig. 1.7 Process of green urea production (Al Afif et al. 2023)

1.10 Myths and Facts Around Biomass

In the past, biomass was surrounded by misconceptions in society. However, these gaps are gradually being filled, and today, awareness is growing. Young generations are moving in a positive direction, and it won't be long before rural communities and farmers work together to create a new society.

In our society, some misconceptions persist. Proponents of biomass argue that it plays a crucial role in achieving net zero, as it uses wood unsuitable for other products that would otherwise be wasted. However, critics point out that biomass involves burning natural materials, which can pollute the atmosphere. Although biomass is a renewable source of energy and heat, it is often misrepresented and misunderstood (Rebea et al. 2023; Al Afif et al. 2023) (Table 1.3).

Table 1.3 Some myths and facts on the biomass (Rebea et al. 2023; Al Afif et al. 2023)

Sr. No	Myths	Facts
1	Earlier it was thought that burning biomass cause environment pollution	Today the illusion has gone away the biomass is clean and green fuel
2	Earlier we thought that biomass would be very expensive	But today we think switching to biomass require minimum investment the industries can easily shift fossil fuel to biomass fuel
3	Earlier it was thought biomass is not cost friendly	But today it has been proved it is cheapest source of energy and eco-friendly
4	Earlier it was thought biomass is not environment friendly	But today by some of the experimental it has been proved biomass reduced GHG. The biomass emission level is less than fossil fuel

1.11 Hydrogen Production from Biomass

Hydrogen can be obtained from biomass, providing a carbon-neutral renewable energy source and an alternative to fossil fuels. It also helps control greenhouse gases. The main methods for obtaining hydrogen include gasification, pyrolysis, and fermentation. Bio-hydrogen, a biofuel produced from agricultural organic matter and waste, is a clean fuel that does not cause pollution (Rebea et al. 2023).

However, this technology presents challenges, particularly in the transportation of hydrogen, which requires special conditions with high pressure and specialized vessels. Gasification is a high-temperature reaction (around 700 °C) involving carbonaceous materials in biomass, with a controlled amount of oxygen and steam as seen in the Fig. 1.8. The water gas reaction produces more hydrogen. The steam generated in the waste heat boiler is used in the process, and the surplus is utilized for power generation in a steam turbine. Although this process is lengthy, the resulting syngas is sufficiently purified (Al Afif et al. 2023).

The chemical reaction is as follow (Al Afif et al. 2023):

$$C_6H_{12}O_6 + O_2 + H_2O \rightarrow CO + CO_2 + H_2 + \text{ other species}$$

And water gas reaction is as follow (Al Afif et al. 2023):

$$CO + H_2O \rightarrow CO_2 + H_2(+\text{ small amount of heat})$$

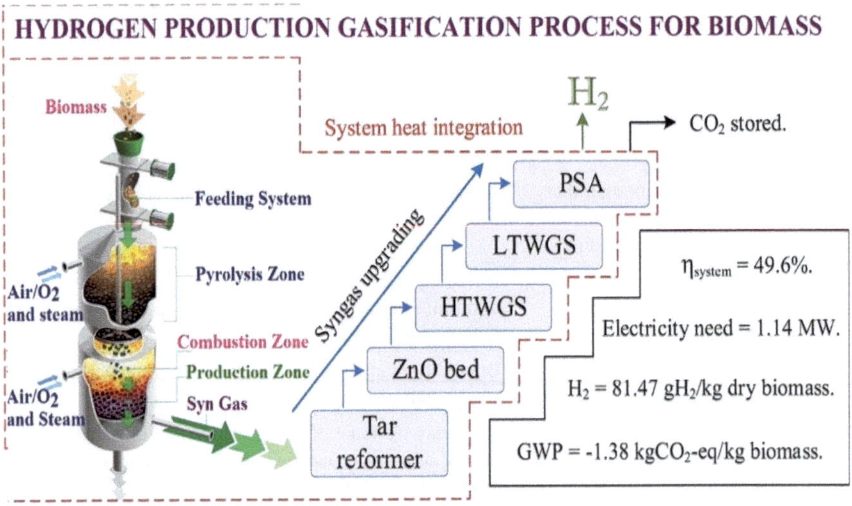

Fig. 1.8 Hydrogen production and gasification process (Al Afif et al. 2023)

1.12 Advantages for Industrial Sustainability

Biomass offers several significant advantages. First, it does not require mining and cannot be depleted like fossil fuels. With the abundance of plants on Earth, biomass can be a primary source of renewable energy, used as a sustainable alternative to fossil fuels. Crops, fast-growing trees, algae, and corn stover are excellent materials for energy development (Rebea et al. 2023).

Biological waste like ash and sludge does not harm the environment; the ash is even an excellent organic fertilizer, and byproducts like bricks can be made. Fast-growing plants such as energy willow or Jerusalem artichoke produce oxygen, improve soil conditions, and provide good biomass. Biochar, produced during pyrolysis, is valuable for agriculture and environmental use. Previously, we discarded black liquor from paper mills, but today it is recycled and used to power the mills. Some industries convert this liquor into syngas to generate electricity (Zhang 2023).

Key Advantages of Biomass (Zhang 2023; Rebea et al. 2023; Al Afif et al. 2023):

- Renewable and Abundant: Biomass is an endless energy source that can be used repeatedly, making it sustainable. It is abundant in many parts of the world, making it readily available.
- Low Cost: Often cheaper than traditional energy sources, biomass is an attractive option for businesses and individuals.
- Carbon-Neutral: Unlike fossil fuels, biomass does not emit CO_2 when burned, making it a more environmentally friendly option.
- Versatile: Biomass can generate electricity, heat boilers, produce transportation fuels like biodiesel, ethanol, and biogas, as well as plastics and fabrics.
- Clean Energy Source: Biomass is a clean, renewable energy source. Its initial energy comes from the sun, and plants or algae can regrow quickly. Burning biomass does not produce the same air pollutants as burning fossil fuels.
- Job Creation: The biomass industry can create jobs in the areas it is used, helping to boost local economies.
- Local Sourcing: Biomass can be sourced locally, reducing the need for long-distance transport of energy resources.
- Waste Reduction: When biomass is burned, its volume reduces significantly in the form of ash, which reduces landfill disposal costs.

1.13 Biomass and Environment

The carbon cycle is an essential process that coordinates the exchange of carbon between the Earth's layers: the atmosphere, hydrosphere, and lithosphere. This cycle has numerous benefits, including regulating sunlight. It occurs primarily through photosynthesis, decomposition, and respiration, influenced by human activities. For example, carbon absorbed by the soil is decomposed by organisms, while plants release carbon-based nutrients into the biosphere through photosynthesis. The carbon

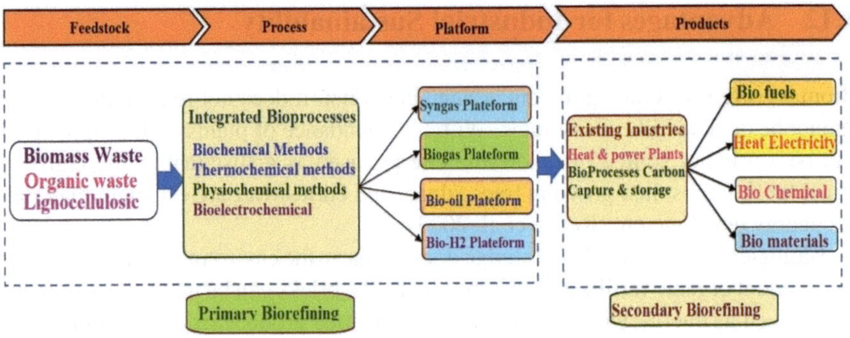

Fig. 1.9 Process of bioenergy (Rebea et al. 2023; Al Afif et al. 2023)

in fossil fuels has been sequestered for millions of years (Fig. 1.9). When these fuels are extracted and burned for energy, the stored carbon is released into the atmosphere without being reabsorbed. In contrast, biomass comes from recently living organisms, and its carbon can continue to circulate in the carbon cycle (Global Bioenergy Statistics 2022; Zhang 2023).

To maintain the carbon cycle, it is crucial to continuously cultivate plants and forests. This process takes time and requires a diverse supply of trees and crops to maintain a healthy environment. Certain trees, such as the peepal (Ficus religiosa), neem, mango, tulsi (basil), bamboo, oak, walnut, mahogany, and cassia, are recognized for their ability to purify the environment by absorbing polluted gases. These trees, along with others like the horse chestnut, black walnut, American sweetgum, ponderosa pine, red pine, white pine, London plane, Hispaniolan pine, Douglas fir, scarlet oak, red oak, Virginia live oak, and bald cypress, are effective at absorbing and storing CO_2. The peepal and mango trees are particularly notable for their air-purifying capabilities. The peepal tree produces significant amounts of oxygen during the day, making it an essential component of urban landscapes for reducing pollution and fostering healthier ecosystems. The roots of the peepal tree also absorb chromate from the soil, contributing to environmental purification. Mango wood, when burned, releases formaldehyde, a gas that kills bacteria and purifies the air (Rebea et al. 2023; Al Afif et al. 2023).

1.14 Disadvantages

Although biomass is renewable, it can become non-renewable if it is not replenished quickly after use. For example, a forest may take hundreds of years to re-establish, which is still much faster than fossil fuels like peat, which can take 900–1000 years to regenerate. Most biomass requires arable land to grow, meaning that land used for biofuel crops like corn and soybeans is unavailable for food production or natural habitat conservation. Some biofuels, such as biodiesel and bioethanol, are not as efficient as gasoline for continuous processes (Al Afif et al. 2023).

The production of methane from human and animal waste can also harm the environment. There is a risk of deforestation when forests are cut down without being replanted at the same rate, as wood is a major source of biomass energy. Biomass has a lower calorific value than coal or natural gas, resulting in the production of large amounts of ash and waste during combustion, which requires frequent removal of ash and cleaning of heating installations. Pellets, chips, and straw require proper storage in dry conditions to maintain their efficiency. In urban areas, it is difficult to plant trees and plants due to the risks associated with power lines. Conventional energy sources are often cheaper and more cost-effective than biomass, limiting its adoption. More research and technology are needed in this field to preserve the future (Al Afif et al. 2023).

1.15 Conclusion

In the face of increasing pollution and the depletion of conventional fossil fuels, biomass emerges as a promising alternative for rural development and environmental sustainability. By adopting techno-enviro-economic solutions, biomass can be transformed into hydrogen through low-carbon processes, contributing to renewable energy systems. This approach addresses the dual challenge of reducing greenhouse gas emissions and managing solid waste effectively. Biochemical, mechanical, and thermochemical methods enable the conversion of biomass into biofuels like ethanol and biodiesel, which are being actively promoted by governments worldwide for blending with diesel. Technologies such as briquetting, which enhances energy density while minimizing moisture content, and advanced emission control systems like electrostatic precipitators and catalytic reduction units further optimize biomass utilization.

Modern advancements in bioenergy have opened doors to replacing fossil fuels, which are nearing extinction. Biomass represents a sustainable energy source capable of supporting societal development, rural upliftment, and economic growth. Its versatility in producing clean energy and creating jobs for farmers and entrepreneurs makes it an essential component of future energy ecosystems. Governments committed to sustainable development must continue to support biomass technologies to ensure a cleaner, greener future.

References

Al Afif R, Bikić S, Radojcin M (2023) Bio energy conversion technologies: a review study and case study. J Process Energy in Agric 27(1). https://doi.org/10.5937/jpea27-43884

Biomass Properties and Handling Bio Renewables Education Laboratory Biomass Properties Summer Academy Instructor's Guide JB/CB 2011

Ciolkosz D, Miller B, Wallace R (2010) Characteristics of biomass as a heating fuel. Renew Altern Energy Fact Sheet 5:1–2

Global Bioenergy Statistics (2022) World bioenergy association

Kalak T (2023) Potential use of industrial biomass waste as a sustainable energy source in the future. Energies 16(4):1783

Rebea K, Machailos S, Hughes KJ, Engham D, Md. Pourkshanian (2023) Energy conversion and management, vol 298. Elsevier, p 117812

Saleem M (2022) Possibility of utilizing agriculture biomass as a renewable and sustainable future energy source. Heliyon 8(2)

Zhang J (2023) Genetic engineering for bio hydrogen production from microalgae. Iscience 26(8):107255

Chapter 2
Pretreatment Techniques for Biomass Waste

Ilham Jioui, Soumia Aboulhrouz, Ikrame Ayouch, Othmane Amadine, and Mohamed Zahouily

Abstract This chapter aims to provide a comprehensive review of the various pretreatment methods applied to biomass waste. These procedures are critical for enhancing the efficiency and effectiveness of converting biomass waste into energy or high-value-added products. The discussion will cover physical, chemical and biological pretreatment techniques. Each technique offers unique advantages and challenges, depending on the type of biomass, the application and economic and environmental factors. In this chapter, we present an overview of these pretreatment methods, highlighting their role in improving treatment efficiency, product yield and the overall sustainability of biomass treatment processes.

2.1 Introduction

Biomass waste refers to organic materials that are typically discarded as by-products of agricultural, industrial, or residential processes. These materials can be transformed into energy or other valuable products through various technologies. There are many types of biomass waste including agriculture residues (straw, stalks, husks…), industrial waste (food waste, paper and textiles), municipal solid waste (food scraps, garden waste, and paper products) and forestry residues (wood chips, sawdust and

I. Jioui (✉) · I. Ayouch · M. Zahouily
Fert-Explorer, College of Agriculture, Mohammed VI Polytechnic University, Ben Guerir, Morocco
e-mail: ilham.jioui@um6p.ma

S. Aboulhrouz
Materials Science, Energy and Nanoengineering (MSN) Department, Mohammed VI Polytechnic University, Ben Guerir, Morocco

O. Amadine
Chemical & Biochemical Sciences Department, Mohammed VI Polytechnic University, Ben Guerir 43150, Morocco

M. Zahouily
Laboratory of Materials, Catalysis & Natural Resources Valorization, Faculty of Science and Technology, Hassan II University, Casablanca, Morocco

A. Hind (ed.), *Biomass Waste: Processing, Valorization, Industrial Applications*,
https://doi.org/10.1007/978-981-95-0991-1_2

other by-products from logging and wood processing industries) (Bonechi et al. 2017). Biomass waste can be exploited to enhance energy production and environmental management in several ways. It can be converted into bioenergy via combustion (Cherubini et al. 2011), gasification (Carpentieri et al. 2005) and anaerobic digestion (Sawatdeenarunat et al. 2015), providing renewable energy for power generation, heating and biofuels for transport. Moreover, biochar, produced by pyrolysis, improves soil health and sequesters carbon (Lee et al. 2013). Organic waste can also be transformed into anaerobic digesters to produce biogas, mainly methane, used as a renewable energy source (Kasinath et al. 2021). However, the valorization of biomass presents major technical challenges due to its complex composition and recalcitrant structure, which hinder its direct exploitation in many conversion processes. Lignocellulosic biomass, for example, consists mainly of cellulose, hemicellulose and lignin, three closely associated polymers that give plant matter its rigidity and strength (Kumar et al. 2016). While this structure confers ecological advantages on biomass, it also complicates access to valuable constituents. To maximize biomass utilization, it is necessary to use pretreatment methods. These processes enable complex structures to be disassembled, improve the availability of key polymers and facilitate subsequent conversion steps (Carrere et al. 2016; Sai Bharadwaj et al. 2023). Depending on the recovery objectives, pretreatment can also contribute to the selective recovery of specific compounds, such as lignin or sugars.

This chapter explores three main pretreatment techniques for optimizing biomass waste (Fig. 2.1). Biological pretreatments use microorganisms or enzymes (Zabed et al. 2019), to break down lignocellulosic components and other complex polymers, making biomass more accessible for processing in an environmentally friendly manner. Physical pretreatments involve mechanical methods (Feng et al. 2024), microwave (Hoang et al. 2021), extrusion (Duque et al. 2017), pyrolysis (Kwon et al. 2020) or freezing (Rooni et al. 2017) to alter biomass structure. Chemical pretreatments utilize acids (Harmsen et al. n.d.), alkalis (Duque et al. 2018), or solvents (Vaidya et al. 2022)to dissolve or modify lignin and hemicellulose, making cellulose more accessible for enzymatic hydrolysis or fermentation, though they require careful management of chemicals and waste (Sai Bharadwaj et al. 2023; Raina et al. 2024).

Each of these pretreatment approaches presents distinct advantages and challenges. The choice of technique often depends on the specific type of biomass, the intended application and the economic and environmental considerations associated with the process. By examining these methods in detail, this chapter aims to provide an overview of how pretreatment technologies can be effectively applied to optimize biomass feedstocks, thereby advancing the sustainability and efficiency of biomass-based processes.

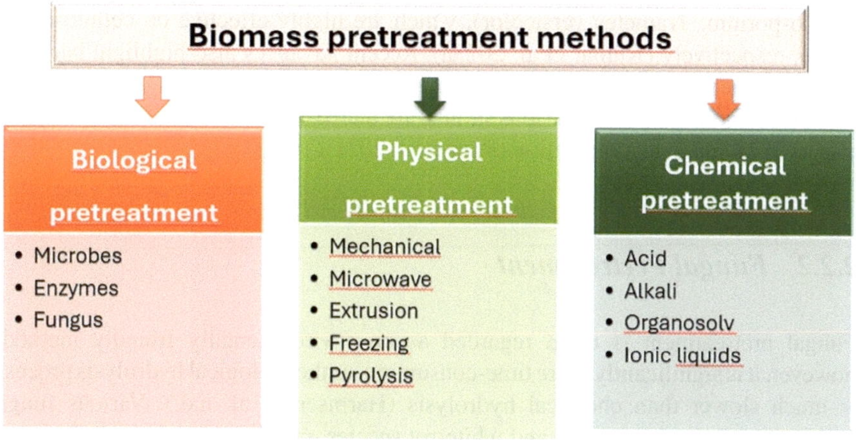

Fig. 2.1 Pretreatment techniques for biomass

2.2 Biological Pretreatment Techniques

The biological pre-treatment of biomass waste uses biological agents to enhance the transformation of biomass into valuable products such as chemicals or biofuels. This approach requires the presence of micro-organisms or enzymes to decompose the complex structures of the biomass facilitating further processing (Brémond et al. 2018). This pretreatment has various advantages, as it does not produce toxic by-products, increases the productivity of the required products, has minimal energy consumption, and is specific to the substrate and reaction process (Karami et al. 2022). However, the process is slow, the space required and the maintenance of optimal conditions for microbial growth are some of the limitations that restrict biological pretreatment processes (Kumar et al. 2023a). The most used biological pretreatment methods are microbial, fungal and enzymatic pretreatment.

2.2.1 *Microbial Pretreatment*

Microbial pretreatment processes for biomass waste exploit the ability of bacteria and other microbes to break down biomass components such as cellulose and lignin into fermentable sugars and other intermediates (Brémond et al. 2018). This improves the efficiency of subsequent steps such as fermentation, where sugars are converted into biofuels or organic acids, or anaerobic digestion, which produces biogas and digestate. Microbial pre-treatment reduces both the use of aggressive chemicals and the impact on the environment and contributes to more sustainable waste management practices. The most prominent microbes are cellulolytic bacteria (Clostridium thermocellum, Paenibacillus campinasensis) and ligninolytic fungi (Phanerochaete chrysosporium, Trametes versicolor), which are highly effective on cellulose and lignin, respectively (Kumar et al. 2023a). Recent advances also highlight bacterial enzymes, such as laccases and peroxidases, for efficient delignification. Microbial consortia, a combination of various strains, are increasingly used for their synergistic effects in biomass decomposition (Raut et al. 2021).

2.2.2 *Fungal Pretreatment*

Fungal pretreatment is often regarded as an environmentally friendly method; however, it is significantly more time-consuming, as the biological hydrolysis process is much slower than chemical hydrolysis (Harmsen et al. n.d.). Various fungi, including soft-rot, brown-rot, and white-rot species, can degrade hemicellulose and lignin with minimal cellulose degradation (Held n.d.). Among these, white-rot fungi, particularly species within the class Basidiomycetes, are the most effective for biological pretreatment. Notably, Phanerochaete chrysosporium stands out due to its high growth rate and superior lignin degradation capability (Saha et al. 2016). Despite advantages such as low energy consumption, no chemical requirements, and environmentally benign conditions, biological pretreatment faces limitations that hinder its commercial viability (Wu et al. 2022). These challenges include the need for large space, extended processing times, and the constant monitoring of microorganism growth.

2.2.3 *Enzymatic Pretreatment*

Enzymatic pretreatment is a specific method performed under relatively mild conditions (50 °C). This technique employs the hydrolytic properties of various enzymes produced by microorganisms, bacteria and fungi (Carrere et al. 2016; Montgomery and Bochmann n.d.). The most commonly studied enzymes are those that hydrolyze lignocellulosic polymers. The efficiency of this pretreatment is highly dependent on the enzymes' ability to access substrate surfaces, making particle size reduction crucial for optimizing bioconversion rates. Biomass is a complex substrate that cannot be completely hydrolyzed by a single enzyme. Enzymatic hydrolysis of biomass results from the synergistic action of three types of enzymes: cellulases (endo 1,4 ß-glucanases and exo 1,4 ß-glucanases), which hydrolyze celluloses into cellobiose, and ß-glucosidases, which hydrolyze cellobiose into glucose (Brémond et al. 2018).

Although this type of pretreatment avoids the use of chemicals, the high cost of enzymes can be a major disadvantage and a limitation of its use (Wu et al. 2022). New research programs are focused on improving the efficiency of enzyme preparation, without increasing production costs.

2.3 Physical Pretreatment Techniques

The physical pretreatment of biomass includes a variety of methods designed to improve its recovery in energy or industrial applications. These include mechanical pretreatments, which reduce particle size and increase specific surface area, thus facilitating subsequent chemical or biological conversion (Gallego-García et al. 2023). The use of microwaves offers rapid, localized heating, promoting the degradation of complex structures such as lignocellulose and improving access to cellular components (Peng et al. 2014). Extrusion, which combines heat, pressure and shear, modifies the physical structure of biomass, making it easier to hydrolyze enzymatically or convert into biofuels (Gallego-García et al. 2023). Freezing uses ice crystals to fracture biomass cells, facilitating extraction of internal constituents such as sugars and proteins (Rooni et al. 2017). Finally, pyrolysis, a thermal process carried out in the absence of oxygen, breaks down biomass at high temperature, producing gases, oils and biochar, which can be used respectively for energy production or as raw materials for other industries (Fahmy et al. 2020). These different methods, often combined, aim to optimize biomass valorization according to its characteristics and processing objectives.

2.3.1 Mechanical Pretreatment

Mechanical size reduction or mechanical comminution or milling is often said to be the first step of pretreatment. Two-roll milling, ball milling, hammer milling, disk milling and colloid milling are several types of milling which are used in biofuel production processes (Gallego-García et al. 2023). The type of the physical pretreatment used is responsible for the final particle size of feedstock achieved, for example, the particle size reduces to 10–30 mm and 0.2–2 mm, after chipping, milling and grinding respectively (Naimi n.d.). Milling desperately consumes high energy which is its most important downside. Wet milling is a more reasonable automatic pretreatment to overcome the energy consumption problem (Yu et al. 2024).

2.3.2 Microwave Irradiation

Microwave pretreatment is a rapid and efficient method for breaking down biomass, using microwave radiation to generate heat and pressure that disrupts the lignocellulosic structure. This process weakens the bonds in cellulose, hemicellulose, and lignin, making them more accessible for enzymatic hydrolysis or chemical conversion (Peng et al. 2014). It offers advantages like fast processing, uniform heating, and improved sugar recovery but can be enhanced further by adding chemicals like acids or alkalis (Diaz et al. 2015). It further improves enzymatic accessibility by increasing the specific surface area, reducing cellulose crystallinity and polymerization, and breaking down lignin (Ethaib et al. n.d.). While promising for biofuel and bioproduct production, challenges like high equipment costs and potential formation of inhibitory by-products need to be addressed to enable large-scale application.

2.3.3 Extrusion

Extrusion pretreatment is a physical pretreatment method used to process biomass waste by applying high shear, pressure, and temperature in a continuous extrusion system. During this process, the biomass is forced through a screw inside an extruder, which disrupts its lignocellulosic structure by reducing particle size, breaking down cellulose crystallinity, and partially hydrolyzing hemicellulose and lignin (Gallego-García et al. 2023). Extrusion alone does not appear to significantly alter the chemical composition of lignocellulosic biomass (Marone et al. 2019). However, the addition of chemicals like acids or alkalis can enhance the solubilization of lignin (Liu et al. 2018), hemicellulose and other non-structural compounds (Han et al. 2020a, b). Reactive extrusion, a specialized method that incorporates a chemical catalyst during the process, combines chemical and physical modifications of the biomass, thereby improving the overall efficiency of the pretreatment. Alkaline extrusion with NaOH has been extensively studied, yielding promising results in recent years (Duque et al. 2018). However, recent research has focused on incorporating more environmentally friendly alternatives, such as ionic liquids (Han et al. 2020a, b) and glycerol (Lu et al. 2018).

This method is energy-efficient, scalable, and effective at processing a wide range of biomass types, but challenges such as high equipment costs and energy consumption need to be optimized for large-scale applications in biofuel and bioproduct production.

2.3.4 Freezing

Freezing pretreatment is an emerging method for processing biomass waste, leveraging the expansion of water during freezing to disrupt the lignocellulosic structure. When biomass is frozen, the formation of ice crystals exerts mechanical stress on cell walls, causing microfractures and loosening the tightly bound matrix of cellulose, hemicellulose, and lignin. This increases the surface area and porosity of the biomass, enhancing its accessibility to enzymes or chemicals in subsequent hydrolysis or fermentation steps (Rooni et al. 2017). Freezing pretreatment is particularly attractive due to its simplicity and lack of chemical additives, making it environmentally friendly. However, its practical application is limited by the energy demands of freezing and the need for optimization to (Lachos-Perez et al. 2023) efficiency for large-scale biofuel and bioproduct production (Rooni et al. 2017).

2.3.5 Pyrolysis

Pyrolysis pretreatment involves heating biomass waste in the absence of oxygen to break it down into smaller, more manageable components, such as biochar (Lee et al. 2013), bio-oil, and syngas (Zhou et al. 2020). Pyrolysis is categorized into slow, fast, and flash types based on temperature, heating rate, and duration. Slow pyrolysis occurs at low temperatures (≤500 °C) with slow heating (0.1–2 °C/s) over hours to days, producing mainly char and valuable chemicals like acetone and methanol (Fahmy et al. 2020). Fast pyrolysis operates at moderate temperatures (400–600 °C), high heating rates (10–200 °C/s), and short durations (0.5–10 s), yielding 50–70 wt% bio-oil and biogas. Flash pyrolysis uses extreme heating rates (up to 2500 °C/s) and very short times (<0.5 s), focusing on maximizing bio-oil yields up to 75–80 wt%. These processes cater to different industrial needs for energy and chemical recovery (Fahmy et al. 2020).

Pyrolysis can enhance biomass utilization by reducing its recalcitrance, improving its energy density, and converting it into value-added intermediates for biofuel and chemical production. While it offers the advantage of producing multiple useful products, challenges such as optimizing process conditions, managing energy requirements, and ensuring the economic viability of large-scale systems must be addressed for widespread adoption.

2.4 Chemical Pre-treatment Techniques

Chemical pretreatment involves the addition of a chemical agent to biomass waste, which acts on the lignocellulosic fractions to facilitate their degradation. Various chemicals have been tested on different substrates. The most studied are reagents for the hydrolysis of organic polymers: acids, bases, organic solvents and ionic liquids (Monlau et al. 2013).

2.4.1 Acidic Pretreatment

The pre-treatment of biomass waste in the presence of acids has been widely studied in order to optimize the conditions for partial or total hydrolysis of biomass. This process is based on breaking down the lignocellulosic complex of plant walls to release more easily biodegradable by-products (Monlau et al. 2013). Various acids are used to perform acid hydrolysis, such as sulfuric acid (H_2SO_4), nitric acid (HNO_3), hydrochloric acid (HCl) or phosphoric acid (H_3PO_4) (R). They are generally used concentrated or diluted (between 0.2% and 2.5% w/w), at temperatures ranging from 130 to 210 °C (Cardona et al. 2010).

Acid pretreatment aims to improve cellulose hydrolysis in order to attack polysaccharides, particularly hemicelluloses, which are easier to hydrolyze than cellulose (Cardona et al. 2010). However, this process results in the formation of various inhibitors such as acetic acid, furfural and 5-hydroxymethylfurfural, which are detrimental to the growth of micro-organisms (Sarkar et al. 2012). The pH range varies from 1 to 4, which requires the use of corrosion-resistant reactors. Other disadvantages are also cited, including their economic cost, which necessitates their recovery after hydrolysis to be recycled and reused in the process (Sun and Cheng n.d.).

2.4.2 *Alkaline Pretreatment*

Alkaline pretreatment is a commonly employed technique, which is usually carried out at room temperature using various bases such as sodium hydroxide (NaOH), potassium hydroxide (KOH), calcium hydroxide ($Ca(OH)_2$) or ammonia (NH_3) (Jędrzejczyk et al. 2019). This process lasts from a few hours to several days and is based on alkaline hydrolysis, which breaks ester bonds in hemi-celluloses and other components such as lignin. This generally leads to an increased hydrolysis of lignin and hemi-celluloses (Harmsen et al. n.d.). Studies have shown the effectiveness of alkaline pretreatments for lignocellulosic biomasses. It was found that pretreatments with calcium oxide and calcium hydroxide offer higher bioconversion yields compared to physical, thermal or biological pretreatments (Bruni et al. 2010). Similarly, pretreatment with 5% sodium hydroxide can increase biogas production from corn cobs by 37% (Zhu et al. 2010). However, these pretreatments can also lead to decreases in biodegradability in biomass processing due to the formation of toxic compounds during saponification. In addition, alkaline pretreatments pose economic and environmental challenges, notably due to the costs of the chemical agents and equipment required and are generally not recommended in agricultural facilities due to the health and environmental risks associated with their handling (Ethaib et al. 2020).

2.4.3 *Organosolv Pretreatment*

Organosolv is a biomass pretreatment method that uses organic solvents or a mixture of organic solvents to solubilize hemicellulose and extract lignin. Commonly used solvents include methanol, ethanol, acetone, ethylene glycol and triethylene glycol (Sai Bharadwaj et al. 2023). In addition, organic acids such as salicylic acid, oxalic acid, and acetylsalicylic acid often serve as catalysts in the organosolv process (Jędrzejczyk et al. 2019). During organosolv treatment of lignocellulosic material with alcohols, internal bonds within lignin and hemicellulose, as well as ether and ester interpolymer bonds, are hydrolyzed. This results in lignin removal and nearly complete solubilization of hemicellulose. The presence of organic acids in the process generates hydrogen ions, which enhance biomass delignification and lignin dissolution. The optimal temperature for the process ranges from 100 to 250 °C, depending on the type of biomass (Vaidya et al. 2022).

The main advantages of organosolv pretreatment compared with other methods are its high conversion yield and the simplicity of recovering and reusing organic solvents at the end of the process (Sannigrahi et al. 2010). However, this pretreatment has certain disadvantages, notably the high cost of organic solvents and the significant investment required for equipment, due to the high volatility of some solvents (Joseph et al. 2023).

2.4.4 Ionic Liquids Pretreatment

Ionic liquids (ILs) are a new class of solvent with low melting point (<100 °C), high polarity, good thermal stability and negligible vapor pressure, consisting entirely of ions (cations and anions) (Zavrel et al. 2009). They reduce the toughness of lignocellulosic biomass by breaking its structure, making it easier to process. ILs can dissolve cellulose by breaking its strong hydrogen bonds, making it easier for enzymes to break it down. They also help remove lignin and hydrolyze hemicellulose, separating these components for individual use. Most ILs used in biomass fractionation are imidazonium salts. Studies show that 1-allyl-3-methylimidazonium chloride (AMIMCl) and 1-butyl-3-methylimidazonium chloride (BMIMCl) can be used effectively as a non-derivatizing solvent for cellulose dissolution at temperatures below 100 °C (Zhang and Lynd 2006).

There are also simpler and cheaper options like protic ionic liquids (Pin et al. 2019) and deep eutectic solvents (Del Mar Contreras-Gámez et al. 2023). However, ILs are expensive, and challenges like recycling, environmental safety, and compatibility with other processes need to be solved. Recent advancements are working on creating cheaper, eco-friendly ILs, combining ILs with other methods, and integrating them into biorefineries to make biomass processing more sustainable and efficient (Jędrzejczyk et al. 2019).

2.5 Comparison of Pretreatment Methods

Comparing biomass waste pretreatment techniques reveals significant differences in mechanisms, efficiency and suitability for different feedstocks. Chemical treatments, such as acid or alkaline pretreatment, effectively remove lignin and improve sugar yields, although they can generate inhibitory by-products and cause severe equipment corrosion issues (Monlau et al. 2013; Bruni et al. 2010). Physical methods, such as mechanical processes, increase the surface area but consume a lot of energy, while thermal processes rapidly disrupt the biomass structure, but can lead to degradation of products. These methods are generally simple, easily scalable, and require

Table 2.1 Advantages and disadvantages of biomass pretreatment techniques

Pretreatment	Advantages	Disadvantages
Biological pretreatement	• Environmentally friendly • Low energy consumption	• Slow process rate • Specific processing conditions (microorganisms and substrates) • Possible high cost of enzymes-assisted
Physical pretreatement	• Simple • Scalable • No or minimal chemicals	• High energy consumption • High equipment cost
Chemical pretreatement	•Efficient lignin removal and improved biodegradability •Can be combined with other methods	• Corrosive chemicals • Generation of inhibitory by-products • Environmental concerns

no chemical reagents. (Jędrzejczyk et al. 2019; Bruni et al. 2010; Feng et al. 2024). Biological pretreatment, which employs microorganisms to degrade lignin and hemicellulose under mild, eco-friendly conditions, is cost-effective and energy-efficient but proceeds slowly and demands specific environmental conditions (Kumar et al. 2023b). The effectiveness of each method depends on the type of biomass, the desired end products and economic considerations. Combined approaches often giving the best results (Table 2.1).

2.6 Conclusion

In conclusion, the success of biomass pretreatment lies in its ability to enhance the accessibility of cellulose and hemicellulose while minimizing inhibitory by-products. The choice of method physical, chemical or biological directly influences downstream processes like fermentation or combustion. Chemical methods, such as acid or alkaline treatments, excel in maximizing sugar yields, physical methods, optimize particle size and handling while biological methods, leveraging microorganisms or enzymes, offer an eco-friendly approach with the potential to reduce energy consumption and avoid chemical residues. Ultimately, effective pretreatment is defined by improved conversion yields, shorter processing times, and adaptability to diverse feedstocks, driving sustainable and cost-effective bioenergy and biochemical production.

References

Bonechi C, Consumi M, Donati A, Leone G, Magnani A, Tamasi G, Rossi C (2017) Biomass. In: Bioenergy systems for the future. Elsevier, pp 3–42. https://doi.org/10.1016/B978-0-08-101031-0.00001-6

Brémond U, de Buyer R, Steyer JP, Bernet N, Carrere H (2018) Biological pretreatments of biomass for improving biogas production: an overview from lab scale to full-scale. Renew Sustain Energy Rev 90 (juillet):583–604. https://doi.org/10.1016/j.rser.2018.03.103

Bruni E, Jensen AP, Angelidaki I (2010) Comparative study of mechanical, hydrothermal, chemical and enzymatic treatments of digested biofibers to improve biogas production. Bioresour Technol 101(22):8713–8717. https://doi.org/10.1016/j.biortech.2010.06.108

Cardona CA, Quintero JA, Paz IC (2010) Production of bioethanol from sugarcane bagasse: status and perspectives. Bioresour Technol 101(13):4754–4766. https://doi.org/10.1016/j.biortech.2009.10.097

Carpentieri M, Corti A, Lombardi L (2005) Life cycle assessment (LCA) of an integrated biomass gasification combined cycle (IBGCC) with CO_2 removal. Energy Convers Manag 46(11–12):1790–1808. https://doi.org/10.1016/j.enconman.2004.08.010

Carrere H, Antonopoulou G, Affes R, Passos F, Battimelli A, Lyberatos G, Ferrer I (2016) Review of feedstock pretreatment strategies for improved anaerobic digestion: from lab-scale research to full-scale application. Bioresour Technol 199(janvier):386–397. https://doi.org/10.1016/j.biortech.2015.09.007

Cherubini F, Peters GP, Berntsen T, Strømman AH, Hertwich E (2011) CO_2 emissions from biomass combustion for bioenergy: atmospheric decay and contribution to global warming: global warming potential of CO_2 from bioenergy. GCB Bioenergy 3(5):413–426. https://doi.org/10.1111/j.1757-1707.2011.01102.x

Del Mar Contreras-Gámez M, Galán-Martín Á, Seixas N, da Costa Lopes AM, Silvestre A, Castro E (2023) Deep eutectic solvents for improved biomass pretreatment: current status and future prospective towards sustainable processes. Bioresour Technol 369(février):128396. https://doi.org/10.1016/j.biortech.2022.128396

Diaz AB, de Souza Moretti MM, Bezerra-Bussoli C, Nunes CD, Blandino A, da Silva R, Gomes E (2015) Evaluation of microwave-assisted pretreatment of lignocellulosic biomass immersed in alkaline glycerol for fermentable sugars production. Bioresour Technol 185(juin):316–323. https://doi.org/10.1016/j.biortech.2015.02.112

Duque A, Manzanares P, Ballesteros M (2017) Extrusion as a pretreatment for lignocellulosic biomass: fundamentals and applications. Renew Energy 114(décembre):1427–1441. https://doi.org/10.1016/j.renene.2017.06.050

Duque A, Manzanares P, González A, Ballesteros M (2018) Study of the application of alkaline extrusion to the pretreatment of eucalyptus biomass as first step in a bioethanol production process. Energies 11(11):2961. https://doi.org/10.3390/en11112961

Ethaib S, Omar R, Kamal SM, Biak DA (n.d.) Microwave-assisted pretreatment of lignocellulosic biomass: a review, n° 2

Ethaib S, Omar R, Siti Mazlina MK, Dayang Radiah AB, Zubaidi SL (2020) Toward sustainable processes of pretreatment technologies of lignocellulosic biomass for enzymatic production of biofuels and chemicals: A Review. BioRes 15(4):10063–10088

Fahmy TYA, Fahmy Y, Mobarak F, El-Sakhawy M, Abou-Zeid RE (2020) Biomass pyrolysis: past, present, and future. Environ Dev Sustain 22(1):17–32. https://doi.org/10.1007/s10668-018-0200-5

Feng Y, Huang J, Zhang J, Yellezuome D, Zhao R, Chen T, Wu J (2024) Mechanical pretreatment of typical agricultural biomass on shape characterization and NO emissions during combustion. Bioresour Technol 397(avril):130477. https://doi.org/10.1016/j.biortech.2024.130477

Gallego-García M, Moreno AD, Manzanares P, Negro MJ, Duque A (2023) Recent advances on physical technologies for the pretreatment of food waste and lignocellulosic residues. Bioresour Technol 369(février):128397. https://doi.org/10.1016/j.biortech.2022.128397

Han S-Y, Park C-W, Endo T, Febrianto F, Kim N-H, Lee S-H (2020a) Extrusion process to enhance the pretreatment effect of ionic liquid for improving enzymatic hydrolysis of lignocellulosic biomass. Wood Sci Technol 54(3):599–613. https://doi.org/10.1007/s00226-020-01170-9

Han S-Y, Park C-W, Kwon G-J, Kim N-H, Kim J-C, Lee S-H (2020b) Ionic liquid pretreatment of lignocellulosic biomass. J Forest Environ Sci 36(2):69–77. https://doi.org/10.7747/JFES.2020.36.2.69

Harmsen P, Huijgen W, Bermudez L, Bakker R (n.d.) Literature review of physical and chemical pretreatment processes for lignocellulosic biomass

Held J (n.d.) Svenskt Gastekniskt Center AB

Hoang AT, Nižetić S, Ong HC, Mofijur M, Ahmed SF, Ashok B, Bui VT, Chau MQ (2021) Insight into the recent advances of microwave pretreatment technologies for the conversion of lignocellulosic biomass into sustainable biofuel. Chemosphere 281(octobre):130878. https://doi.org/10.1016/j.chemosphere.2021.130878

Jędrzejczyk M, Soszka E, Czapnik M, Ruppert AM, Grams J (2019) Physical and chemical pretreatment of lignocellulosic biomass. In: Second and third generation of feedstocks. Elsevier, pp 143–196. https://doi.org/10.1016/B978-0-12-815162-4.00006-9

Jafari Naimi L (n.d.) Experiments and modeling of size reduction of switchgrass in a laboratory rotary knife mill

Joseph P, Opedal MT, Moe ST (2023) The O-factor: using the H-factor concept to predict the outcome of Organosolv Pretreatment. Biomass Convers Biorefinery 13(8):6727–6736. https://doi.org/10.1007/s13399-021-01667-8

Karami K, Karimi K, Mirmohamadsadeghi S, Kumar R (2022) Mesophilic aerobic digestion: an efficient and inexpensive biological pretreatment to improve biogas production from highly-recalcitrant pinewood. Energy 239(janvier):122361. https://doi.org/10.1016/j.energy.2021.122361

Kasinath A, Fudala-Ksiazek S, Szopinska M, Bylinski H, Artichowicz W, Remiszewska-Skwarek A, Luczkiewicz A (2021) Biomass in biogas production: pretreatment and codigestion. Renew Sustain Energy Rev 150(octobre):111509. https://doi.org/10.1016/j.rser.2021.111509

Kumar A, Gautam A, Dutt D (2016) Biotechnological transformation of lignocellulosic biomass in to industrial products: an overview. Adv Biosci Biotechnol 07(03):149–168. https://doi.org/10.4236/abb.2016.73014

Kumar HK, Mohana NC, Amrutha HC, Rakshith D, Harini BP, Satish S (2023a) Biomass conversion through optimization of cellulase from chryseobacterium junjuense Bp17 and their utility in bioethanol production. Energy 283(novembre):129187. https://doi.org/10.1016/j.energy.2023.129187

Kumar JA, Sathish S, Prabu D, Renita AA, Saravanan A, Deivayanai VC, Anish M, Jayaprabakar J, Baigenzhenov O (2023b) Agricultural waste biomass for sustainable bioenergy production: feedstock, characterization and pre-treatment methodologies. Chemosphere 331(août):138680. https://doi.org/10.1016/j.chemosphere.2023.138680

Kwon G, Bhatnagar A, Wang H, Kwon EE, Song H (2020) A review of recent advancements in utilization of biomass and industrial wastes into engineered biochar. J Hazard Mater 400(décembre):123242. https://doi.org/10.1016/j.jhazmat.2020.123242

Lachos-Perez D, Martins-Vieira JC, Missau J, Anshu K, Siakpebru OK, Thengane SK, Morais AR, Tanabe EH, Bertuol DA (2023) Review on biomass pyrolysis with a focus on bio-oil upgrading techniques. Analytica 4(2):182–205. https://doi.org/10.3390/analytica4020015

Lee Y, Ryu C, Park YK, Jung JH, Hyun S (2013) Characteristics of biochar produced from slow pyrolysis of Geodae-Uksae 1. Bioresour Technol 130(février):345–350. https://doi.org/10.1016/j.biortech.2012.12.012

Liu Z, Zhang M, Yang CH (2018) Dual extrusion 3D printing of mashed potatoes/strawberry juice gel. LWT 96(octobre):589–596. https://doi.org/10.1016/j.lwt.2018.06.014

Lu H, Zhang L, Liu C, He Z, Zhou X, Ni Y (2018) A novel method to prepare lignocellulose nanofibrils directly from bamboo chips. Cellulose 25(12):7043–7051. https://doi.org/10.1007/s10570-018-2067-x

Marone A, Trably E, Carrère H, Prompsy P, Guillon F, Joseph-Aimé M, Barakat A, Fayoud N, Bernet N, Escudié R (2019) Enhancement of corn stover conversion to carboxylates by extrusion and biotic triggers in solid-state fermentation. Appl Microbiol Biotechnol 103(1):489–503. https://doi.org/10.1007/s00253-018-9463-x

Monlau F, Aemig Q, Trably E, Hamelin J, Steyer JP, Carrere H (2013) Specific inhibition of biohydrogen-producing clostridium sp. After dilute-acid pretreatment of sunflower stalks. Int J Hydrog Energy 38(28):12273–12282. https://doi.org/10.1016/j.ijhydene.2013.07.018

Montgomery LF, Bochmann G (n.d.) Pretreatment of feedstock for enhanced biogas production

Peng H, Chen H, Qu Y, Li H, Xu J (2014) Bioconversion of different sizes of microcrystalline cellulose pretreated by microwave irradiation with/without NaOH. Appl Energy 117(mars):142–148. https://doi.org/10.1016/j.apenergy.2013.12.002

Pin TC, Nakasu PY, Mattedi S, Rabelo SC, Costa AC (2019) Screening of protic ionic liquids for sugarcane bagasse pretreatment. Fuel 235(janvier):1506–1514. https://doi.org/10.1016/j.fuel.2018.08.122

Raina N, Boonmee R, Kirdponpattara S, Narasingha M, Sriariyanun M, Phitsuwan P, Chuetor S (2024) Process performance evaluation of different chemical pretreatments of lignocellulosic biomass for bioethanol production. Ind Crops Prod 211(mai):118207. https://doi.org/10.1016/j.indcrop.2024.118207

Raut MP, Pandhal J, Wright PC (2021) Effective pretreatment of lignocellulosic co-substrates using barley straw-adapted microbial consortia to enhanced biomethanation by anaerobic digestion. Bioresour Technol 321(février):124437. https://doi.org/10.1016/j.biortech.2020.124437

Rooni V, Raud M, Kikas T (2017) The freezing pre-treatment of lignocellulosic material: a cheap alternative for nordic countries. Energy 139(novembre):1–7. https://doi.org/10.1016/j.energy.2017.07.146

Saha BC, Qureshi N, Kennedy GJ, Cotta MA (2016) Biological pretreatment of corn stover with white-rot fungus for improved enzymatic hydrolysis. Int Biodeterior Biodegrad 109(avril):29–35. https://doi.org/10.1016/j.ibiod.2015.12.020

Sai Bharadwaj AS, Dev S, Zhuang J, Wang Y, Yoo CG, Jeon BH, Aggarwal S, Park SH, Kim TH (2023) Review of chemical pretreatment of lignocellulosic biomass using low-liquid and low-chemical catalysts for effective bioconversion. Bioresour Technol 368(janvier):128339. https://doi.org/10.1016/j.biortech.2022.128339

Sannigrahi P, Miller SJ, Ragauskas AJ (2010) Effects of organosolv pretreatment and enzymatic hydrolysis on cellulose structure and crystallinity in loblolly pine. Carbohyd Res 345(7):965–970. https://doi.org/10.1016/j.carres.2010.02.010

Sarkar N, Ghosh SK, Bannerjee S, Aikat K (2012) Bioethanol production from agricultural wastes: an overview. Renew Energy 37(1):19–27. https://doi.org/10.1016/j.renene.2011.06.045

Sawatdeenarunat C, Surendra KC, Takara D, Oechsner H, Khanal SK (2015) Anaerobic digestion of lignocellulosic biomass: challenges and opportunities. Bioresour Technol 178(février):178–186. https://doi.org/10.1016/j.biortech.2014.09.103

Sun Y, Cheng J (n.d.) Hydrolysis of lignocellulosic materials for ethanol production: a review

Vaidya AA, Murton KD, Smith DA, Dedual G (2022) A review on organosolv pretreatment of softwood with a focus on enzymatic hydrolysis of cellulose. Biomass Convers Biorefinery 12(11):5427–5442. https://doi.org/10.1007/s13399-022-02373-9

Wu Z, Peng K, Zhang Y, Wang M, Yong C, Chen L, Qu P, Huang H, Sun E, Pan M (2022) Lignocellulose dissociation with biological pretreatment towards the biochemical platform: a review. Mater Today Bio 16(décembre):100445. https://doi.org/10.1016/j.mtbio.2022.100445

Yu T, Jing S, Jiaxin L, Aixia W, Mengzi N, Xue G, Lili W, Liya L, Fengzhong W, Litao T (2024) Effects of milling methods on rice flour properties and rice product quality: a review. Rice Sci 31(1):33–46. https://doi.org/10.1016/j.rsci.2023.11.002

Zabed HM, Akter S, Yun J, Zhang G, Awad FN, Qi X, Sahu JN (2019) Recent advances in biological pretreatment of microalgae and lignocellulosic biomass for biofuel production. Renew Sustain Energy Rev 105(mai):105–128. https://doi.org/10.1016/j.rser.2019.01.048

Zavrel M, Bross D, Funke M, Büchs J, Spiess AC (2009) High-throughput screening for ionic liquids dissolving (ligno-)cellulose. Bioresour Technol 100(9):2580–2587. https://doi.org/10.1016/j.biortech.2008.11.052

Zhang Y-H, Lynd LR (2006) A functionally based model for hydrolysis of cellulose by fungal cellulase. Biotechnol Bioeng 94(5):888–898. https://doi.org/10.1002/bit.20906

Zhou N, Zhou J, Dai L, Guo F, Wang Y, Li H, Deng W et al (2020) Syngas production from biomass pyrolysis in a continuous microwave assisted pyrolysis system. Bioresour Technol 314(octobre):123756. https://doi.org/10.1016/j.biortech.2020.123756

Zhu JY, Pan X, Zalesny RS (2010) Pretreatment of woody biomass for biofuel production: energy efficiency, technologies, and recalcitrance. Appl Microbiol Biotechnol 87(3):847–857. https://doi.org/10.1007/s00253-010-2654-8

Chapter 3
Sustainable Technologies for Biomass Conversion: Last Trends

Naoual Ouchitachne, Houda Elboukhari, Soumia Aboulhrouz, A. M. Abourriche, and Mina Oumam

Abstract In response to growing environmental concerns and the urgent need for renewable energy sources, biomass has emerged as a versatile and sustainable option for energy production. This chapter explores recent advancements in sustainable biomass conversion technologies, emphasizing innovative processes that efficiently transform biomass into electricity, fuels, heat, and biochemical products while minimizing environmental impact. As a renewable resource, biomass is crucial in the shift toward sustainable energy, offering alternatives to fossil fuels and contributing to the reduction of greenhouse gas emissions. Additionally, this chapter addresses the challenges associated with the high pretreatment costs of biomass and the necessity for process optimization to achieve maximum efficiency. It explores innovations and new techniques aimed at enhancing the economic viability of these technologies, such as improved conversion methods and strategies to maximize the utilization of biomass components. By providing a comprehensive overview, the chapter highlights biomass as a sustainable energy solution and emphasizes the importance of advanced technologies in propelling the sector towards a cleaner, more efficient future.

Keywords Biomass conversion · Renewable energy · Sustainable technologies · Circular economy · Greenhouse gas reduction

---

N. Ouchitachne · M. Oumam
Laboratory of Engineering and Materials LIMAT, Faculty of Science Ben M'Sik, Hassan II University, Casablanca, Morocco

H. Elboukhari · A. M. Abourriche
Laboratoire de Chimie Analytique et Moléculaire (LCAM), Faculty of Science Ben M'Sik, Hassan II University, Casablanca, Morocco

S. Aboulhrouz (✉)
Materials Science, Energy and Nanoengineering (MSN) Department, Mohammed VI Polytechnic University, Ben Guerir, Morocco
e-mail: soumia.aboulhrouz@um6p.ma

A. Hind (ed.), *Biomass Waste: Processing, Valorization, Industrial Applications*,
https://doi.org/10.1007/978-981-95-0991-1_3

3.1 Introduction

Biomass stands as a renewable and versatile energy source, playing a pivotal role in addressing climate change, reducing reliance on fossil fuels, and advancing sustainable energy solutions. Its potential to enhance energy security and foster rural economic development has intensified the pursuit of innovative biomass transformation technologies. These advancements aim to meet the world's escalating energy demands while preserving ecological integrity (Goyal et al. 2008). Sustainable biomass conversion into thermal energy, biofuels, and chemicals offers multiple environmental and economic benefits. This approach not only provides a renewable energy source but also significantly reduces greenhouse gas emissions, promotes the circular use of resources, and minimizes waste by valorizing agricultural and industrial by-products. By integrating biomass conversion into existing industrial processes, we can enhance energy efficiency and contribute to a more sustainable and resilient energy system (Ibitoye et al. 2023). Biomass feedstocks are derived from various natural waste sources, including landscaping debris, forest maintenance residues, purpose-grown energy crops, and agricultural by-products. These materials serve as renewable resources for producing energy, fuels, and chemicals, contributing to sustainable energy solutions (Fig. 3.1). Observing our environment reveals an abundance of natural and sustainable biomass energy sources. Utilizing these resources can facilitate a transition toward a greener and more ecologically friendly lifestyle.

In recent years, advancements in biomass conversion technologies were oriented to greater efficiency of energy conversion, lower emissions of pollutants, and better economic feasibility for fostering wider use. Traditional methods of biomass conversion, such as direct combustion and basic gasification, have been largely supplemented or replaced by sophisticated approaches that can promise clean, efficient, and

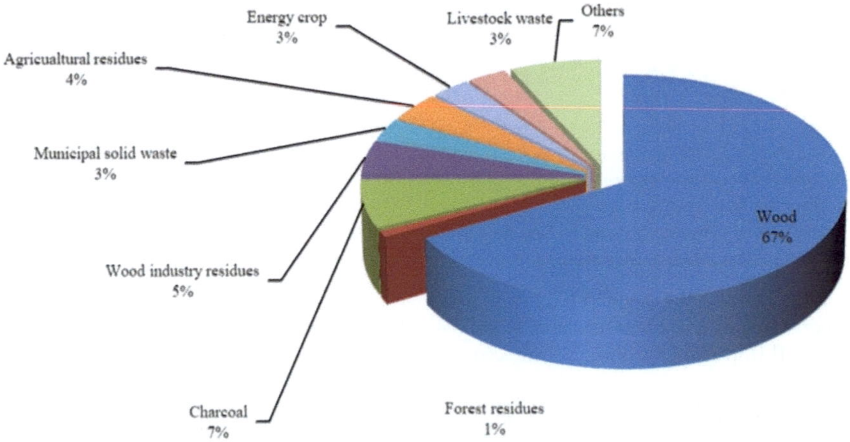

Fig. 3.1 Percentage of biomass sources in bioenergy combination (Ibitoye et al. 2023)

economically viable energy generation. The integration of such techniques further ensured the optimal use of biomass in these processes, thus allowing industries to reap the fullest application of biomass as a green energy source (Juneja 2024). This chapter provides a comprehensive review of contemporary sustainable biomass conversion technologies, emphasizing both established and emerging methods applicable to diverse energy production forms. It delves into the principles, innovations, and potential environmental impacts associated with advanced thermomechanical, thermochemical, and biochemical conversion processes. Key techniques such as pyrolysis, gasification, liquefaction, plasma technology, anaerobic digestion, and fermentation are examined for their efficacy in transforming biomass into fuels. Additionally, the chapter presents real-world case studies demonstrating the implementation of these technologies across various sectors, highlighting their benefits and practical applications. Serving as a comprehensive guide, it elucidates the pathways, challenges, and opportunities inherent in the sustainable utilization of biomass.

3.2 Sustainable Conversion Technologies for Thermal Energy from Biomass

Global efforts to mitigate climate change underscore the urgent need for a rapid transition from fossil fuels to renewable energy sources (Elum and Momodu 2017). This shift is essential to reduce greenhouse gas emissions and limit global warming to 1.5 °C above pre-industrial levels, as outlined in the Paris Agreement (Matthews and Wynes 2022; Meinshausen et al. 2022). Achieving this target necessitates substantial reductions in greenhouse gas (GHG) emissions. Key renewable resources, including wind, solar energy, and biomass, play a fundamental role in this endeavor. The European Union's Green Deal amplifies these objectives by aiming to make Europe the first climate-neutral continent by 2050, with an interim goal of reducing GHG emissions by 55% by 2030 compared to 1990 levels (Wolf et al. 2021).

Biomass is the most widely available versatile of all renewable energy sources (Barot 2022). It broadly consists of a wide range of organic materials, mainly agricultural residues in the form of stalks and husks from crops, forestry by-products including wood chips and sawdust, industrial wastes resulting from manufacturing processes, and municipal organic waste in the form of food scraps and yard clippings (Millati et al. 2019). These other sources of biomass represent a huge resource, in most instances almost unused, with great potential for energy conversion.

Biomass-to-energy conversion meets the increasing demand for renewable energy and helps very significantly in waste management. Rather than allowing organic materials to decompose in landfills or be incinerated; both of which contribute to greenhouse gas emissions; biomass can be processed through various methods to produce clean energy. These methods include combustion, gasification, pyrolysis, and biochemical processes such as anaerobic digestion. Biomass offers a dual advantage in combating climate change: it not only reduces waste but also serves as a reliable energy source. Unlike intermittent renewables like solar and wind, biomass can generate a consistent, dispatchable electricity supply, making it integral to the energy transition, particularly in regions abundant in agricultural and forestry residues. In the European Union, biomass significantly contributes to renewable energy, especially in heating, where it accounts for approximately 93% of renewable energy used for heating and cooling (Malico et al. 2019).

Biomass energy resources are obtained using various technologies that vary in terms of efficiency, scalability, and environmental concerns. These include thermochemical processes, thermal and mechanical conversion, and biochemical conversion including direct conversion, gasification, pyrolysis, and hydrothermal liquefaction (Fig. 3.2). All of these processes are environmentally friendly and can convert biomass into thermal energy, electricity, and biofuels (Zhang and Zhang 2019). These technologies contribute to the reduction in the environmental footprint of energy production by using waste materials, hence reducing dependency on fossil fuels and contributing to the circular economy models. Biomass energy systems also enable the development and job creation in rural settings—wherever biomass feedstocks are available. Sustainable management of biomass resources, inclusive of reforestation and agricultural waste recycling, further increases the environmental and economic benefits of this source of renewable energy. This means that countries are capable of meeting their targets on GHG reductions, as well as increased resilience and diversification of energy infrastructures, by developing biomass in the context of wider energy Systems (Dasappa 2011).

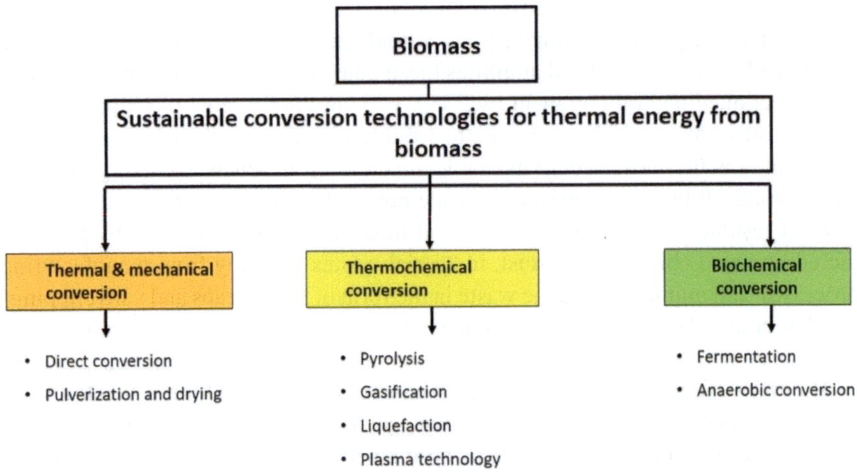

Fig. 3.2 Summary of biomass conversion technologies

3.2.1 Thermal and Mechanical Conversion

The thermal and mechanical conversion of biomass is one of the key areas in converting organic material into useful energy or other value-added products (Luo and Zhou 2022). These methods are essential in the leveraging of biomass as a renewable resource through the provision of alternative means to fossil fuels in such a way that sustainability is fostered. These processes use heat or mechanical forces to break down biomass into usable forms, either directly for energy production or as an intermediate in other industrial uses. A further understanding of the principles and mechanisms of these methods for the conversion of biomass is required for efficiency optimization in energy production, reduction in environmental impacts, and enhancement of overall biomass value in energy systems (Luo et al. 2019).

Thermal conversion of biomass is a process in which heat is applied to change the physical and chemical structure of biomass, and one typical example of such a process is direct combustion used to produce heat, especially in the form of electricity. This old technique is widely used but demands optimization of the combustion conditions to attain maximum energy efficiency and low emissions (Zhang et al. 2010). Mechanical conversion, on the other hand, is done by processes such as pulverization and drying to prepare the biomass either by increasing its surface area or reducing its moisture content. Both improve combustion and enhance overall energy output by reducing losses from moisture and enhancing subsequent thermal or chemical conversion steps (Esteban and Carrasco 2006).

3.2.1.1 Direct Conversion

The first conversion technique is direct conversion technologies may be considered as one fundamental process of biomass thermal and mechanical conversions, wherein the biomass is directly transformed into usable energy by combustion, through chemical energy that is stored in the biomass. Being one of the oldest and most conventional methods of energy production from biomass, direct conversion principally consists of the combustion of organic material to produce heat, which might then be utilized either directly or indirectly in electrical generation, heating, or within any industrial application. In contrast, direct conversion entails the release of energy stored in biomass through combustion in the presence of oxygen (Zhao et al. 2017). It is usually carried out under controlled conditions, such as in a boiler or furnace, where the transferring of heat either to water or any other medium is done, and the resultant steam produced may also be used for driving turbines for electrical production or heating applications in the system of residential or industrial types. Efficiency in direct combustion mainly depends on factors such as biomass type, moisture content, and the technology of combustion applied. One of the most critical parameters in direct conversion is biomass moisture content (Mandø 2013). Biomass with high levels of moisture reduces the energy efficiency in the combustion process since high energy is used to vaporize the water content before it gets combusted completely.

Due to this fact, pre-drying biomass is many times a necessary step simply to increase the energy yield during combustion. Biomass with low moisture contents, such as wood pellets or agricultural residues, usually offers higher combustion efficiency compared to raw and untreated biomass. Biomass fuels may vary in form, such as wood, agricultural wastes like straw or husk, and organic residues from various industries that can be combustible (Gebreegziabher et al. 2013). However, each biomass type is thought to possess different combustion characteristics. The woody biomass has shown generally higher calorific values with lower ash contents, hence its higher efficiency in energy production. Herbaceous biomass, however, generally displays a lower energy density and higher ash content than wood; this could create problems in combustion systems, namely slagging and fouling of the boilers. Several different combustion technologies have been developed to try and make this direct conversion more efficient and much more viable. The most common practice is grate-fired combustion, where biomass is combusted on a grate that allows controlled air feeding, hence promoting complete combustion and lowering the resulting emissions (Faaij 2006).

Among advanced technologies, fluidized-bed combustion enables biomass to be burned in a bed of suspended particles, which improves control of combustion parameters, temperature and air distribution, as it increases efficiency and reduces pollutant emissions, notably NOx and particulates (Fig. 3.3). The study by Li and Kuo (2020). In a fluidized-bed combustion system, for example, efficiency can be around 10% higher than in conventional grate systems. This is because the technology maintains temperature and distributes oxygen appropriately to ensure complete combustion, with a minimum of unburnt residue (Li and Kuo 2020).

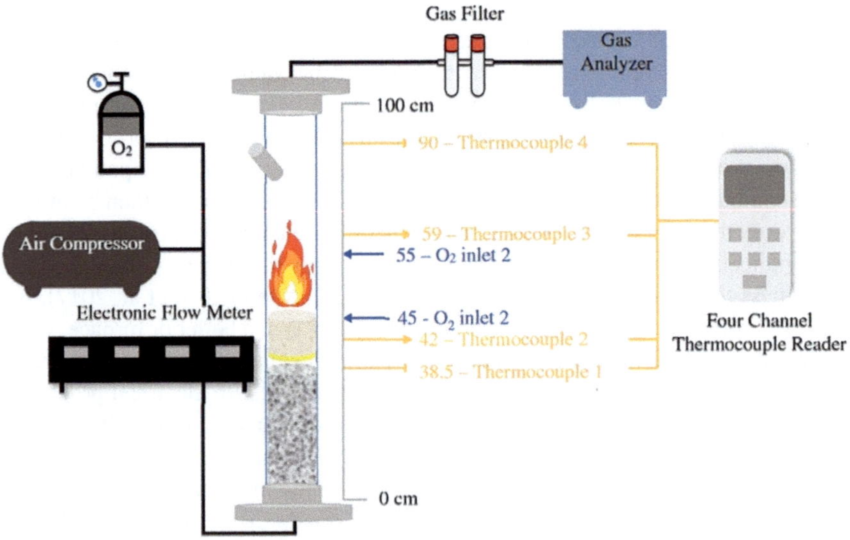

Fig. 3.3 Schematic of lab-scale fluidized-bed combustor (Li and Kuo 2020)

The biomass should be sourced from materials that are regenerated over time, such as forests or crops, so that the overall carbon footprint of biomass combustion can be lower compared to that from fossil fuels. Because this reduces the warming effect, by as much as 20%, direct conversion is highly feasible for greenhouse gas emissions reduction, especially in the case of coal or oil replacement in energy production. However, the combustion route has its own set of drawbacks. Biomass combustion, even though from a renewable energy source, will also contribute to air pollution if not handled appropriately. It could emit particulate matter, CO, and VOCs during burning due to which air quality and human health would be affected. Emission control technologies like electrostatic precipitators and fabric filters that capture particulates reduce harmful emissions, and thus mitigate these impacts, fitted into modern combustion systems. Over the past decades, direct conversion technologies have seen improvement in efficiency. As a superior form of co-firing, biomass is combusted together with coal at old power plants for easy transition to renewable energy but still leveraging on existing infrastructure (Gil and Rubiera 2019). Co-firing reduces demand for coal and lowers carbon emissions while guaranteeing reliability in energy generation (Agbor et al. 2014). Apart from this, the research into increasing the efficiency of the direct combustion systems is being done in the form of better boiler design, better control over the combustion conditions, and improved integration with combined heat and power systems. In conclusion, Biomass energy conversion through direct conversion is a wide-based and very simple means of extracting energy from biomass, and there is immense potential in this area for sustainable energy generation. This represents a practical way of utilizing renewable biomass resources, ranging from wood to agricultural waste, for heat and electricity production with reduced greenhouse gas emissions compared to conventional systems based on fossil fuels. It is to be expected that direct combustion methods will further increase their efficiency and environmental performance in pace with developing technology, thus securing its niche in the global transition toward renewable energy sources. Table 3.1 describes the process of direct biomass conversion technologies.

3.2.1.2 Pulverization and Drying

The second biomass thermal and mechanical conversion technology: Pulverization and drying are two of the fundamental processes included in the thermal and mechanical conversion of biomass. However, both bear different yet interrelated functions. Pulverization involves the mechanical reduction of biomass into smaller, finer particles, while drying consists of the extraction of excess moisture from the biomass material. In other words, these processes put the biomass in a better condition for thermal conversion by providing the feedstock with an enhanced quality best suited for subsequent thermal treatment such as combustion, pyrolysis, or gasification (Faaij 2006). Pulverization is an important mechanical process in biomass conversion, whereby the main objective is the size reduction of whole, large, heterogeneous biomass particles to fine and uniform powders. Such a size reduction is very

Table1 Overview of the key steps in direct conversion of biomass into energy, highlighting the transformation of thermal energy into electricity

Process stage	Description
Biomass collection	Biomass, mainly from wood, agricultural waste, or organic residues, is collected from various sources
Biomass preparation	The biomass is treated in such a way-pre-treatment, for example, may involve drying or size reduction-that it becomes more efficient for combustion
Combustion in furnace	Biomass is burned in a combustion chamber, releasing heat energy
Heat transfer to boiler	Heat from combustion is transferred to a boiler, which uses it to produce steam from water
Steam generation	The high-pressure steam produced in the boiler is directed to a steam turbine
Turbine operation	The steam turbine rotates as steam flows through it, converting thermal energy into mechanical energy
Electricity generation	The turbine drives a generator that converts mechanical energy into electrical energy
Heat recovery	Several systems incorporate heat recovery processes that further enhance efficiency by putting the residual heat to other use, such as district heating
Emission control	The air pollutants coming out of the combustion process are filtered or treated before release into the environment

critical, as we have already mentioned, in improving the efficiency of subsequent thermal conversion operations: combustion, pyrolysis, or gasification. Biomass in its raw form is constituted of particulates that are highly varied in size and shape, hence leading almost invariably to non-uniformity in thermal reactions and poor efficiency in energy conversion. Pulverization ensures that the size of particles is reduced, increasing the surface-to-volume area. This will allow much better heat transfer during thermal conversion, improving rates of reactions and making sure there is more uniformity in heat treatment. Pulverization is important because of its ability to enhance even heat distribution during thermal processes. There would be increased uniformity of temperature pickup and transfer with finer pulverization of biomass, hence ensuring that in any combustion, gasification, or pyrolysis process, the whole biomass sample reacts uniformly. It ensures that no single portion of the biomass gets overheated more than others underreacted, which leads to incomplete combustion and loss in proper energy conversion and combustion efficiency (Kobayashi et al. 2008). However, pulverization efficiency is influenced by several critical factors, such as the type of biomass being processed, its moisture content, and the specific equipment used for pulverization. Different biomass types require different levels of energy input in their effective pulverization. For example, soft biomass, like agricultural residues of straw or grass origin, would generally be easier to pulverize due to lower density and lignin content. In contrast, the more lignin and cellulose a biomass contains the energy consumption to break down the biomass into fines is considerably higher (Kobayashi et al. 2008). Biomass moisture content is another critical factor. The more moisture biomass has, the harder it is to be pulverized because the added water increases cohesion between particles and can lead to clogging or reduced mill efficiency. Since such would require pre-treatment methods like drying before pulverization, it can allow an increase in effectiveness due to a reduction of moisture levels (Guo et al. 2015). Biomass pulverization is done using various mills, which offer different advantages based on the peculiarities of the biomass in question (Tumuluru and Wright 2010). Taking this as an example, hammer mills have been in wide use for the pulverization of soft biomass materials. Biomass mills commonly employ a pulverization mechanism using fast-moving hammers that smash the biomass into smaller particles. In the process, hammer mills are effective at relatively fine particle sizes and are commonly adopted owing to their simplicity and inexpensiveness (Fig. 3.4). Ball mills utilize rotating cylinders that are loaded with steel or ceramic balls for grinding biomass into fine powders. These kinds of mills turn out to be fairly effective for harder materials like woody biomass, which requires higher mechanical forces to bring about size reduction. Other options include roller mills, which crush biomass between rotating rollers. They provide a controlled and uniform particle size, therefore suitable for applications calling for tight particle size distribution (Oyedeji et al. 2020).

Pulverization drastically affects thermal efficiency. For example, in combustion, pulverized biomass presents a much larger surface area to oxygen, leading to more complete and efficient combustion (Yin and Yan 2016). This acts not only to increase the energy yield but also to reduce the formation of pollutants commonly associated with incomplete combustion, such as carbon monoxide and particulate matter. The finely pulverized biomass allows for faster kinetics of the reaction to take place during biomass gasification, hence better production of a syngas-the mixture containing hydrogen and carbon monoxide with other gases used either directly as fuel or as feedstock for further chemical synthesis. Pyrolysis refers to thermal decomposition in an oxygen-free atmosphere. Pulverization can favorably influence this process

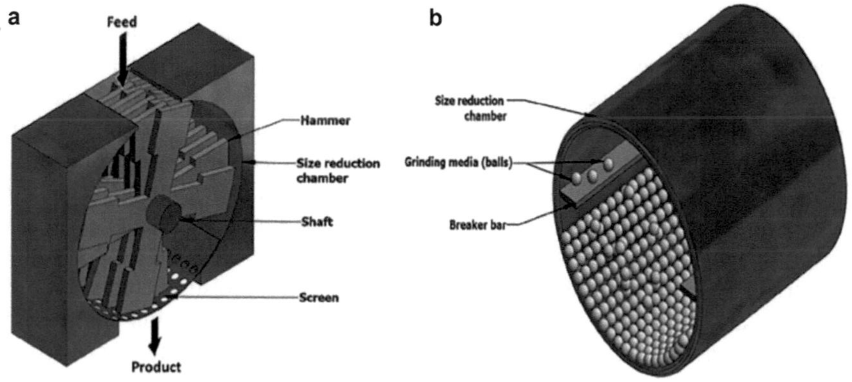

Fig. 3.4 Illustration of a representative components of **a** hammer mill and **b** ball mill (Oyedeji et al. 2020)

as well. The finer and more homogenous the particles, the more homogeneous the heat distribution within the biomass for higher yields of bio-oil and biochar, both of which are considered very important by-products of the pyrolysis process (Pandey et al. 2022). Although pulverization has its advantages, there are also several challenges with its implementation for industrial-scale biomass conversion. One of the major challenges is the energy intensity of the pulverization process. Reduction of biomass into fine powders, maybe those that contain substantial quantities of woody biomass or high moisture content materials, is quite energy intensive. This could increase the operational cost of the whole process and decrease the net energy gain from the biomass conversion process (Jiang et al. 2017). In this context, pulverization has been considered as an important process for both mechanical and thermal conversion of biomass because better combustions, pyrolysis, and gasification would be significantly enhanced by increasing the heat transfer rate, reaction rates, and efficiency of conversion. However, pulverization has several challenges because of its energy-intensive nature and also some handling difficulties of fine biomass powders. A further consequence of biomass being a developing renewable energy source is that pulverization technology will continue to innovate toward its maximum role within a sustainable energy system. Drying is a very important stage in the thermal and mechanical conversion of biomass and plays a significant role in the quality and efficiency of other thermal treatments of biomass, namely combustion, gasification, pyrolysis and liquefaction. Biomass is generally harvested or collected in its raw form, with a high moisture content (Pang and Mujumdar 2010). This means that excessive water removal is required to give the biomass a higher calorific value, lower weight and better storage capacity. The chapter also covers drying in thermal and mechanical biomass conversions, focusing on the importance of moisture reduction, the types of drying technologies, their operational mechanism and their application in sustainable biomass conversion. Biomass thermal conversion necessarily involves the process of drying, which directly affects the yield of energy, as well as the efficiency of the process. High moisture biomass results in energy loss because most of the input energy is used for the evaporation of water rather than useful thermal output. Through the reduction of moisture, biomass improves in calorific value, becoming more efficient for energy production and thermal process stabilization. It is also easier to handle, transport, and even store, due to the reduced weight of dried biomass and reduced vulnerability to microbial degradation that normally compromises quality during storage (Luk et al. 2013). This moisture reduction during combustion results in higher and more stable combustion temperatures, again increasing the quantity of energy produced, at the same time reducing harmful emissions. Tucho et al. have shown that eucalyptus chips dried to around 10% moisture burn more completely and produce fewer harmful emissions than those with a moisture content of 50% (Tucho et al. 2023). In addition, drying of biomass before gasification improves the quality of syngas, as excessive moisture dilutes the energy density of syngas and increases water vapor content, which eventually reduces thermal efficiency. In this regard, Makwana et al. (2023) have shown in a review of studies on rice husk gasification that pre-dried biomass produces a syngas with a higher hydrogen concentration, improving its suitability for power generation and fuel applications. Drying methods for biomass are generally classified into two classes, namely: natural drying method and mechanical drying method, which are selected according to the biomass type, desired moisture level, and operational requirements. Natural drying, or air drying, is one of the oldest and most energy-effective methods of drying. The principle behind this method consists of spreading biomass in open-air yards or simple shelters to let it dry naturally due to sunlight and ambient air. Mainly suitable for small-scale applications, as well as those where time is ample, natural drying is the most cost-effective one since it requires only a minimal amount of equipment and external energy input. It is particularly helpful in those regions that are hot and dry because the dry air and high temperatures promote the evaporation of moisture from biomass. Natural drying is appropriate for the kind of materials like agricultural residues which do not require rapid processing. The process however is so at the mercy of climatic parameters that the method is simply impossible to carry out in highly humid or rainfall regions (Labbé et al. 2018). Therefore, natural drying remains a good option for small-scale biomass drying in arid and warm regions. In as much as it heavily relies on weather conditions, and the drying time is generally quite long, it is still cost-effective and involves only the use of natural resources without the capital and operational costs as in the mechanical drying technologies (Susanto et al. 2024). The second type of drying is Mechanical drying methods become necessary for the rapid, uniform, and controlled reduction of moisture in biomass, particularly in an industrial setting. Unlike natural drying dependent on prevailing external conditions, the mechanical systems allow for uniform drying through the application of controlled heat and airflow. This makes them ideal for large-scale biomass processing. Control to such a level is important for operations where moisture stability in biomass is called for, for instance, pellet making, biofuel processing, and biochemical uses. Luk et al. have shown that the moisture content of biomass can be reduced by 5–10% through mechanical drying, optimizing the yield of energy from thermal conversion processes and reducing weight and volume for efficient transport and storage (Luk et al. 2013). The main choices for drying biomass are rotary dryers, flash dryers, and belt dryers. The rotary dryer, in particular, is one of the most commonly used mechanical drying technologies in biomass processing. Rotary dryers use cylindrical, rotating drums which create a continuous high-capacity drying environment. Rotation of the drum presents the biomass to a stream of heated air flowing either parallel or counter to feed direction, thus allowing fast and uniform removal of moisture. As such, the design of rotary dryers allows for the intake of huge volumes of biomass from the feedstock, including wood chips, bark, agricultural residues, and other forms of bulk feedstock, thereby making them indispensable in industries dealing with wood pellet manufacture, biofuel production, and agricultural processing (Fagernäs et al. 2010; Havlík and Dlouhý 2020). In the rotary dryer, heated air generated from either combustion chambers or heat exchangers passes through in either co-current or counter-current directions to the biomass. Co-current design, where the air and biomass are moving in the same direction, provides for rapid initial drying, particularly suited to biomass with higher initial moisture. A counter-current design in which air and biomass flow in opposite directions offers controlled drying rates, and focused capacity for specified levels of low moisture in biomass. Research by Del Giudice et al. (2019)

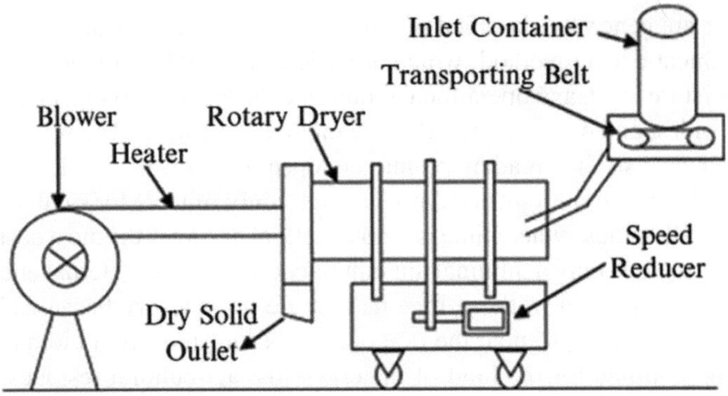

Fig. 3.5 Rotary dryer (John et al. 2024)

Indicates that biomass drying removes moisture from biomass, resulting in a significant reduction in the amount of energy normally required to produce energy from biomass. This moisture is removed in such a way that its calorific value increases with the use of a rotary dryer (Fig. 3.5). For example, wood chips, which initially had a high moisture content, were efficiently dried: 17% in the case of poplar, and up to 31% for vine and locust chips. Moisture reduction improves fuel quality and reduces transport costs. This rotary drying process is therefore particularly useful for small-scale biomass processing and waste heat recovery from power plants (Del Giudice et al. 2019).

The second type is the belt dryers, whereby biomass is conveyed through a drying chamber into which heated air is either blown over or channeled through biomass on a conveyor belt. Biomass, during passage along the belt, provides for gentle drying suited to low-density materials such as sawdust, straw, and other fine biomass in progressive steps. It results in minimal particle breakdown because of slow and controlled motion, hence the preservation of biomass quality in terms of structural integrity. This is of great importance in processing applications, such as biochar production, where retaining the physical structure of the feedstock improves finished product quality (Shirinbakhsh and Amidpour 2017). Of the most important advantages of belt dryers is the possibility of accurate temperature control, and minimizing thermal degradation, which may be common in other more aggressive drying systems. The moisture control in the belt dryers is done progressively to obtain uniform consistency of the end product (Friso 2020). This process is quite vital in agricultural waste materials and sawdust meant for bioenergy or bioproducts, in which consistent moisture is of the essence in the respective combustions and pelletization. The study of Vassilev et al., gave evidence that belt dryers are highly functional when dealing with biomass that has high initial moisture content. This is because the drying can achieve gradual material moisture reduction without exposing the biomass to excessive heat. In such a perspective, the drying of high-moisture sawdust with belt dryers has ensured better combustion efficiency towards energy application since it managed to avoid charring or scorching which normally reduces the calorific value of the biomass. The flexibility and gentle drying action of belt dryers make them especially advantageous for industrial biomass applications that demand high throughput without compromising material integrity. The ability to accurately control moisture content and retain biomass properties gave prominence to the technology of belt dryers in several sectors related to renewable energy and sustainable processing of biomass (Vassilev et al. 2012). Other forms of mechanical dryers for biomass processing include the flash dryer (Fig. 3.6). Flash drying rapidly exposes biomass particles to a stream of hot air, at high speed, evaporating the moisture almost instantly. This technique is quite suitable for small or finely ground biomass particles that require rapid drying, such as wood chips or agricultural residues (John et al. 2024).

There are other types of mechanical dryer, such as vacuum dryers, superheated steam dryers and so on. Each method has unique advantages and disadvantages, with the choice generally based on the specific characteristics of the biomass, such as moisture content, particle size, thermal sensitivity and intended application. To conclude this section, pulverisation and drying are the two main processes for preparing biomass for thermal and mechanical conversion. Reducing the particle size and

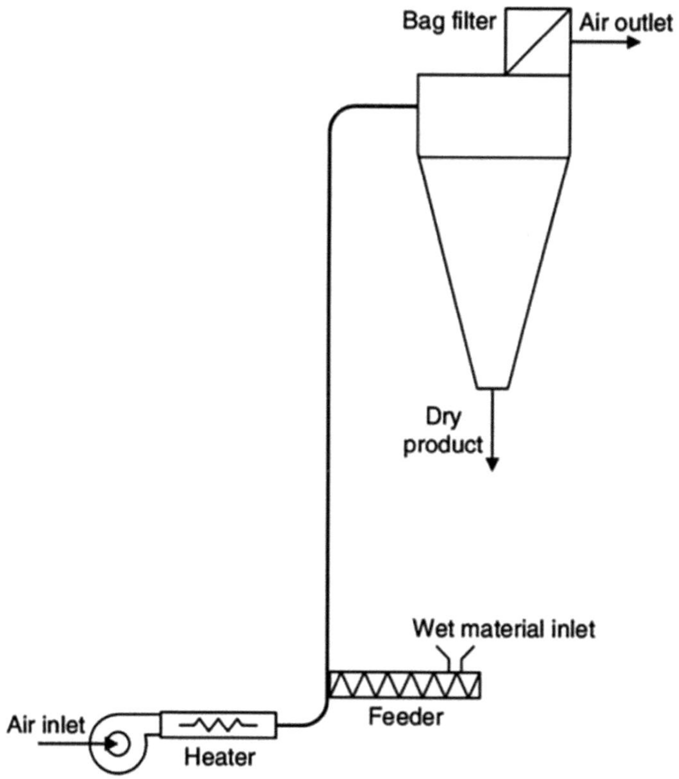

Fig. 3.6 Flash dryer (John et al. 2024)

moisture content of the material significantly improves the efficiency of biomass conversion technologies. On the other hand, the energy demand of these processes must be carefully managed to ensure the overall energy efficiency of biomass use. Future innovations in pulverisation and drying technologies are likely to be guided by a few key factors: minimising energy consumption per tonne of biomass processed, improving biomass handling properties and better integration into an industrial-scale biomass conversion system (Tshikovhi and Motaung 2023).

3.2.2 *Thermochemical Conversion*

Thermochemical conversion of biomass is a key process for transforming organic matter into energy and high-value chemical product. This process uses heat to break down biomass into simpler molecules through various thermochemical mechanisms such as pyrolysis, gasification, and hydrothermal liquefaction (Nguyen and Trinh 2022). Unlike biochemical processes, which rely on microorganisms or enzymes and are often limited by slow reaction rates or the need for expensive catalysts. In fact, thermochemical conversions occur at high temperatures. These high temperatures enable the rapid and complete decomposition of organic matter, particularly lignocellulosic biomass such as wood and agricultural residues, which are typically difficult to degrade through biological means.

Through this process, biomass is converted into three main types of products: gases, liquids, and solids. Regarding the gaseous products, synthesis gas (or syngas) is produced, which mainly consists of carbon monoxide (CO), hydrogen (H_2), and methane (CH_4). This gas can be used for electricity generation, as feedstock in chemical processes, or in the synthesis of liquid fuels (Khaskhachikh et al. 2021). As for the liquids, bio-oils produced from thermal decomposition can be refined and used as substitutes for fossil fuels in the energy and chemical sectors (Demirbas 2006). Finally, biochar is produced as a solid carbon-rich residue, which can be used as a soil amendment to improve water retention and fertility, or as an adsorbent to purify gases and liquids (Rao et al. 2016).

The advantages of thermochemical conversion include great flexibility in utilizing various types of biomasses and the ability to simultaneously produce gases, liquids, and solids from a single source. Furthermore, these processes allow for the maximization of by-product valorization, especially biochar, which can play a key role in carbon sequestration and soil enhancement.

3.2.2.1 Pyrolysis

Pyrolysis is a crucial thermochemical process for converting biomass into valuable products. It occurs in the absence of oxygen at temperatures typically ranging from 300 to 700 °C. This leads to the complete thermal decomposition of the polymers present in biomass, including cellulose, hemicellulose, and lignin (Papari and

Hawboldt 2015). One of the primary advantages of pyrolysis is its flexibility, allowing the production of a wide range of products used in the energy, chemical, and agricultural sectors, offering a promising alternative to fossil fuels. Moreover, pyrolysis is often used as the first step in more complex processes such as gasification and combustion (Babu 2008). As a result of this decomposition, pyrolysis produces three main types of products: biochar, bio-oil, and syngas. For Biochar, which is a carbon-rich solid residue, it is produced from the carbonization of biomass. It's predominantly produced during slow pyrolysis, which is characterized by relatively low temperatures (300–450 °C) and slow heating rates, promoting the formation of a solid residue rich in carbon (Shahrun et al. 2024).

Biochar plays an important role in various environmental, economic, and industrial aspects. It sustainably enhances soil properties by increasing its water retention capacity and improving its fertility, which boosts crop productivity and reduces the water irrigation and the need for chemical fertilizers. Additionally, it contributes to long-term carbon sequestration, thereby reducing carbon dioxide emissions and mitigating the effects of climate change (Wu et al. 2022). Furthermore, biochar is used in the purification of water and air from industrial pollutants due to its high surface area and porosity, which enable it to adsorb contaminants such as heavy metals, pesticides, and organic pollutants. This makes it a sustainable solution for purifying water sources (Ong et al. 2019). In addition, biochar has potential applications in renewable energy systems, where it can be used as a solid fuel due to its high energy content (Kan et al. 2016). Furthermore, biochar has been explored as a component in building materials such as concrete and asphalt, enhancing the durability and insulation properties of construction materials while contributing to a lower carbon footprint (Babu 2008).

Simultaneously, bio-oil, a liquid rich in organic compounds such as phenols, acids, ketones, and alcohols, is another key product of pyrolysis. Bio-oil has strong energy potential, but its high oxygen content makes it chemically unstable, often requiring additional refining to improve its properties (Negahdar et al. 2016). One of its emerging uses is in transportation fuels, where bio-oil can be upgraded to a renewable alternative to diesel and gasoline through hydrogenation or cracking processes (Wu et al. 2022). Moreover, bio-oil can be used as a feedstock for chemicals, where it serves as a precursor for the production of valuable bio-based chemicals such as resins and adhesives (Ong et al. 2019). Its utilization in heating systems is also gaining attention, as it can be combusted directly for the production of heat or electricity in industrial boilers.

Due to its complex composition, bio-oil is suitable for biorefineries, where it can be fractionated into different chemical components for diverse applications (Kan et al. 2016). Fast pyrolysis, carried out at high heating rates (up to 1000 °C/s), is the most favorable process for maximizing bio-oil yield, with production rates reaching up to 75% (Carrasco Díaz et al. 2024). In addition to biochar and bio-oil, pyrolysis generates syngas, which consists primarily of CO, carbon dioxide (CO_2), H_2, and CH_4. Syngas, also known as bio-gas, has a broad range of applications due to its versatility. It is often used for electricity generation via gas turbines or internal combustion engines. Additionally, syngas serves as a feedstock for producing synthetic fuels through the Fischer-Tropsch process, where it is converted into liquid hydrocarbons such as synthetic diesel and jet fuel. Moreover, syngas can be used in chemical synthesis, providing the base materials for the production of methanol, ammonia, and other industrial chemicals. Gas production is particularly favored at temperatures above 600 °C, where the complete thermal decomposition of volatile compounds increases gas yields (Zeng et al. 2017).

The product yield in pyrolysis depends on several key operational parameters. The first and most important factor is the temperature of the process. Higher temperatures (above 500 °C) favor gas and bio-oil production, while lower temperatures (300–400 °C) lead to increased biochar production (Wu et al. 2022). The heating rate also significantly influences the outcome: faster heating rates result in a rapid thermal breakdown, maximizing bio-oil yield, whereas slower pyrolysis, with lower heating rates, promotes the formation of biochar (Wang et al. 2020). Another critical factor is the residence time, which refers to the duration during which the biomass is maintained at high temperatures. Longer residence times allow for a more complete conversion of volatile compounds, resulting in higher gas and bio-oil production. Additionally, the chemical composition of the biomass used plays a decisive role in the distribution of products. For instance, biomass rich in cellulose and hemicellulose (such as agricultural residues) produces more bio-oil, while lignocellulosic biomass (such as wood) tends to produce more biochar due to the resistant structure of lignin, which decomposes more slowly.

Recent technological advancements have improved the efficiency of the pyrolysis process through the use of catalysts (Table 3.2). The addition of alkaline and alkaline earth metals (AAEMs) in catalytic pyrolysis systems not only increases gas and bio-oil yields but also reduces biochar formation (Ong et al. 2019). These catalysts also enhance the deoxygenation of bio-oil, improving its calorific value and chemical stability (Kabir and Hameed 2017). Furthermore, reactor design plays a crucial role in optimizing the pyrolysis process. Fluidized bed and fixed bed reactors are commonly used to maximize heat transfer and improve the uniform mixing of biomass particles. Fluidized bed reactors, in particular, provide precise temperature control and better homogeneity in the mixing process, resulting in improved product quality (Kan et al. 2016).

3.2.2.2 Gasification

Gasification is a complex thermochemical process that converts carbon-based organic materials, such as biomass, coal, and solid waste, into a gaseous mixture called syngas. This synthetic gas primarily consists of CO, H_2, CO_2, and CH_4, along with traces of other light hydrocarbons (Hardianto et al. 2015). Unlike combustion, which requires a significant amount of oxygen, gasification operates in an oxygen-limited environment, preventing complete combustion and promoting syngas formation rather than pure carbon dioxide (Kumar et al. 2009; Gomaa et al. 2020). Syngas can be used for electricity generation, synthetic fuel production through processes

Table 3.2 Effects of various catalysts on pyrolysis product

Catalyst type	Effect on bio-oil yield	Effect on syngas yield	Effect on biochar yield	Additional benefits	References
Bentonite	Increases	–	Enhances BET surface area	Improves heavy metal removal	Dai et al. (2017)
ZSM-5 zeolite	Enhances aromatics	–	–	Increases phenols and hydrocarbons	Wang et al. (2011, 2012)
NiCu/ biochar	Significantly increases	–	–	Enhances Fischer-Tropsch synthesis	Luo et al. (2022)
Fe-loaded biochar	–	Promotes hydrogen	–	Reduces tar formation	Zhaoyan et al. (2022)
Calcite	Decreased	Decreased	Increased	–	Merdun and Sezgin (2018)
Dolomite	Decreased	Decreased	Increased	–	Merdun and Sezgin (2018)
Zeolite	Decreased	Increased	Decreased	Enhanced tar yield, improved hydrocarbon content in bio-oil	Merdun and Sezgin (2018), Wijayanti et al. (2021)
Oil shale ash	Enhanced by 5%	Increased	Reduced	Improved oil quality, reduced activation energy, increased short-chain hydrocarbons	Huang et al. (2017)
HZSM-5	Decreased	Increased	Decreased	Enhanced H_2 and CO content in syngas	Saad et al. (2015)

such as Fischer-Tropsch synthesis, or as a feedstock for various chemical industries (Kumar et al. 2009).

The gasification process generally follows four key stages: drying, pyrolysis, partial oxidation, and reduction. These stages convert organic matter into gas at high temperatures, typically between 800 and 1200 °C, in a low-oxygen or steam environment (Molino et al. 2016). The first stage, drying, occurs at temperatures below 200 °C. Biomass can contain up to 30% moisture, and this step is essential to remove water. Without drying, the remaining moisture would reduce the efficiency of gasification by consuming heat for evaporation, thereby decreasing the overall energy yield of the process (Wu et al. 2022). Pyrolysis, which occurs at around 200–600 °C, breaks down the organic compounds into three main fractions: tars, char, and volatile gases (Saleh and Sudarmanta 2018). The remaining carbon-rich char is further transformed during the reduction stage (Ren et al. 2020). The oxidation stage takes place between 800 and 1200 °C, where a small amount of oxygen or air is introduced to trigger partial combustion. This combustion generates the heat necessary to drive the endothermic stages of pyrolysis and reduction (Kumar et al. 2009). In the reduction stage, the gases produced during pyrolysis and oxidation react with the remaining char to produce CO and hydrogen H_2. This involves reactions such as the Boudouard reaction and steam reforming, which generate a high-energy fuel gas (Gomaa et al. 2020).

The products of gasification include several elements, each with specific applications. The main product is syngas, primarily composed of carbon monoxide CO and hydrogen H_2, with minor amounts of CH_4 and carbon dioxide CO_2. Syngas can be used as fuel for electricity generation, to power engines or turbines, or be converted into chemicals such as methanol or ammonia (De Tissera et al. 2019). Char, a solid byproduct, consists mainly of incombustible carbon and ash, and it can be used as a biofuel or soil amendment to enhance soil fertility (Ren et al. 2020). Tars are heavy hydrocarbons produced during pyrolysis. Although they can foul equipment, solutions are being explored to break them down into simpler hydrocarbons or use them in other chemical processes (Wu et al. 2022). Ash, composed of inorganic residues, can be utilized in industrial or agricultural applications depending on its chemical composition.

One of the main advantages of gasification is its ability to convert a wide range of feedstocks, including biomass waste and municipal solid waste, into useful energy, thereby reducing greenhouse gas emissions (Wu et al. 2022). Unlike combustion, which requires large amounts of oxygen and primarily produces CO_2, gasification generates a mixture of combustible gases, such as CO, H_2, and CH_4, which can be directly used in gas turbines or internal combustion engines or converted into high-value chemicals such as methanol and liquid fuels (Laurence and Ashenafi 2012). Additionally, gasification offers better energy efficiency compared to the direct combustion of biomass. It allows for electricity generation from gas turbines or the use of syngas in combined cycles, increasing the overall efficiency of plants (Breault 2010). For instance, by combining a gas cycle with a steam cycle, gasification plants can achieve significantly higher thermal efficiencies than conventional coal-fired power plants (Ren et al. 2020). Moreover, gasification holds great potential for regions with limited (Molino et al. 2016). Another significant advantage is the reduction of pollutant emissions, such as sulfur oxides (SOx) and nitrogen oxides (NOx). By gasifying fuels like biomass, it is possible to capture and purify harmful emissions before they are released into the atmosphere, making gasification a more environmentally friendly solution than traditional combustion.

Despite its many advantages, gasification presents technical challenges. One of the main obstacles to large-scale implementation is the formation of tars, which are heavy hydrocarbons produced during the pyrolysis of biomass. These tars can clog equipment, leading to high maintenance costs and reduced efficiency (Palma 2013). Reducing tar production is an active area of research, and the use of catalysts in the gasification process has proven promising for enhancing tar cracking and improving syngas quality (Ren et al. 2020). Catalysts, such as alkali metals or metal oxides, help accelerate the chemical reactions needed to fully convert heavy hydrocarbons into lighter gases. Additionally, they lower the required temperature for these reactions, thereby improving overall energy efficiency (Wu et al. 2022). However, managing the deactivation of catalysts and their regeneration presents another significant technical challenge, requiring ongoing research to make the process more robust and economical (Ren et al. 2020). The moisture content of biomass is another major challenge. Biomass that is too wet decreases process efficiency, as a large amount of energy is consumed simply to evaporate the water present in the material. This often requires pre-treating the biomass by drying it, adding an additional step to the process and increasing costs (Breault 2010).

The efficiency of gasification depends on many factors, including the nature of the feedstock, operating conditions, and the type of reactor used. Better understanding these parameters is essential to optimize the production of high-quality syngas and minimize unwanted byproducts. The composition and nature of the biomass or carbonaceous material used directly impact the efficiency of gasification. For example, the carbon content, moisture, ash, and volatile matter in the materials affect not only the syngas yield but also the formation of byproducts like tars (Pereira et al. 2012). Biomass with high moisture content requires more energy for water evaporation, reducing the overall process efficiency. Temperature is a critical factor that influences both the composition of syngas and the overall process efficiency. Higher temperatures favor the conversion of carbon into gases like CO and H_2 through reactions such as the Boudouard reaction ($C + CO_2 \leftrightarrow 2CO$) and steam reforming ($C + H_2O \leftrightarrow CO + H_2$) (Kumar et al. 2009). However, excessively high temperatures can degrade reactors and increase energy costs (Gomaa et al. 2020). Typically, an optimal temperature range of 800–1200 °C is suitable for most gasification processes, although this can vary depending on the type of biomass and reactor used (Breault 2010). The ratio between the gasifying agent (air, oxygen, steam, or carbon dioxide) and the feedstock is another critical factor affecting the process efficiency. Excess oxygen results in complete combustion, reducing syngas production and increasing the formation of CO_2, an undesirable byproduct (Wu et al. 2022). Conversely, insufficient oxygen or steam may not provide enough heat to maintain endothermic reactions, leading to incomplete carbon conversion.

The type of reactor used in the gasification process also has a significant impact on efficiency. Fixed-bed reactors are simple and inexpensive to design, but they tend to produce more tars and offer lower energy yields compared to fluidized-bed reactors or advanced gasification systems (Jaiswal et al. 2020). Fluidized-bed reactors ensure better mixing of gasifying agents and biomass, promoting more complete carbon conversion and better temperature control within the reactor. In addition, solar gasification technologies are being developed to use concentrated solar energy as a heat source, reducing the reliance on fossil fuels and increasing overall energy efficiency (Boujjat et al. n.d.).

3.2.2.3 Liquefaction

Biomass liquefaction is a thermochemical process designed to convert organic materials such as agricultural residues, algae, and lignocellulosic waste into liquid biofuels, primarily in the form of bio-oil (Jena and Das 2011). This process is especially effective for wet biomass, as it eliminates the need for pre-drying, unlike other technologies such as pyrolysis. Hydrothermal liquefaction (HTL) takes place in an aqueous environment at temperatures between 250 and 400 °C and pressures ranging from 5 to 25 MPa (Jatoi et al. 2022). Under these conditions, water serves as both a solvent and a reactant, aiding in the breakdown of biomass into bio-oil (Chan et al. 2018). This method is particularly advantageous for wet biomass, such as algae and sewage sludge, because it does not require pre-drying, thus reducing energy costs. For example, a bio-oil yield of 43% was achieved from microalgae under optimized conditions at 300 °C and 9 MPa (Dandamudi et al. 2019). On the other hand, direct liquefaction (LD) uses organic solvents like methanol, ethanol, or acetone at temperatures between 250 and 350 °C (Li et al. 2008a). The solvents aid in breaking down the complex lignocellulosic structures, thereby converting the biomass into bio-oil. Adding metallic catalysts, such as nickel (Ni) or copper (Cu), further improves bio-oil quality by enhancing deoxygenation reactions. For instance, liquefying lignin with a nickel catalyst yielded a bio-oil output of 29.69% at 350 °C (Aysu and Durak 2015).

The liquefaction process relies on thermochemical mechanisms that break down biomass into bio-oil and other products (Fig. 3.7). In hydrothermal liquefaction, the process begins with hydrolysis, where water at high temperature and pressure breaks the chemical bonds in polymers like cellulose and lignin (Chacon-Parra et al. 2021). The resulting intermediates, such as sugars and phenols, are further converted into bio-oil through various chemical reactions. The presence of supercritical water in HTL accelerates these reactions while allowing better solubilization of intermediate products (Jensen et al. 2017). In direct liquefaction, the primary mechanism is solvolysis, where organic solvents assist in breaking down the biomass polymers, increasing bio-oil yields (Li et al. 2008b). The addition of catalysts helps to accelerate dehydration and decarboxylation reactions, removing water and CO_2 from the intermediates, thereby improving the energy quality of the bio-oil (De Filippis et al. 2016). For example, in an algal liquefaction study, the oxygen content of the bio-oil was reduced to about 10% following catalytic treatment, improving its stability (Nkosi et al. 2021).

The main product of liquefaction is bio-oil, a liquid biofuel with a high energy density, typically ranging between 30 and 36 MJ/kg (Chen et al. 2019). This bio-oil is a renewable fuel that can be used to produce electricity, power engines or turbines, or be converted into chemicals such as methanol or ammonia. In addition to bio-oil, liquefaction produces gases such as CH_4, CO_2, and H_2. Although their proportion is generally lower than that of bio-oil, these gases can be valorized to produce energy or used in industrial processes (Nkosi et al. 2021).

Soil gradation, significantly influences the undrained shear strength and liquefaction resistance of soils, with larger grain sizes generally providing higher stability (Hazout et al. 2017). Relative density and confining pressure are also essential, as higher values enhance the cyclic stress behavior and reduce liquefaction potential in granular materials. For example, an increase in confining pressure from 50 to 100 kPa decreases the cyclic stress ratio (CSR) from 0.30 to 0.12, indicating greater resistance

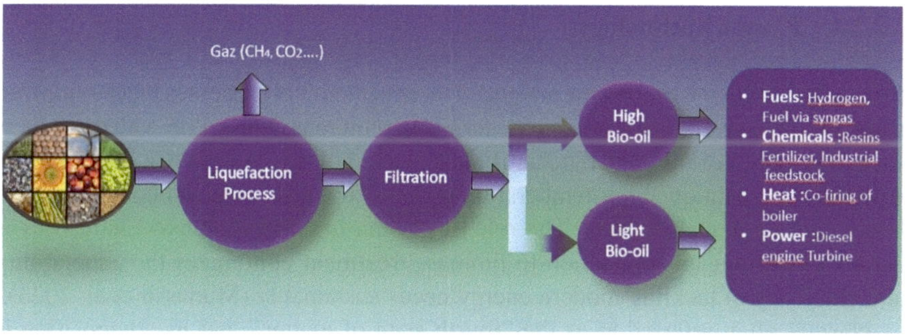

Fig. 3.7 Biomass liquefaction process: bio-oil production and its applications

at higher pressures (Hoque et al. 2017). Additionally, a higher groundwater table and deeper sand layers negatively correlate with liquefaction risk, as these conditions reduce the likelihood of pore water pressure buildup during seismic events. Seismic parameters such as magnitude, epicentral distance, and earthquake duration have a direct impact on liquefaction, with stronger and longer events increasing susceptibility (Papathanassiou et al. 2005). The fines content in sandy soils influences both relative density and the uniformity coefficient, which in turn affects liquefaction susceptibility. Soils with higher fines content are generally more prone to liquefaction under seismic loads. The combined influence of fines content and the uniformity coefficient is significant in determining liquefaction resistance, as both parameters must be considered together to accurately predict soil behavior under seismic conditions (Monkul et al. 2021).

Liquefaction offers several advantages over other thermochemical processes such as pyrolysis or gasification. One of the primary benefits is its ability to process wet biomass without the need for pre-drying, which significantly reduces energy costs. Furthermore, liquefaction provides flexibility in the types of biomass that can be processed, including diverse feedstocks such as algae, agricultural residues, and forest waste (Dandamudi et al. 2019). Additionally, the process can be optimized to produce high-quality bio-oils that require minimal post-processing treatment compared to other technologies. Despite its advantages, liquefaction presents some technical challenges. The quality of the bio-oil produced, particularly in terms of its high oxygen content and acidic compounds, often requires additional treatments, such as hydrodeoxygenation, to make it suitable as a high-quality fuel (Baloch et al. 2018). Moreover, the use of organic solvents in direct liquefaction increases operational costs and necessitates efficient recycling systems to recover and reuse these solvents. Additionally, managing the catalysts, which can deactivate due to the accumulation of ash and carbon deposits, poses further challenges, especially in terms of catalyst regeneration (Aysu and Durak 2015).

3.2.2.4 Plasma Technology

Plasma technology is among the advanced thermochemical processes that use ionized gases for the conversion of organic materials like biomass into value-added products such as syngas, bio-oils, and biochar is plasma technology, as shown in Fig. 3.8. This will also offer a unique high-temperature process supporting cleaner conversion and lesser environmental impacts. The role of plasma in energy-intensive industries is hereby emerging; its application to biomass treatment epitomizes the adaptability and efficiency of tackling modern energy needs sustainably (Mudassir et al. 2023).

Plasma, often referred to as the "fourth state of matter," and its composition is very different from that of solids, liquids, and gases. Plasma as a new phase of matter would logically appear as the continuation of other states, obtained by successive heat addition. Whereas solids, liquids, and gases are ordinary phases produced by phase transitions (Burm 2012), as shown in Fig. 3.9.

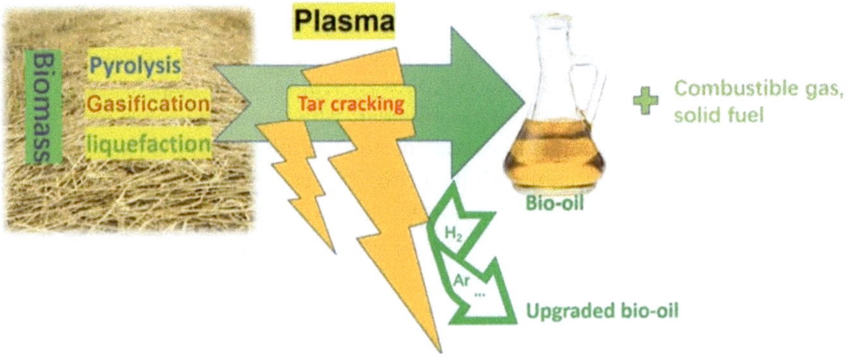

Fig. 3.8 Application of plasma in the thermal conversion of biomass (Shao et al. 2022)

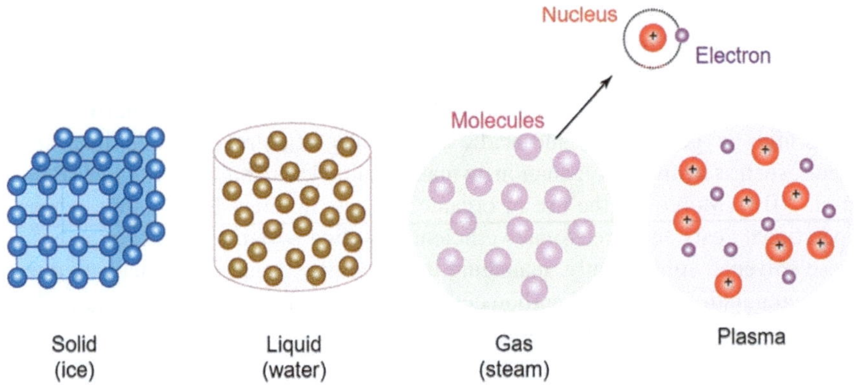

Fig. 3.9 Representation of four states of matter (Wu and Şahin 2018)

Plasma is a high-energy mixture of electrons, ions, neutral particles, and excited species. In biomass conversion, this mixture will break down complex organic compounds with a high degree of efficiency at temperatures higher than in conventional thermal approaches (de Oliveira et al. 2022). Plasma is produced by an electrical discharge in gases, which ionizes the gas and produces a highly reactive medium capable of sustaining very high temperatures (10^4 K or more) depends on the degree of ionization, as shwon in Fig. 3.10. Such a temperature value is considered important in biomass applications since it is able to decompose lignocellulosic structures rapidly into simpler molecules with extremely minimal residue. The high energy and reactivity of the plasma result in a cleaner conversion process that significantly reduces unwanted by-product formation, tar being among them. Muvhiiwa et al. demonstrated that a temperature of 1000 °C optimises biomass conversion with 82% efficiency and reduces tar formation. Ranges of 800–900 °C are also recommended for efficient conversion (Muvhiiwa et al. 2018). Plasma reactors, which are essential for directing this high-energy environment towards specific biomass conversion processes, fall into two main categories: thermal plasma systems and non-thermal plasma systems. Thermal plasma reactors are designed to achieve high temperatures through thermal equilibrium, where the plasma maintains a uniform temperature for electrons, ions and neutral molecules. Thermal plasma is produced at high pressure (>10 kPa) using direct current (DC) or alternating current (AC) sources, radio frequencies (RF) or microwaves, at temperatures between 2,000 and 20,000 K (Schriver et al. 1998). This breaks down the most stable chemical bonds in biomass, facilitating the almost complete conversion of organic matter into simpler compounds such as synthesis gas (a mixture of hydrogen and carbon monoxide) and bio-oil. This intense heat not only speeds up reaction rates, but also minimises the production of undesirable by-products such as tar, a persistent problem in conventional gasification processes. One of the main advantages of thermal plasma gasification is that it produces a high-purity, hydrogen-rich synthesis gas. The absence of oxygen in the plasma reactors minimizes the formation of carbon dioxide, while the reactivity product is in the direction of carbon monoxide and hydrogen. Impurities such as tars, sulfur, and nitrogen oxides can contaminate the syngas during conventional gasification, which complicates its further use in fuel applications. The impurities in the feedstocks undergo dissociation at very high temperatures in the thermal plasma process and produce clean gas, which can be useable for fuel cell applications or in the production of synthetic fuel (Ma et al. 2020).

The process uses steam as the main gasification agent, together with the gas forming the plasma and the heat carrier. The results show that the composition of the gas produced is mainly hydrogen (43.86% by volume) and carbon monoxide (30.93% by volume), with a H_2/CO ratio of 1.42 and a net calorific value (NCV) of 10.23 MJ/Nm^3. The process achieved a carbon conversion efficiency of 100% and an energy conversion efficiency of 48.83%, requiring 1.78 kWh of electrical energy to produce 1 kg of syngas (Tamošiūnas et al. 2023). In addition to syngas, thermal plasma reactors are also very effective in generating bio-oils and biochar by pyrolysis. Pyrolysis in a thermal plasma environment decomposes biomass in the absence of oxygen which may produce liquid bio-oil suitable for conversion into

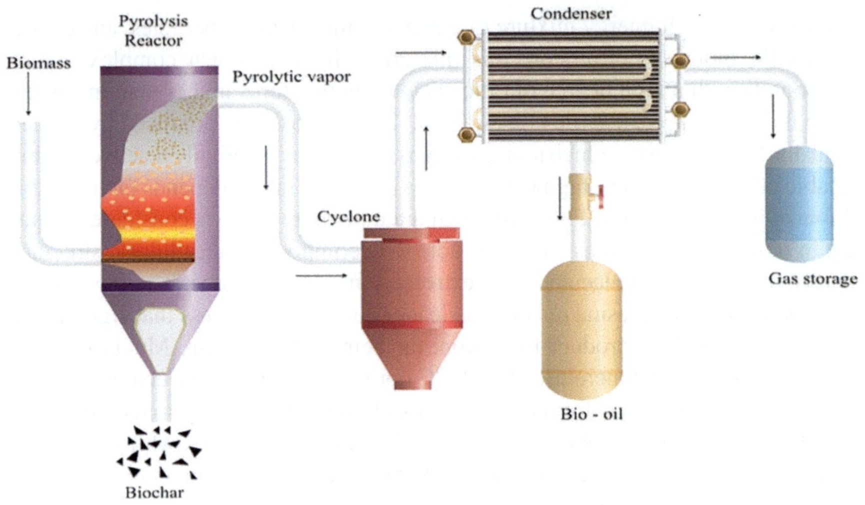

Fig. 3.10 Schematic diagram of biomass pyrolysis to biochar and bio-oil (Yogalakshmi et al. 2022)

synthetic fuels and chemicals. It has been proved that this process produces bio-oils with high yields and low contaminants since the high temperatures suppress char formation and secondary reactions that degrade oil quality (Panwar 2024). For example, a study by Yogalakshmi et al. (2022) investigated plasma thermal pyrolysis of lignocellulosic biomass as another promising method in which renewable plant biomass is converted into highly valued products, namely bio-oil, biochar, and pyrolysis gases. Influenced by temperature, residence time, among other issues, these processes enhance the production of energy-rich bio-oils, yet all these oils require further treatment for the complete removal of water and oxygen. The resulting bio-oils represent a complex mixture of carbonyls, furans, and phenols providing high energetic content but needing further adjustment to make it feasible as fuel sources.

In conclusion, thermal plasma technology is under evolution, while the developing applications related to biomass conversion in clean energy production and waste management show its competence to reach sustainability. Further, as this technology continues to mature and becomes more economically viable, thermal plasma reactors could become one of the cornerstones for sustainable bioenergy infrastructure that will minimize waste and provide cleaner energy sources in diverse industrial contexts (Kalak 2023).

The second class of plasma reactors includes non-thermal plasma systems. Non thermal plasma (NTP), sometimes also called cold plasma, is usually generated under conditions in which the electrons can acquire high kinetic energies while the bulk gas temperature remains comparatively low, often below 100 °C (Palma et al. 2020). The property allows non-thermal plasma to drive only the selective chemical reactions without heating the whole biomass sample. In contrast to thermal plasma, where all the components are exposed to high temperatures and are fully decomposed, the selective energy distribution in NTP will allow only for specific changes in biomass, such as depolymerization of lignin or extraction of particular compounds. This thus constitutes an ideal treatment when partial protection of one biomass component is required, especially in enhancing efficiency in downstream processing (Pipliya et al. 2023). One of the promising applications of NTP is selective degradation of lignin, a complex lignocellulosic biomass polymer. It forms strong bonds with cellulose and hemicellulose. Due to the linear, rigid structure, this resistant material against degradation restricts access to cellulose during biofuel production most of the time, reducing process efficiency. An experiment carried out by Mei et al. (2020) employed NTP to cause selective depolymerization of lignin in lignocellulosic biomass to realize the value of some valuable chemicals and biofuels. Grounds for the experiments using three biomass types-sawdust, corncobs, and rice straw-have been made in a plasma-assisted liquefaction (PEL) process. The results indicate that the energy yield is largely controlled by the biomass chemical composition, more precisely, the lignin and ash content. For instance, sawdust with its high lignin content and low ash content gave a very high yield for liquid products. Besides yielding the same products, PEL is unique in several ways when compared to traditional liquefaction techniques. It achieves high yields in a matter of minutes at relatively low temperatures and does not need supplementary heat input. Acidic, alcoholic, and phenolic compounds have been identified by chemical analysis in the liquid products, testifying to effective lignin destruction. In general, based on the results of this work, PEL has emerged as a promising, energy-efficient technology that provides fast and effective conversion of different types of biomass into renewable fuels and chemicals (Mei et al. 2020). The main advantages of NTP in biomass treatment include selective reactivity, very minimal consumed energy, and the ability to preserve some biomass structures while enhancing the destruction of others. This specificity is useful in applications that want the integrity of cellulose, hemicellulose, or other valuable components preserved, as seen in biofuel and chemical extraction. Moreover, since NTP normally operates at lower temperatures, energy input for most thermal methods is lower, which would increase the economic feasibility of the biomass treatment process. However, several problems need to be solved for NTP systems before scaling up for industrial applications. Advanced reactor designs are required to maintain plasma stability at large-scale levels and to provide uniform treatment for each batch of biomass. Second, the equipment for generating plasma, such as DBD reactors, can be very costly, although continuous research is being conducted to reduce such costs and render the technology more scalable (Shao et al. 2022).

3.2.3 Biochemical Conversion

Biochemical conversion of biomass plays a crucial role in the transition toward a circular and sustainable economy. This process relies on the use of microorganisms and enzymes to transform biomass into high-value products such as biofuels, biogas, and biochemicals (De Paula et al. 2019). Unlike thermochemical conversion,

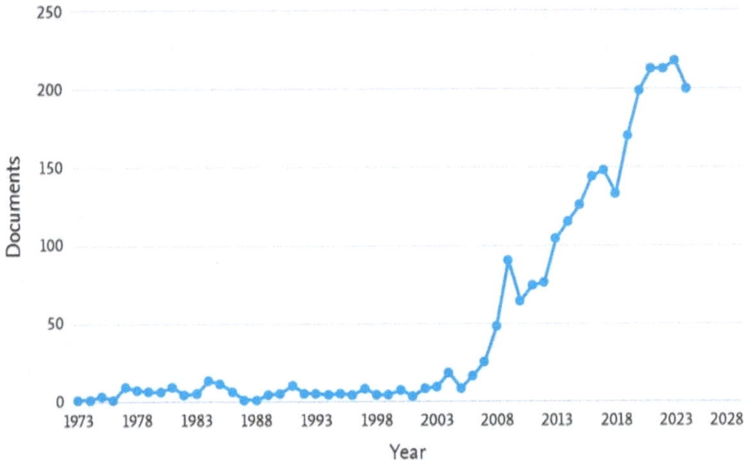

Fig. 3.11 Trends in scientific publications on biochemical conversion of biomass (1973–2023)

biochemical conversion is more environmentally friendly, requiring less energy and producing fewer greenhouse gas (GHG) emissions. Biomass, particularly agricultural residues, lignocellulosic material, and organic waste, represents an abundant and underutilized renewable resource (Osman et al. 2023). The world production of biomass is estimated at 146 billion metric tons annually, primarily from wild plant growth a significant portion of which can be used for energy production (Balat 2006). The Fig. 3.11 shows an exponential increase in scientific publications on biochemical conversion of biomass since 2005. By 2024, more than 230 scientific papers were published on this topic, reflecting the growing importance of this technology in the renewable energy sector. This growth can be attributed to several factors, including increasing concerns about climate change, the depletion of fossil fuel resources, and rising investments in biotechnology research.

3.2.3.1 Fermentation

Fermentation is a crucial biological process for converting organic matter into biofuels and high-value chemicals. This process relies on several methods, such Dark fermentation wich is a process where microorganisms convert sugars into hydrogen and organic acids, such as acetic and butyric acids, without the presence of light (Bastidas-Oyanedel et al. 2015). This process is driven by a complex community of microorganisms, including bacteria like *Clostridium* and *Enterobacterium*, which are capable of breaking down various organic substrates, including industrial wastes and lignocellulosic biomass (Dzulkarnain et al. 2022). By optimizing key factors such as pH, temperature, substrate concentration, hydraulic retention time, inoculum selection, and hydrogen removal, researchers improved the efficiency of dark fermentation, achieving a hydrogen yield of 2.15 mol per mol of glucose consumed, significantly higher than non-optimized conditions (Kucharska et al. 2018). On the other hand, fermentation of lignocellulosic residues often requires pretreatment to break down the complex structure of cellulose and lignin before microorganisms can effectively ferment the sugars (Fu et al. 2024). Combining pretreatment methods, such as steam explosion followed by extrusion, has shown significant efficacy. A study revealed that this combination resulted in 84% glucan recovery and 91% hemicellulose recovery, increasing biofuel yields by 1.4 times compared to methods that use only steam explosion (Oliva et al. 2017). Solid-state fermentation is particularly effective for producing enzymes and biofuels from solid biomass, such as lignocellulosic residues. This method relies on fungi that produce enzymatic cocktails capable of breaking down the biomass and releasing fermentable sugars (Taherzadeh-Ghahfarokhi et al. 2019). For example, the use of fungi such as *Trichoderma reesei* and *Aspergillus niger* has been shown to maximize ethanol yield. In one study, a yield of 84% of the theoretical maximum was achieved from sugarcane bagasse, demonstrating the effectiveness of this approach (Pirota et al. 2014). This process also simplifies production by allowing the use of the entire fermentation medium, thus reducing costs related to purchasing commercial enzymes. A more innovative method is syngas fermentation, which utilizes gases produced from biomass gasification, such as CO and H_2, to produce ethanol and other high-value chemicals (Fig. 3.12). This technology enables the use of industrial residues or organic waste as raw materials (Wang et al. 2015). A study demonstrated that syngas fermentation can produce 480 g of ethanol per cubic meter of treated gas, with energy efficiency reaching 70%, though improvements are still needed to stabilize cultures and increase yields (Phillips et al. 2017).

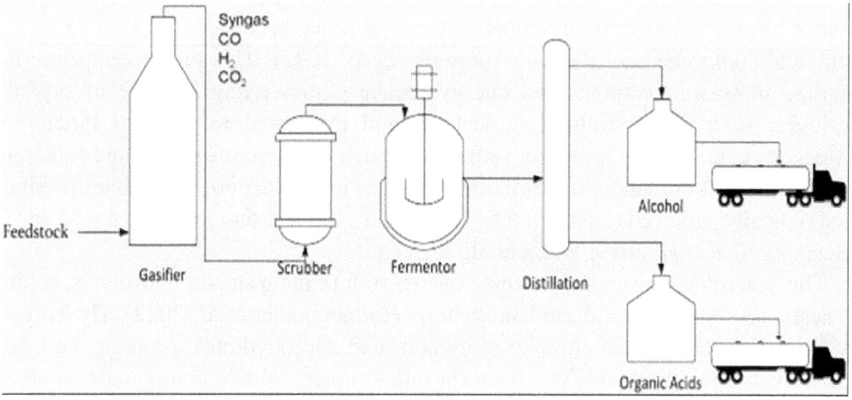

Fig. 3.12 Flowchart of syngas fermentation, showing conversion of feedstock to biofuels and chemicals (Phillips et al. 2017)

The products generated by fermentation processes offer a wide range of applications across various industries. Bioethanol, primarily used in the transportation sector, is a renewable fuel that can be blended with gasoline to reduce greenhouse gas emissions. It is also used as an industrial solvent in pharmaceuticals and cosmetics (Ferreira et al. 2019). Organic acids, such as lactic acid and acetic acid, are key fermentation products used in the food and pharmaceutical industries and in the production of biodegradable plastics, such as PLA (polylactide), a material increasingly favored for eco-friendly packaging (Chidi et al. 2018). Lastly, enzymes, produced through solid-state fermentation, play a crucial role in the paper, food, and biofuel industries, facilitating the breakdown of biomass into fermentable sugars (Pirota et al. 2014).

The environmental benefits of fermentation processes are significant. For example, producing bioethanol from lignocellulosic biomass can reduce CO_2 emissions by 60–90% compared to fossil fuels, depending on the type of biomass used and the treatment conditions (Osman et al. 2021). Moreover, using agricultural, industrial, and municipal waste as raw materials allows for better waste management while providing a renewable energy source. However, several challenges remain. The high costs of pretreatments, particularly steam explosion, remain a major obstacle to the economic viability of large-scale fermentation processes (Du et al. 2022). While steam explosion is effective at breaking down lignocelluloses, it requires significant energy investment, thus increasing the overall process costs. Additionally, the production of fermentation inhibitors, such as phenolic compounds resulting from lignin degradation, can reduce the efficiency of fermenting microorganisms (Du et al. 2022).

3.2.3.2 Anaerobic Conversion

Anaerobic conversion, a biochemical process involving the breakdown of organic matter by microorganisms in oxygen-free conditions, produces CH_4 and CO_2, the principal components of biogas (Andreides et al. 2022). This process is primarily applied in waste treatment and energy recovery, converting a range of organic residues, such as agricultural, industrial, and municipal waste, into bioenergy (Meegoda et al. 2018). Optimizing conditions such as pH, temperature, and substrate concentration is crucial for maximizing methane yield, with optimal methane production typically achieved at a pH of 6.5–7.5 and a mesophilic temperature around 35 °C (Scamardellaa et al. 2019; Wang et al. 2024).

The anaerobic conversion process consists of four main stages: hydrolysis, acidogenesis, acetogenesis, and methanogenesis (Induchoodan et al. 2022). Hydrolysis is the initial step, where complex polymers like carbohydrates, proteins, and fats are broken down by hydrolytic bacteria into simpler, soluble compounds, such as sugars, amino acids, and fatty acids (Zhang et al. 2016). Hydrolysis is often the rate-limiting step in anaerobic digestion, especially for lignocellulosic biomass due to lignin's resistance to microbial attack. Studies report that acidification pretreatment can improve hydrolysis efficiency by up to 40%, with optimal conditions typically at temperatures of 30–50 °C and pH levels of 5–7 (Sicchieri et al. 2025). Prehydrolysis at higher temperatures (around 35 °C) further reduces the lag phase and boosts efficiency, with ideal hydrolysis occurring between 25 and 30 °C, especially for substrates like cellulose (Zhang et al. 2016). Following hydrolysis, acidogenesis occurs, where acidogenic bacteria convert the breakdown products into volatile fatty acids (VFAs), such as acetic, propionic, and butyric acids, as well as hydrogen and CO_2 (Liu et al. 2012). Acidogenesis efficiency can vary significantly based on substrate composition, with lipid-rich substrates producing higher VFAs. However, excessive VFA accumulation can decrease pH and inhibit microbial growth, necessitating careful process monitoring to maintain pH within optimal ranges (Yu et al. 2016). Acidification of food waste, for example, has shown to increase acidogenesis efficiency by 21–239%, depending on microbial strain and substrate quality (Grana et al. 2024). Acetogenesis then takes place, where acetogenic bacteria convert VFAs into acetic acid, hydrogen, and additional CO_2. This stage is particularly sensitive to VFA and hydrogen levels, as high concentrations can inhibit bacterial function. Studies indicate that hydrogen pressure should remain below 10^{-4} atm to support effective acetogenesis and maintain process stabilit (Xu et al. 2024). In lipid-rich waste, acetogenesis involves β-oxidation, where long-chain fatty acids are broken down into acetate, essential for the subsequent methanogenesis phase (Rodríguez-Méndez et al. 2017). The final stage, methanogenesis, is carried out by methanogenic archaea that utilize acetic acid, hydrogen, and CO_2 to produce methane and water (Zhao et al. 2021). Methanogenesis is highly pH-sensitive, requiring stable conditions above pH 6.5, and is temperature-dependent, with optimal performance observed around 35–40 °C in mesophilic conditions. This stage can achieve methane yields up to 60% of the theoretical potential, with process retention times between 5 and 16 days in batch reactors (Domingues et al. 2024).

Anaerobic digestion produces biogas as a primary product, which is widely used in heating, electricity generation, and as a renewable vehicle fuel (Noori et al. 2024). The secondary product, digestate, is nutrient-rich and can be applied as an organic fertilizer, enhancing soil fertility and promoting sustainable agriculture by reducing the need for synthetic fertilizers. Studies indicate that solid-phase digestate is particularly beneficial for long-term soil structure, while liquid-phase digestate provides readily available nutrients, supporting varied agricultural applications (Knoop et al. 2017).

Several factors influence anaerobic digestion efficiency, including pH, temperature, substrate concentration, and inoculum selection. Pretreatment methods like acidification, mechanical disintegration, and thermal hydrolysis have been shown to enhance substrate accessibility, significantly increasing methane yields by 21–239%, particularly in lignocellulosic biomasses that are otherwise difficult to degrade (Mirmohamadsadeghi et al. 2021). Computational models, including Artificial Neural Networks (ANNs), have also successfully optimized variable selection, with lignin content and particle size shown to be key parameters in biogas yield prediction and digestate quality assessment (Xu et al. 2024). Anaerobic conversion is a process with significant environmental and economic benefits, as it helps reduce methane emissions, a potent greenhouse gas, by capturing it from organic waste decomposition and using it as biofuel, thereby aiding in combating climate change. The process also produces a byproduct known as digestate, which enhances soil health and reduces dependency on chemical fertilizers, thus decreasing water pollution caused by eutrophication. Additionally, digestate application as a fertilizer improves soil health, reduces dependency on chemical fertilizers, and mitigates eutrophication in water systems. However, challenges remain, particularly in handling the cost of pretreatment methods for lignocellulosic materials and managing phenolic inhibitors produced from lignin breakdown (Koszel et al. 2016). Detoxification processes, such as enzymatic or microbial treatments, can alleviate these inhibitory effects, though they add complexity and cost to the overall (Mazzanti et al. 2025).

3.3 Conclusion and Future Perspectives

Biomass conversion represents a critical pathway for developing renewable energy sources and high-value products within a circular economy. Thermal and mechanical preparation technologies, such as milling and drying, streamline initial processing steps, thereby optimizing the quality and reactivity of biomass. In particular, thermochemical processes like pyrolysis, gasification, liquefaction, and plasma treatment allow for diversified and targeted conversions of biomass. Pyrolysis produces biochar, bio-oil, and gas, contributing to energy production while offering a means of sequestering carbon in solid form within soils. Gasification yields a flexible syngas used for heat, electricity, and as a basis for synthetic fuels. Liquefaction, on the other hand, transforms biomass into liquid fuels, often integrated into bioenergy systems, while plasma, though an emerging technology, shows potential for rapid and highly energetic conversions of organic waste.

On the biochemical side, fermentation and anaerobic digestion provide viable alternatives for transforming organic waste into biofuels. Fermentation enables the valorization of organic substrates for the production of ethanol and hydrogen, while capturing and reducing greenhouse gas emissions. Anaerobic digestion, widely applied in agricultural and municipal waste treatment, generates biogas for energy applications and produces digestate, a nutrient-rich by-product used as a natural fertilizer in agriculture, thereby supporting sustainable farming practices.

Even with these advances, challenges persist, particularly the energy and economic costs associated with pretreatments, especially for lignocellulosic biomass. Innovations such as acidification, microbial inhibitor detoxification, and the integration of advanced pretreatment systems are underway to overcome these barriers. Artificial Neural Network (ANN) models and other modeling tools also optimize biogas yields by adjusting variables such as particle size and substrate composition to refine overall process performance.

Future perspectives include the development of integrated systems that combine thermochemical and biochemical technologies to maximize biomass valorization.

Additionally, enhancing pretreatments and conversion processes, as well as stabilizing fermentation and methanogenesis methods, will contribute to the economic viability of these technologies. Overall, biomass conversion represents a pivotal strategy for a transition to renewable energy and a sustainable economic model, reducing waste and emissions while offering significant environmental and industrial benefits.

References

Agbor E, Zhang X, Kumar A (2014) A review of biomass co-firing in North America. Renew Sustain Energy Rev 40:930–943

Andreides D, Fliegerova KO, Pokorna D, Zabranska J (2022) Biological conversion of carbon monoxide and hydrogen by anaerobic culture: prospect of anaerobic digestion and thermochemical processes combination. Biotechnol Adv 58:107886

Aysu T, Durak H (2015) Thermochemical conversion of Datura stramonium L. by supercritical liquefaction and pyrolysis processes. J Supercrit Fluids 102:98–114

Babu BV (2008) Biomass pyrolysis: a state-of-the-art review. Biofuels Bioprod Biorefining: Innov Sustain Econ 2(5):393–414

Balat M (2006) Biomass energy and biochemical conversion processing for fuels and chemicals. Energy Sour Part A 28(6):517–525

Baloch HA, Nizamuddin S, Siddiqui MT, Riaz S, Jatoi AS, Dumbre DK, Mubarak NM, Srinivasan MP, Griffin GJ (2018) Recent advances in production and upgrading of bio-oil from biomass: a critical overview. J Environ Chem Eng 6(4):5101–5118

Barot S (2022) Biomass and bioenergy: resources, conversion and application. In: Renewable energy for sustainable growth assessment, pp 243–262

Bastidas-Oyanedel J-R, Bonk F, Thomsen MH, Schmidt JE (2015) Dark fermentation biorefinery in the present and future (Bio) chemical industry. Rev Environ Sci Bio/Technol 14:473–498

Boujjat H, Rodat S, Abanades S (n.d.) Solar-hybrid thermochemical gasification of wood particles and solid recovered fuel in a continuously-fed prototype reactor. Energies 13(2020):5217

Breault RW (2010) Gasification processes old and new: a basic review of the major technologies. Energies 3(2):216–240

Burm KTAL (2012) Plasma: the fourth state of matter. Plasma Chem Plasma Process 32:401–407

Chacon-Parra A, Lewis D, van Eyk P (2021) Elemental nitrogen balance, reaction kinetics and the effect of ethanol on the hydrothermal liquefaction of soy protein. Chem Eng J 425:130576

Chan YH, Quitain AT, Yusup S, Uemura Y, Sasaki M, Kida T (2018) Liquefaction of palm kernel shell in sub-and supercritical water for bio-oil production. J Energy Inst 91(5):721–732

Chen W-H, Lin Y-Y, Liu H-C, Chen T-C, Hung C-H, Chen C-H, Ong HC (2019) A comprehensive analysis of food waste derived liquefaction bio-oil properties for industrial application. Appl Energy 237:283–291

Chidi BS, Bauer FF, Rossouw D (2018) Organic acid metabolism and the impact of fermentation practices on wine acidity: a review. S Afr J Enol Vitic 39(2):1–15

Dai L, Fan L, Liu Y, Ruan R, Wang Y, Zhou Y, Zhao Y, Yu Z (2017) Production of bio-oil and biochar from soapstock via microwave-assisted co-catalytic fast pyrolysis. Bioresour Technol 225:1–8

Dandamudi KP, Raj TM, Markovski JS, Lammers P, Deng S (2019) Hydrothermal liquefaction of green microalga Kirchneriella sp. under sub-and super-critical water conditions. Biomass Bioenerg 120:224–228

Dasappa S (2011) Potential of biomass energy for electricity generation in sub-saharan Africa. Energy Sustain Dev 15(3):203–213

Del Giudice A, Acampora A, Santangelo E, Pari L, Bergonzoli S, Guerriero E, Petracchini F, Torre M, Paolini V, Gallucci F (2019) Wood chip drying through the using of a mobile rotary dryer. Energies 12(9):1590
Demirbas MF (2006) Current technologies for biomass conversion into chemicals and fuels. Energy Sour Part A: Recov Utiliz Environ Effects 28(13):1181–1188
De Filippis P, De Caprariis B, Scarsella M, Petrullo A, Verdone N (2016) Biocrude production by hydrothermal liquefaction of olive residue. Int J Sustain Dev Plan 11(5):700–707
de Oliveira LB, de Azevedo AR, Marvila MT, Pereira EC, Fediuk R, Vieira CM (2022) Durability of geopolymers with industrial waste. Case Stud Construct Mater 16:e00839
De Paula RG, Antoniêto AC, Ribeiro LF, Srivastava N, O'Donovan A, Mishra PK, Gupta VK, Silva RN (2019) Engineered microbial host selection for value-added bioproducts from lignocellulose. Biotechnol Adv 37(6):107347
Díaz C, Alejandra LA, Brodu N, Montes-Jiménez V, Taouk B (2024) Upgrading of pyrolysis bio-oil by catalytic hydrodeoxygenation, a review focused on catalysts, model molecules, deactivation, and reaction routes. Molecules 29(18):4325
De Tissera S, Köpke M, Simpson SD, Humphreys C, Minton NP, Dürre P (2019) Syngas biorefinery and syngas utilization. Biorefineries 247–280
Domingues PS, Sousa HP, Oliveira NS, Ribeiro AM, Ferreira A (2024) Determination of the biochemical methane potential of swine hydrolyzate. Renew Energy 235:121287
Du Y-H, Wang M-Y, Yang L-H, Tong L-L, Guo D-S, Ji X-J (2022) Optimization and scale-up of fermentation processes driven by models. Bioengineering 9(9):473
Dzulkarnain EL, Nadia JO, Audu WR, Dagang ZW, Abdul-Wahab MF (2022) Microbiomes of biohydrogen production from dark fermentation of industrial wastes: current trends, advanced tools and future outlook. Bioresour Bioprocess 9(1):16
Elum ZA, Momodu AS (2017) Climate change mitigation and renewable energy for sustainable development in Nigeria: a discourse approach. Renew Sustain Energy Rev 76:72–80
Esteban LS, Carrasco JE (2006) Evaluation of different strategies for pulverization of forest biomasses. Powder Technol 166(3):139–151
Faaij A (2006) Modern biomass conversion technologies. Mitig Adapt Strat Glob Change 11:343–375
Fagernäs L, Brammer J, Wilén C, Lauer M, Verhoeff F (2010) Drying of biomass for second generation synfuel production. Biomass Bioenergy 34(9):1267–1277
Ferreira RM, Mota MJ, Lopes RP, Sousa S, Gomes AM, Delgadillo I, Saraiva JA (2019) Adaptation of Saccharomyces cerevisiae to high pressure (15, 25 and 35 MPa) to enhance the production of bioethanol. Food Res Int 115:352–359
Friso D (2020) Conveyor-belt dryers with tangential flow for food drying: mathematical modeling and design guidelines for final moisture content higher than the critical value. Inventions 5(2):22
Fu X, Qiao J, Xu Z, Xu C, Li X (2024) Deep eutectic solvent cocktail enhanced the pretreatment efficiency of lignocellulose. Ind Crops Prod 210:118040
Gebreegziabher T, Oyedun AO, Hui CW (2013) Optimum biomass drying for combustion–a modeling approach. Energy 53:67–73
Gil MV, Rubiera F (2019) Coal and biomass cofiring: fundamentals and future trends. In: New trends in coal conversion. Elsevier, pp 117–40
Gomaa MR, Mustafa RJ, Al-Dmour N (2020) Solar thermochemical conversion of carbonaceous materials into syngas by co-gasification. J Clean Prod 248:119185
Goyal HB, Seal D, Saxena RC (2008) Bio-fuels from thermochemical conversion of renewable resources: a review. Renew Sustain Energy Rev 12(2):504–517
Grana M, Riboli G, Tatangelo V, Mantovani M, Gandolfi I, Turolla A, Ficara E (2024) Anaerobic valorization of sewage sludge pretreated through hydrothermal carbonization: volatile fatty acids and biomethane production. Bioresour Technol 412:131279
Guo Z, Chen X, Xu Y, Liu H (2015) Study of flow characteristics of biomass and biomass–coal blends. Fuel 141:207–213

Hardianto T, Amalia AR, Suwono A, Riauwindu P. (2015) Study of Indonesia low rank coal utilization on modified fixed bed gasification for combined cycle power plant. IOP Conf Ser: Mater Sci Eng 88:12042 (IOP Publishing)
Havlík J, Dlouhý T (2020) Indirect dryers for biomass drying—Comparison of experimental characteristics for drum and rotary configurations. ChemEngineering 4(1):18
Hazout L, Zitouni Z-A, Belkhatir M, Schanz T (2017) Evaluation of static liquefaction characteristics of saturated loose sand through the mean grain size and extreme grain sizes. Geotech Geol Eng 35:2079–2105
Hoque MM, Ansary MA, Yasin SJ (2017) Effects of relative density and effective confining pressure on liquefaction resistance of sands. In: The 19th international conference on soil mechanics and geotechnical engineering. Seoul, pp 1027–1030
Huang L, Zhang YM, Zhang L, Zhang XC, Sun GG (2017) Effects of shale ash and FCC catalyst on adjusting secondary reaction of volatiles in oil shale pyrolysis. CIESC J 68(10):3770
Ibitoye SE, Mahamood RM, Jen TC, Loha C, Akinlabi ET (2023) An overview of biomass solid fuels: biomass sources, processing methods, and morphological and microstructural properties. J Bioresour Bioprod
Induchoodan TG, Haq I, Kalamdhad AS (2022) Factors affecting anaerobic digestion for biogas production: a review. In: Advanced organic waste management, pp 223–233
Jaiswal R, Furuvik NCIS, Thapa RK, Moldestad BME (2020) A CPFD model to investigate the influence of feeding positions in a gasification reactor. Int J Energy Prod Manag 5(3):223–233
Jatoi AS, Shah AA, Ahmed J, Rehman S, Sultan SH, Shah AK, Raza A, Mubarak NM, Hashmi Z, Usto MA (2022) Hydrothermal liquefaction of lignocellulosic and protein-containing biomass: a comprehensive review. Catalysts 12(12):1621
Jena U, Das KC (2011) Comparative evaluation of thermochemical liquefaction and pyrolysis for bio-oil production from microalgae. Energy Fuels 25(11):5472–5482
Jensen CU, Rodriguez Guerrero JK, Karatzos S, Olofsson G, Iversen SB (2017) Fundamentals of Hydrofaction^{TM}: renewable crude oil from woody biomass. Biomass Convers Biorefinery 7(4):495–509
Jiang J, Wang J, Zhang X, Wolcott M (2017) Characterization of micronized wood and energy-size relationship in wood comminution. Fuel Process Technol 161:76–84
John N, Fathima PS, Harsha VS, Paul NM, Nisha P (2024) Physical conversion of biomass: dewatering, drying, size reduction, densification, and separation. In: Handbook of biomass. Springer, pp 987–1013
Juneja P (2024) Biomass waste conversion technologies for its utilization and energy generation in India: a perspective. IOP Conf Ser: Earth Environ Sci 1285:12010 (IOP Publishing)
Kabir G, Hameed BH (2017) Recent progress on catalytic pyrolysis of lignocellulosic biomass to high-grade bio-oil and bio-chemicals. Renew Sustain Energy Rev 70:945–967
Kalak T (2023) Potential use of industrial biomass waste as a sustainable energy source in the future. Energies 16(4):1783
Kan T, Strezov V, Evans TJ (2016) Lignocellulosic biomass pyrolysis: a review of product properties and effects of pyrolysis parameters. Renew Sustain Energy Rev 57:1126–1140
Khaskhachikh VV, Larina OM, Sychev GA, Ya Gerasimov G, Zaichenko VM (2021) Pyrolitic methods of the thermal processing of solid municipal waste. High Temp 59(2):373–383
Knoop C, Dietrich N, Heinrich M, Dornack C, Raab T (2017) Degradation of soil amendments produced from municipal organic waste digestate during a two-year field study. In: 25th European biomass conference and exhibition, Stockholm, pp 1720–1723
Kobayashi N, Guilin P, Kobayashi J, Hatano S, Itaya Y, Mori S (2008) A new pulverized biomass utilization technology. Powder Technol 180(3):272–283
Koszel M, Kocira A, Lorencowicz E (2016) The evaluation of the use of biogas plant digestate as a fertilizer in alfalfa and spring wheat cultivation. Fresenius Environ Bull 25:3258–3264
Kucharska K, Hołowacz I, Konopacka-Łyskawa D, Rybarczyk P, Kamiński M (2018) Key issues in modeling and optimization of lignocellulosic biomass fermentative conversion to gaseous biofuels. Renew Energy 129:384–408

Kumar A, Jones DD, Hanna MA (2009) Thermochemical biomass gasification: a review of the current status of the technology. Energies 2(3):556–581
Labbé R, Carey P, Trincado G, Thiers O (2018) Secado natural de biomasa forestal: efecto de la altura y cobertura del apilado en la provincia de Valdivia, Chile. Bosque (Valdivia) 39(3):449–456
Laurence LC, Ashenafi D (2012) Syngas treatment unit for small scale gasification-application to IC engine gas quality requirement
Li X, Hu H, Jin L, Hu S, Wu B (2008a) Approach for promoting liquid yield in direct liquefaction of Shenhua Coal. Fuel Process Technol 89(11):1090–1095
Li X, Hu S, Jin L, Hu H (2008b) Role of iron-based catalyst and hydrogen transfer in direct coal liquefaction. Energy Fuels 22(2):1126–1129
Li Y-H, Kuo W-C (2020) The study of optimal parameters of oxygen-enriched combustion in fluidized bed with optimal torrefied woody waste. Int J Energy Res 44(9):7416–7434
Liu Y, Zhang Y, Quan X, Li Y, Zhao Z, Meng X, Chen S (2012) Optimization of anaerobic acidogenesis by adding Fe0 powder to enhance anaerobic wastewater treatment. Chem Eng J 192:179–185
Luk HT, Lam TYG, Oyedun AO, Gebreegziabher T, Hui CW (2013) Drying of biomass for power generation: a case study on power generation from empty fruit bunch. Energy 63:205–215
Luo X, Ma C, Chen Z, Zhang X, Niu N, Li J, Liu S, Li S (2019) Biomass-derived solar-to-thermal materials: promising energy absorbers to convert light to mechanical motion. J Mater Chem A 7(8):4002–4008
Luo Z, Zhou J (2022) Thermal conversion of biomass. In: Handbook of climate change mitigation and adaptation. Springer, pp 965–1021
Luo W, Wang T, Zhang S, Zhang D, Dong H, Song M, Zhou Z (2022) Catalytic co-pyrolysis of herb residue and polypropylene for pyrolysis products upgrading and diversification using nickel-X/biochar and ZSM-5 (X= iron, cobalt, copper). Biores Technol 349:126845
Ma W, Chu C, Wang P, Guo Z, Lei S, Zhong L, Chen G (2020) Hydrogen-rich syngas production by DC thermal plasma steam gasification from biomass and plastic mixtures. Adv Sustain Syst 4(10):2000026
Makwana J, Dhass AD, Ramana PV, Sapariya D, Patel D (2023) An analysis of waste/biomass gasification producing hydrogen-rich syngas: a review. Int J Thermofluids 100492
Malico I, Pereira RN, Gonçalves AC, Sousa AMO (2019) Current status and future perspectives for energy production from solid biomass in the European industry. Renew Sustain Energy Rev 112:960–977
Mandø M (2013) Direct combustion of biomass. In: Biomass combustion science, technology and engineering. Elsevier, pp 61–83
Matthews HD, Wynes S (2022) Current global efforts are insufficient to limit warming to 1.5 C. Science 376(6600):1404–1409
Mazzanti G, Demichelis F, Fino D, Tommasi T (2025) A closed-loop valorization of the waste biomass through two-stage anaerobic digestion and digestate exploitation. Renew Sustain Energy Rev 207:114938
Meegoda JN, Li B, Patel K, Wang LB (2018) A review of the processes, parameters, and optimization of anaerobic digestion. Int J Environ Res Public Health 15(10):2224
Mei D, Liu S, Wang S, Zhou R, Zhou R, Fang Z, Zhang X, Cullen PJ, Ostrikov KK (2020) Plasma-enabled liquefaction of lignocellulosic biomass: balancing feedstock content for maximum energy yield. Renew Energy 157:1061–1071
Meinshausen M, Lewis J, McGlade C, Gütschow J, Nicholls Z, Burdon R, Cozzi L, Hackmann B (2022) Realization of Paris Agreement pledges may limit warming just below 2 C. Nature 604(7905):304–309
Merdun H, Sezgin İV (2018) Products distribution of catalytic co-pyrolysis of greenhouse vegetable wastes and coal. Energy 162:953–963
Millati R, Cahyono RB, Ariyanto T, Azzahrani IN, Putri RU, Taherzadeh MJ (2019) Agricultural, industrial, municipal, and forest wastes: an overview. In: Sustainable resource recovery and zero waste approaches, pp 1–22

Mirmohamadsadeghi S, Karimi K, Azarbaijani R, Yeganeh LP, Angelidaki I, Nizami A-S, Bhat R, Dashora K, Vijay VK, Aghbashlo M (2021) Pretreatment of lignocelluloses for enhanced biogas production: a review on influencing mechanisms and the importance of microbial diversity. Renew Sustain Energy Rev 135:110173

Molino A, Chianese S, Musmarra D (2016) Biomass gasification technology: the state of the art overview. J Energy Chem 25(1):10–25

Monkul MM, Kendir SB, Tütüncü YE (2021) Combined effect of fines content and uniformity coefficient on cyclic liquefaction resistance of silty sands. Soil Dyn Earthq Eng 151:106999

Mudassir MA, Batool M, Kousar S, Makarem MA, Razia ET, Meshksar M, Murtaza M, Tariq K, Din MA, Bodlah MA, Rahimpour MR (2023) Plasma-assisted hydrodeoxygenation of bio-oils. Fuel Process Technol 250:107872

Muvhiiwa RF, Sempuga B, Hildebrandt D, Van Der Walt J (2018) Study of the effects of temperature on syngas composition from pyrolysis of wood pellets using a nitrogen plasma torch reactor. J Anal Appl Pyrol 130:159–168

Negahdar L, Gonzalez-Quiroga A, Otyuskaya D, Toraman HE, Liu L, Jastrzebski JT, Van Geem KM, Marin GB, Thybaut JW, Weckhuysen BM (2016) Characterization and comparison of fast pyrolysis bio-oils from pinewood, rapeseed cake, and wheat straw using 13C NMR and comprehensive GC × GC. ACS Sustain Chem Eng 4(9):4974–4985

Nguyen AQ, Trinh LTP (2022) Thermochemical conversion of cellulose and hemicellulose. In: Biomass utilization: conversion strategies. Springer, 107–131

Nkosi N, Muzenda E, Gorimbo J, Belaid M (2021) Developments in waste tyre thermochemical conversion processes: gasification, pyrolysis and liquefaction. RSC Adv 11(20):11844–11871

Noori MT, Das I, Yadav AK (2024) Biogas upgradation using integrated anaerobic digestion and microbial electrochemical technologies. In: Emerging trends and advances in microbial electrochemical technologies. Elsevier, pp 273–292

Oliva JM, Negro MJ, Manzanares P, Ballesteros I, Chamorro MÁ, Sáez F, Ballesteros M, Moreno AD (2017) A sequential steam explosion and reactive extrusion pretreatment for lignocellulosic biomass conversion within a fermentation-based biorefinery perspective. Fermentation 3:1–15

Ong HC, Chen W-H, Farooq A, Gan YY, Lee KT, Ashokkumar V (2019) Catalytic thermochemical conversion of biomass for biofuel production: a comprehensive review. Renew Sustain Energy Rev 113:109266

Osman AI, Mehta N, Elgarahy AM, Al-Hinai A, Al-Muhtaseb AA, Rooney DW (2021) Conversion of biomass to biofuels and life cycle assessment: a review. Environ Chem Lett 19:4075–4118

Osman AI, Lai ZY, Farghali M, Yiin CL, Elgarahy AM, Hammad A, Ihara I, Al-Fatesh AS, Rooney DW, Yap PS (2023) Optimizing biomass pathways to bioenergy and biochar application in electricity generation, biodiesel production, and biohydrogen production. Environ Chem Lett 21(5):2639–2705

Oyedeji O, Gitman P, Qu J, Webb E (2020) Understanding the impact of lignocellulosic biomass variability on the size reduction process: a review. ACS Sustain Chem Eng 8(6):2327–2343

Palma CF (2013) Modelling of tar formation and evolution for biomass gasification: a review. Appl Energy 111:129–141

Palma V, Cortese M, Renda S, Ruocco C, Martino M, Meloni E (2020) A review about the recent advances in selected nonthermal plasma assisted solid-gas phase chemical processes. Nanomaterials 10(8):1596

Pandey PH, Ukarde TM, Mahale JS, Vasishta A, Pawar HS (2022) Syngas production via biomass gasification. In: Biomass, biofuels, biochemicals. Elsevier, pp 211–261

Pang S, Mujumdar AS (2010) Drying of woody biomass for bioenergy: drying technologies and optimization for an integrated bioenergy plant. Drying Technol 28(5):690–701

Panwar NL (2024) Pyrolysis technologies for biochar production in waste management: a review. Clean Energy 8(4):61–78

Papari S, Hawboldt K (2015) A review on the pyrolysis of woody biomass to bio-oil: focus on kinetic models. Renew Sustain Energy Rev 52:1580–1595

Papathanassiou G, Pavlides S, Christaras B, Pitilakis K (2005) Liquefaction case histories and empirical relations of earthquake magnitude versus distance from the broader aegean region. J Geodyn 40(2–3):257–278

Pereira EG, Silva JND, De Oliveira JL, Machado CS (2012) Sustainable energy: a review of gasification technologies. Renew Sustain Energy Rev 16(7):4753–4762

Phillips JR, Huhnke RL, Atiyeh HK (2017) Syngas fermentation: a microbial conversion process of gaseous substrates to various products. Fermentation 3(2):28

Pipliya S, Kumar S, Babar N, Srivastav PP (2023) Recent trends in non-thermal plasma and plasma activated water: effect on quality attributes, mechanism of interaction and potential application in food & agriculture. Food Chem Adv 2:100249

Pirota RDPB, Delabona PS, Farinas CS (2014) Simplification of the biomass to ethanol conversion process by using the whole medium of filamentous fungi cultivated under solid-state fermentation. BioEnergy Res 7:744–752

Rao S, Lu Y, Huang F, Cai Y-X, Cai K-Z (2016) A review of researches on effects of biochars on soil microorganisms. J Ecol Rural Environ 32(1):53–59

Ren J, Liu Y-L, Zhao X-Y, Cao J-P (2020) Biomass thermochemical conversion: a review on tar elimination from biomass catalytic gasification. J Energy Inst 93(3):1083–1098

Rodríguez-Méndez R, Le Bihan Y, Béline F, Lessard P (2017) Long chain fatty acids (LCFA) evolution for inhibition forecasting during anaerobic treatment of lipid-rich wastes: case of milk-fed veal slaughterhouse waste. Waste Manag 67:51–58

Saad A, Ratanawilai S, Tongurai C (2015) Catalytic cracking of pyrolysis oil derived from rubberwood to produce green gasoline components. BioResources 10(2):3224–3241

Saleh AR, Sudarmanta B (2018) Experimental investigation on multi-stage downdraft gasification: influence of air ratio and equivalent ratio to the gasifier performance. AIP Conf Proc 1983 (AIP Publishing)

Scamardellaa D, De Crescenzoa C, Marzocchellaa A, Molinob A, Chianesea S, Savastanoc V, Tralicec R, Karatzaa D, Musmarraa D (2019) Simulation and optimization of pressurized anaerobic digestion and biogas upgrading using aspen plus. Chem Eng 76

Schriver D, Ashour-Abdalla M, Richard RL (1998) On the origin of the ion-electron temperature difference in the plasma sheet. J Geophys Res Space Phys 103(A7):14879–14895

Shahrun MS, Rahman MHA, Baharom NA, Jumat F, Saad MJ, Mail MF, Zawawi NZ, Suherman FHS (2024) Design of a pyrolysis system and the characterisation data of biochar produced from coconut shells, carambola pruning, and mango pruning using a low-temperature slow pyrolysis process. Data Brief 52:109997

Shao S, Ye Z, Sun J, Liu C, Yan J, Liu T, Li X, Zhang H, Xiao R (2022) A review on the application of non-thermal plasma (NTP) in the conversion of biomass: catalyst preparation, thermal utilization and catalyst regeneration. Fuel 330:125420

Shirinbakhsh M, Amidpour M (2017) Design and optimization of solar-assisted conveyer-belt dryer for biomass. Energy Equipm Syst 5(2):85–94

Sicchieri IM, Bortoloti TC, de Quadros F, Bortoloti MA, Teixeira A, de Souza H, Nogueira P, Araki K, Fernandes F, Kuroda EK (2025) Biochars from waste as an additive material in anaerobic co-digestion: characterization and application in batch and semi-continuous systems. Fuel 381:133342

Susanto EE, Saptoro A, Kumar P, Tiong ANT, Putranto A, Suherman S (2024) 7E+ Q analysis: a new multi-dimensional assessment tool of solar dryer for food and agricultural products. Environ Dev Sustain 26(7):16363–16385

Taherzadeh-Ghahfarokhi M, Panahi R, Mokhtarani B (2019) Optimizing the combination of conventional carbonaceous additives of culture media to produce lignocellulose-degrading enzymes by trichoderma reesei in solid state fermentation of agricultural residues. Renew Energy 131:946–955

Tamošiūnas A, Gimžauskaitė D, Aikas M, Uscila R, Snapkauskienė V, Zakarauskas K, Praspaliauskas M (2023) Biomass gasification to syngas in thermal water vapor arc discharge plasma. Biomass Convers Biorefinery 13(18):16373–16384

Tshikovhi A, Motaung TE (2023) Technologies and innovations for biomass energy production. Sustainability 15(16):12121
Tucho BG, Etefa HF, Kumar V, Raba GA, Efa MT, Dejene FB (2023) Effect of wood moisture content on the performance of wood burning cook stoves. Int J Sustain Eng 16(1):1–10
Tumuluru JS, Wright CT (2010) A review on biomass densification technologie for energy application
Vassilev SV, Baxter D, Andersen LK, Vassileva CG, Morgan TJ (2012) An overview of the organic and inorganic phase composition of biomass. Fuel 94:1–33
Wang L, Lei H, Ren S, Bu Q, Tang J, Zhang Q, Ruan R (2011) Microwave assisted pyrolysis of douglas fir pellets over ZSM-5 zeolite catalysts. In: 2011 Louisville, Kentucky, August 7-10, 2011. American Society of Agricultural and Biological Engineers, p 1
Wang L, Lei H, Ren S, Bu Q, Liang J, Wei Y, Liu Y, Lee GS, Chen S, Tang J, Zhang Q (2012) Aromatics and phenols from catalytic pyrolysis of douglas fir pellets in microwave with ZSM-5 as a catalyst. J Anal Appl Pyrolys 98:194–200
Wang F, Zhang Y, Xie H, Peng Y, Song A (2015) Microorganism screening for ethanol production using gasification gas from agricultural residue. Trans Chinese Soc Agric Eng 31(7):221–226
Wang H, Hu H, Yang Y, Liu H, Tang H, Xu S, Li A, Yao H (2020) Effect of high heating rates on products distribution and sulfur transformation during the pyrolysis of waste tires. Waste Manag 118:9–17
Wang X, Ming X, Chen M, Han X, Li X, Zhang D (2024) Effect of acidification pretreatment on two-phase anaerobic digestion of acidified food waste. Waste Manag 190:208–216
Wijayanti W, Sasongko MN, Musyaroh M (2021) Effect of zeolite on the catalytic cracking of tar yields mahogany wood pyrolysis. J Appl Eng Sci 19(4):989–995
Wolf S, Teitge J, Mielke J, Schütze F, Jaeger C (2021) The European green deal—More than climate neutrality. Intereconomics 56:99–107
Wu Y, Şahin S (2018) 3.13 fusion energy production
Wu Y, Wang H, Li H, Han X, Zhang M, Sun Y, Fan X, Tu R, Zeng Y, Xu CC, Xu X (2022) Applications of catalysts in thermochemical conversion of biomass (pyrolysis, hydrothermal liquefaction and gasification): a critical review. Renew Energy 196:462–481
Xu X, Su X, Wang X, Qin G, Liu W (2024) Bioaugmentation with targeted recombinant functional consortia to improve lignocellulosic biowaste co-anaerobic digestion performance. Chem Eng J 499:156151
Yin C, Yan J (2016) Oxy-fuel combustion of pulverized fuels: combustion fundamentals and modeling. Appl Energy 162:742–762
Yogalakshmi KN, Sivashanmugam P, Kavitha S, Yukesh Kannah R, Rajesh Banu J (2022) Lignocellulosic biomass-based pyrolysis: a comprehensive review. Chemosphere 286:131824
Yu B, Zhang D, Dai X, Lou Z, Yuan H, Zhu N (2016) The Synthetic effect on volatile fatty acid disinhibition and methane production enhancement by dosing FeCl3 in a sludge thermophilic anaerobic digestion system. RSC Adv 6(25):21090–21098
Zeng J, Xiao R, Zhang H, Wang Y, Zeng D, Ma Z (2017) Chemical looping pyrolysis-gasification of biomass for high H_2/CO syngas production. Fuel Process Technol 168:116–122
Zhang L, Gao R, Naka A, Hendrickx TLG, Rijnaarts HHM, Zeeman G (2016) Hydrolysis rate constants at 10–25 °C can be more than doubled by a short anaerobic pre-hydrolysis at 35 °C. Water Res 104:283–291
Zhang J, Zhang X (2019) The thermochemical conversion of biomass into biofuels. In: Biomass, biopolymer-based materials, and bioenergy. Elsevier, 327–368
Zhang L, Xu CC, Champagne P (2010) Overview of recent advances in thermo-chemical conversion of biomass. Energy Convers Manag 51(5):969–982
Zhao X, Liu W, Deng Y, Zhu JY (2017) Low-temperature microbial and direct conversion of lignocellulosic biomass to electricity: advances and challenges. Renew Sustain Energy Rev 71:268–282

Zhao L, Wang X-T, Chen K-Y, Wang Z-H, Xu X-J, Zhou X, Xing D-F, Ren N-Q, Lee D-J, Chen C (2021) The underlying mechanism of enhanced methane production using microbial electrolysis cell assisted anaerobic digestion (MEC-AD) of proteins. Water Res 201:117325
Zhaoyan HU, Jingxin Z, Yiliang HE (2022) Catalytic pyrolysis of polypropylene plastics and product properties with Fe-loaded sludge biochar. Chem Ind Eng Progr 42(2):631

Chapter 4
Biomass Conversion: Sustainable and Industrial Applications

Karim Dânoun, Soumia Aboulhrouz, and Mohamed Zahouily

4.1 Introduction

The uninterrupted exploitation of fossil fuels has highlighted the urgent need for the advancement of renewable alternatives. A considerable number of countries have earmarked financial resources for the investigation of sustainable options for finite resources. Among these alternatives, biomass has emerged as a pivotal solution due to its diverse applications in bioenergy, biofuels, and biomaterials (Das et al. 2024; Stolarski et al. 2024; Paul et al. 2024; Poornima et al. 2024; Ashokkumar et al. 2024). In the contemporary era, the conversion of biomass is associated with reduced environmental impacts, particularly in light of the considerable damage caused by the utilization of fossil fuels in recent decades (Poornima et al. 2024; Jiang et al. 2022; Saleem 2022; Zhang et al. 2023; Wei et al. 2024). In response to these global concerns, governments have established a series of agreements and regulations with the objective of mitigating environmental pollution and promoting the use of renewable energy sources. Biomass plays a pivotal role in a multitude of national strategies, as evidenced by initiatives such as the European Commission's "Biomass Action Plan" and the U.S. Department of Energy's Multi-Year Plan. The principal aim of the aforementioned initiative is to curtail CO_2 emissions in accordance with the stipulations of the Kyoto Protocol, thereby addressing the challenges posed by global

K. Dânoun (✉)
Department of Chemical & Biochemical Sciences - Green Process Engineering (CBS), Mohammed VI Polytechnic University (UM6P), Ben Guerir, Morocco
e-mail: karim.danoun@UM6P.ma

S. Aboulhrouz
Department of Materials Science and Nanoengineering (MSN), Mohamed VI Polytechnic University (UM6P), Ben Guerir, Morocco

M. Zahouily
Laboratory of Biotechnology, Agri-Food, Materials and Environment, FST Mohammadia University Hassan II, Casablanca, Morocco

A. Hind (ed.), *Biomass Waste: Processing, Valorization, Industrial Applications*,
https://doi.org/10.1007/978-981-95-0991-1_4

warming (Sikkema et al. 2021). The latter places particular emphasis on policies pertaining to agriculture and energy, with a focus on the conversion and valorization of renewable resources for the production of fuels, biodegradable plastics, and other valuable materials through research and public engagement. The current estimate of the quantity of usable biomass on Earth is approximately 5 billion tons, distributed across both terrestrial and submerged habitats. The estimated potential production capacity of this biomass is 33,000 exajoules (EJ), which is approximately 100 times the global annual energy consumption (Renewable Energy Statistics 2022). Despite the production of over 56 million terajoules per year, or 1.23 million tons of oil equivalent, only 15% of this biomass is utilized. The proportion of biomass used for renewable energy sources varies significantly on a global scale. In industrialized countries, biomass accounts for approximately 11% of total energy consumption, while in developing countries, this figure can reach as much as 50% (Saleem 2022). To illustrate, the United States employs biomass for approximately 3% of its energy requirements, amounting to an annual consumption of approximately 3 million terajoules. In Europe, biomass accounts for approximately 4% of total energy consumption, which equates to approximately 50 million tons of oil equivalent annually. It is noteworthy that some Nordic countries exceed the European average, with biomass contributing up to 20% of their total energy supply (Liu et al. 2024, 2025). The extensive utilization of biomass demonstrates its potential as a renewable energy source, which can be transformed into a variety of useful forms of energy through different conversion processes. These processes can be broadly categorised into three primary types: thermal, biochemical, and chemical conversion (Begum et al. 2024; Asaad et al. 2024; Saravanan et al. 2023). The selection of the most appropriate conversion method is dependent upon a number of factors, including the specific type, quantity and characteristics of the biomass feedstock, as well as the desired energy form or end-use requirements (Fig. 4.1). Furthermore, environmental regulations, policy frameworks, economic conditions, and project-specific considerations exert a considerable influence on this decision-making process. Ultimately, the required energy form and the availability of suitable feedstocks are the primary determinants in selecting the most appropriate conversion route, facilitating the effective utilisation of biomass to meet energy needs across diverse regions. In connection with all these things, the objective of this chapter is to provide a comprehensive overview of the various biomass conversion processes, which are essential for transforming raw biomass into valuable energy and materials. This chapter will examine the sustainable applications of biomass, with particular emphasis on its role in renewable energy production, biofuels and biodegradable materials. Furthermore, the chapter will examine the industrial applications of biomass, investigating how different sectors can utilize this resource for energy and product development. In addition, the chapter will address the challenges associated with the implementation of biomass conversion technologies, including economic, logistical, and regulatory barriers that may impede their widespread adoption. Finally, the chapter will present a comprehensive overview of future directions for biomass utilization, considering advancements in technology, policy developments, and emerging research areas.

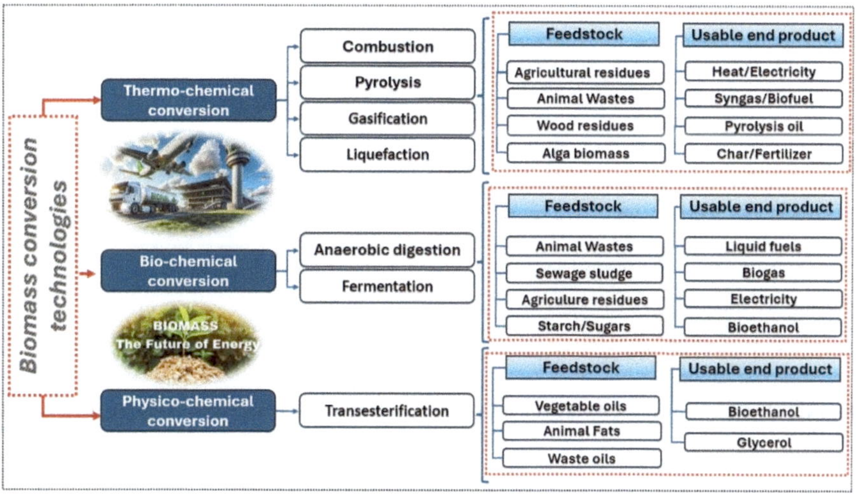

Fig. 4.1 Main conversion technologies and their corresponding products

By providing this holistic view, the chapter aims to inform stakeholders about the potential of biomass as a key player in the transition to sustainable energy systems.

4.2 Biomass Conversion Technologies

This section provides a comprehensive account of the three principal processes for converting biomass energy: thermochemical, biochemical, and physicochemical conversions. Each process employs distinctive mechanisms to transform biomass into a range of usable energy forms, thereby enhancing its role in renewable energy strategies.

4.2.1 Thermo-Chemical Conversion

In thermochemical conversion processes, energy is generated through the application of heat in conjunction with chemical reactions. This approach exploits the thermal decomposition of biomass to produce a range of energy forms and by-products. The four principal thermochemical conversion processes are combustion, pyrolysis, gasification and liquefaction.

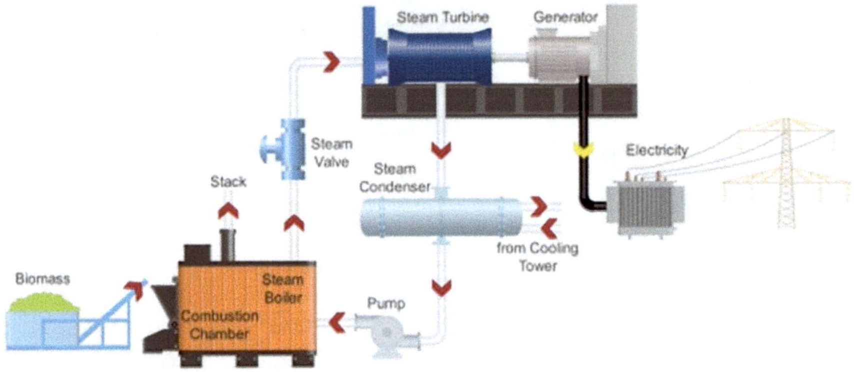

Fig. 4.2 Biomass conversion via combustion process. Reproduced with permission from Ref. Tursi (2019)

4.2.1.1 Combustion

Biomass combustion technology is acknowledged as the most established and pervasive thermochemical conversion process. This method is particularly well-established for the generation of both electricity and heat. The combustion of biomass encompasses three principal stages: drying, pyrolysis and reduction (Poornima et al. 2024; Bisen et al. 2025; Chormare et al. 2023). During these stages, the release of unstable gases and the combustion of solid char occur. It is noteworthy that in excess of 70% of the total heat produced during combustion is derived from the burning of these unstable gases. Biomass combustion plants that use wood and forest residues as fuel typically generate between 20 and 50 megawatts of electricity (MWe), achieving electrical efficiencies of about 25–30%. Recently, innovations like fluidized bed technologies and improved gas cleaning methods have significantly boosted these systems performance (Wu et al. 2022). As a result, current electricity production levels can reach between 50 and 80 MWe, with efficiencies rising to as high as 30–40%. The combustion process is applicable to a range of biomass forms, but its efficacy is optimised when the moisture content is below 50%. The combustion of biomass is an extremely efficient process, resulting in the production of a substantial amount of heat. This makes it an economically viable option. However, the process itself is complex, involving a continuous cycle of solid–gas and solid–solid reactions. The combustion of a substance requires a combination of air, heat and fuel in order to sustain the reaction. A typical diagram of the biomass combustion process is illustrated in Fig. 4.2.

4.2.1.2 Pyrolysis

Pyrolysis is a conversion process that transforms specific types of biomass into three principal products: liquid bio-oil, solid charcoal, and gaseous combustible gas. This occurs through partial combustion at approximately 500 °C in an oxygen-free environment (Lebaka 2013). The elevated temperatures facilitate the vaporization of the biomass's volatile components, resulting in gases that can be condensed into liquid form through liquefaction. The resulting liquid fuel can be stored and utilized for various heating and electricity generation applications. Additionally, the pyrolysis process yields other combustible products, including charcoal, gas, and a variety of valuable chemicals (Amenaghawon et al. 2021). As previously stated, pyrolysis systems operating at temperatures below 600 °C are classified as simple, whereas industrial plants often reach temperatures between 600 and 1000 °C, resulting in a gas with a higher hydrogen content. The advanced processes require the implementation of more sophisticated systems and management strategies. The pyrolysis process comprises four principal stages, each occurring at a different temperature. The initial stage of the process entails the drying of the incoming biomass at temperatures between 100 and 120 °C. Subsequently, at 275 °C, the outlet gases, predominantly nitrogen (N_2) and carbon dioxide (CO_2), along with acetic acid and methanol, undergo distillation. The third step comprises exothermic reactions occurring between 280 and 350 °C. These reactions result in the breaking of weaker chemical bonds, thereby releasing complex mixtures of substances. These include ketones, aldehydes, phenols, and gases such as CO_2, carbon monoxide (CO), methane (CH_4), ethane (C_2H_6), and hydrogen (H_2). Ultimately, at 350 °C, all volatile compounds are removed via evaporation, resulting in a higher proportion of H_2 and CO, while the remaining carbon forms charcoal residues. The condensed liquids produced during pyrolysis, such as tar and pyroleptic acid, can be separated and subsequently purified. The approximate yields from 100 kg of dry wood are as follows: 30 kg of charcoal, 14 m^3 of gas (with a calorific value of 10.4 MJ/m^3), 7.6 L of wood oil and light tar, 1.4 L of methyl alcohol, 3 kg of pitch, and 5.3 L of acetic acid, among other products (Poornima et al. 2024). Figure 4.3 illustrates the standard diagram of the biomass pyrolysis process. The pyrolytic process can be conducted in several ways, each of which affects the production of bio-oil, syngas, and carbonaceous residues. (i) The oldest pyrolysis method is carbonisation, which occurs at temperatures between 300 and 500 °C, yielding primarily solid vegetable coal. (ii) Conventional pyrolysis, conducted at approximately 500 °C, yields three fractions in roughly equal proportions, though with longer reaction times. (iii) Fast pyrolysis occurs at temperatures between 500 and 650 °C, facilitating gasification reactions rapidly and resulting in a liquid fraction that constitutes 70–80% of the biomass. Typically, fast pyrolysis yields approximately 60% bio-oil, 20% bio-coal, and 20% gas. (iv) Lastly, flash pyrolysis is performed at temperatures exceeding 650 °C, with contact times of less than one second, resulting in the production of gaseous fractions with an efficiency of up to 80%.

4.2.1.3 Gasification

The gasification process is a thermochemical method that converts solid carbon materials, specifically solid biomass, into synthesis gas (syngas), which primarily consists

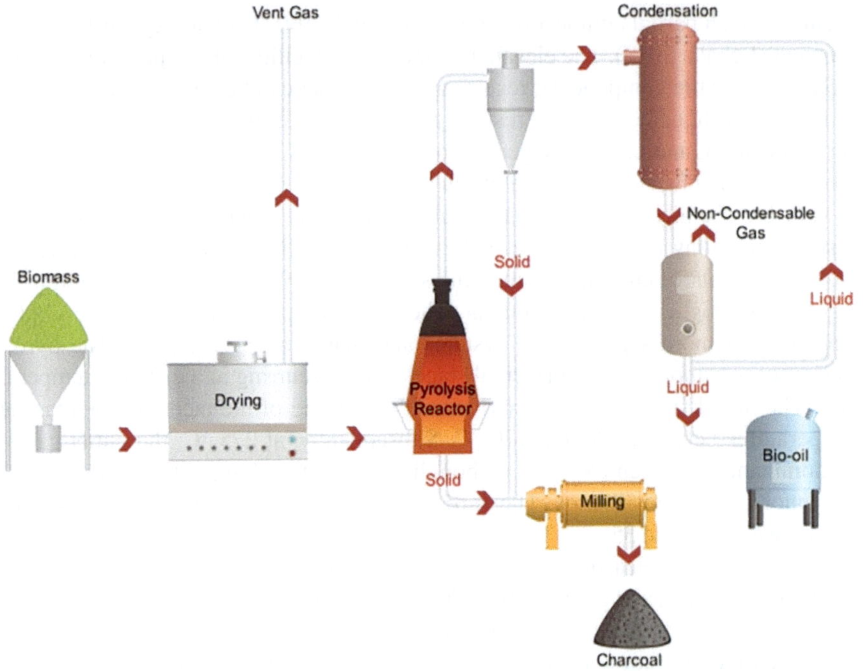

Fig. 4.3 Biomass conversion via pyrolysis process. Reproduced with permission from Ref. Tursi (2019)

of carbon monoxide (CO), hydrogen (H_2), and nitrogen (N_2). The syngas serves as a fuel for electricity generation and is also a precursor for a variety of products in the petrochemical and refining industries, including methanol, ammonia, and synthetic gasoline (Matsumoto 2024; Shahabuddin et al. 2020). The development of gasification technologies can be traced back to the early decades of the twentieth century. Although a variety of treatment methods exist, they can be characterised by a number of common features. The prevailing technology entails the partial oxidation of solid biomass at elevated temperatures, resulting in syngas with varying ratios of CO and H_2, which collectively account for over 85% of the total gas volume. The remaining components typically include carbon dioxide (CO_2) and methane (CH_4) (Aghamiri and Lahijani 2024; Lombardelli et al. 2024). This process is conducted within a sealed chamber under conditions of air suction or at reduced pressure relative to atmospheric pressure (Fig. 4.4). Subsequently, the syngas produced undergoes a purification process, during which impurities are removed and trace elements are recovered. This purification step is particularly advantageous as it effectively eliminates pollutants, especially those containing high levels of heteroatoms or impurities that are challenging to treat through alternative methods. For instance, sulphur can frequently be recovered in its elemental form, while purified syngas can be employed for chemical production and/or electricity generation (Mondal et al. 2011). The process of biomass gasification can be divided into four principal stages: drying, pyrolysis,

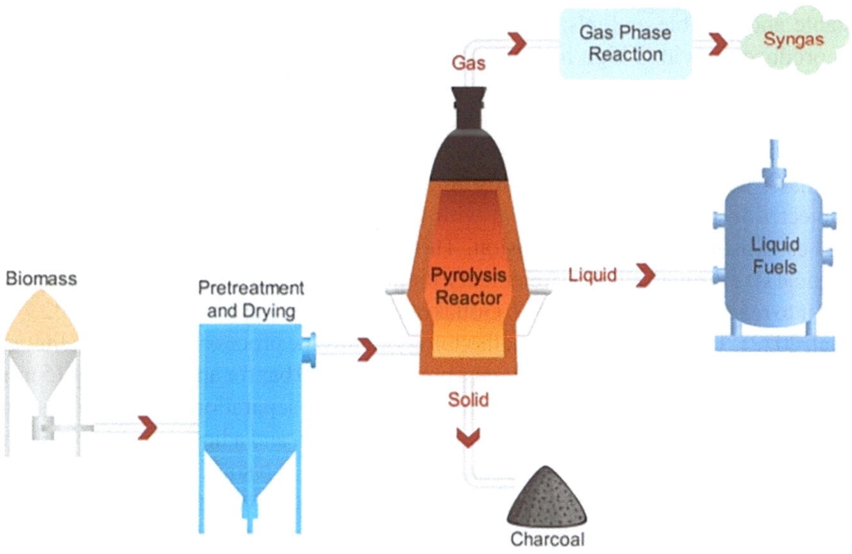

Fig. 4.4 Biomass conversion via gasification. Reproduced with permission from Ref. Tursi (2019)

oxidation, and reduction. The drying phase is responsible for the removal of moisture from the biomass. Subsequently, pyrolysis occurs, whereby thermal decomposition gives rise to the liberation of volatile compounds. The subsequent oxidation stage involves the partial combustion of the biomass, which generates heat and facilitates the conversion of the remaining solid carbon into gas. Subsequently, in the reduction phase, the generated gases undergo further reaction to form syngas. The composition of the produced gas includes both combustible and non-combustible fractions, which vary depending on the type of biomass used and the operational conditions of the gasification process. In general, the calorific value of the generated gas ranges from 4.5 to 6 MJ/m^3, representing approximately 10–50% of the calorific value of natural gas (Akia et al. 2014). Therefore, this technology offers a viable solution to produce energy from biomass, despite the potential for the generation of considerable quantities of pollutants as by-products, including tar, ammonia, hydrogen sulphide and hydrochloric acid.

4.2.1.4 Liquefaction

Liquefaction is a process that converts biomass into valuable liquid fuels by using water at moderate temperatures, typically between 280 and 370 °C, and under high pressures of 10–25 MPa. The process yields a liquid biogranulate that closely resembles crude oil, along with other by-products in gaseous, liquid, and solid forms (López Barreiro et al. 2013; Bast et al. 2023; Bhushan et al. 2020). The resulting products exhibit a high heating value and low oxygen content, rendering them chemically stable and suitable as fuels. The principal objective of the liquefaction process is to produce oil with a high hydrogen-to-carbon (H/C) ratio. Liquefaction processes can be classified into two main categories, distinguished by the type of raw material employed. The first category encompasses processes designed for lignocellulosic biomass, which is typically in a dry state. The second category is reserved for algal biomass, which is characteristically in a wet state. In both instances, biomass is subjected to preliminary treatment with the objective of eliminating contaminants and establishing a stable suspension. This frequently necessitates the reduction of particle size through alkaline treatments, thereby facilitating the pumping of the material into reactors (Yang et al. 2019; Gollakota et al. 2018; Ravichandran et al. 2022; Ni et al. 2022). The most prevalent methodology entails the processing of lignocellulosic biomass at approximately 350 °C and a pressure of 150 bar for approximately 15 min. Under these conditions, the biomass undergoes phase separation, yielding a gaseous phase comprising CO_2 and other gases, a solid residue, and a liquid phase containing biocrude. The solid residues can be employed directly as a source of biofuel or fertilizer, while the aqueous phase can be recycled within the plant for water-consuming processes or for anaerobic digestion. The resulting biocrude, which is characterized by a low oxygen content, requires further refinement before it can be utilized on a commercial scale (Fig. 4.5).

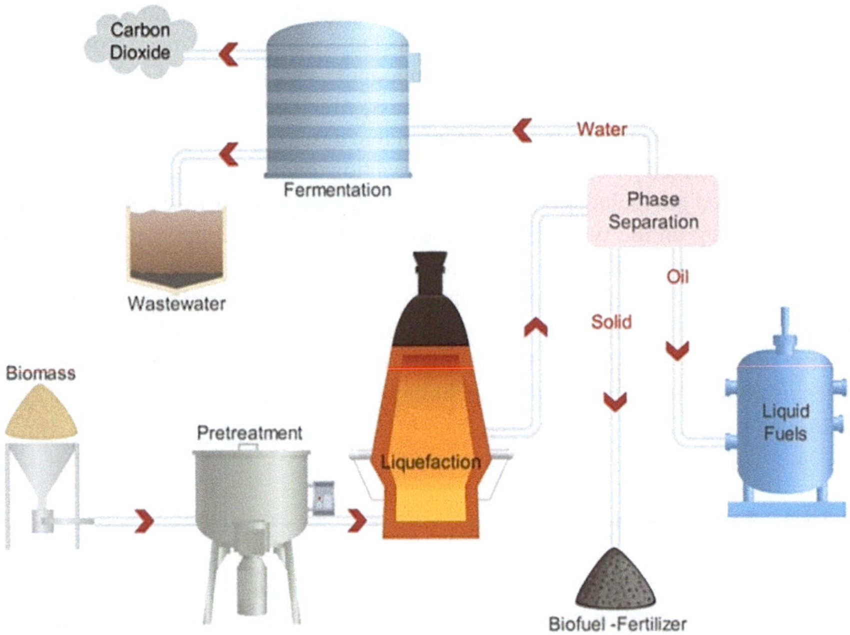

Fig. 4.5 Biomass conversion via liquefaction process. Reproduced with permission from Ref. Tursi (2019)

4.2.2 *Biochemical Conversion of Biomass*

Biochemical conversion processes facilitate the breakdown of biomass into accessible carbohydrates, which can subsequently be transformed into liquid fuels, biogas, and various bio-products through the action of biological agents, including bacteria and enzymes. The most employed biochemical technologies in this context are anaerobic digestion and fermentation (Brethauer and Studer 2015).

4.2.2.1 Anaerobic Digestion

Anaerobic digestion is a multi-step biological process that plays a pivotal role in effective waste management and the generation of renewable energy (Zamani et al. n.d.). The process comprises four principal stages: hydrolysis, acidogenesis, acetogenesis and methanogenesis. Throughout these stages, a series of chemical reactions occur via natural metabolic pathways, facilitated by microorganisms in an oxygen-free environment. These reactions facilitate the breakdown of complex organic macromolecules into simpler compounds, resulting in the production of biogas, a mixture of methane, carbon dioxide, and trace amounts of other gases, as well as digestate. The most common feedstocks for anaerobic digestion include sewage sludge, agricultural residues, municipal solid waste (MSW), and animal manure. A schematic representation of the anaerobic digestion process is provided in Fig. 4.6. The energy content of the biogas produced by this process is 20–40% lower than the heating value of the original raw materials. This method is particularly well-suited to the processing of organic waste with a moisture content of between 80 and 90%. A notable advantage of this process is that the resulting biogas can be employed directly as an ignition gas for engines and turbines. The overall conversion efficiency of anaerobic digestion is estimated to be approximately 21%. Furthermore, residual heat from the engines and turbines can be captured and reused through a heat exchanger.

4.2.2.2 Fermentation

The fermentation of organic materials is a biochemical process that involves the conversion of simple sugars (both hexoses and pentoses) into ethanol and carbon dioxide under anaerobic conditions (Peinemann and Pleissner 2020; Kovalev et al. 2023). This process is primarily facilitated by microorganisms, such as yeasts, which play a crucial role in the fermentation process. The most utilized microorganism for this process is *Saccharomyces cerevisiae.* The feedstocks for fermentation can be classified into three categories: sugars, starches and lignocellulosic substrates. The theoretical yield from 100 g of hexoses or pentoses is approximately 51.14 g of ethanol and 48.86 g of carbon dioxide. In addition to ethanol and carbon dioxide, the fermentation process also yields by-products such as glycerol and carboxylic acids. The efficiency and quality of the fermentation process are influenced by various

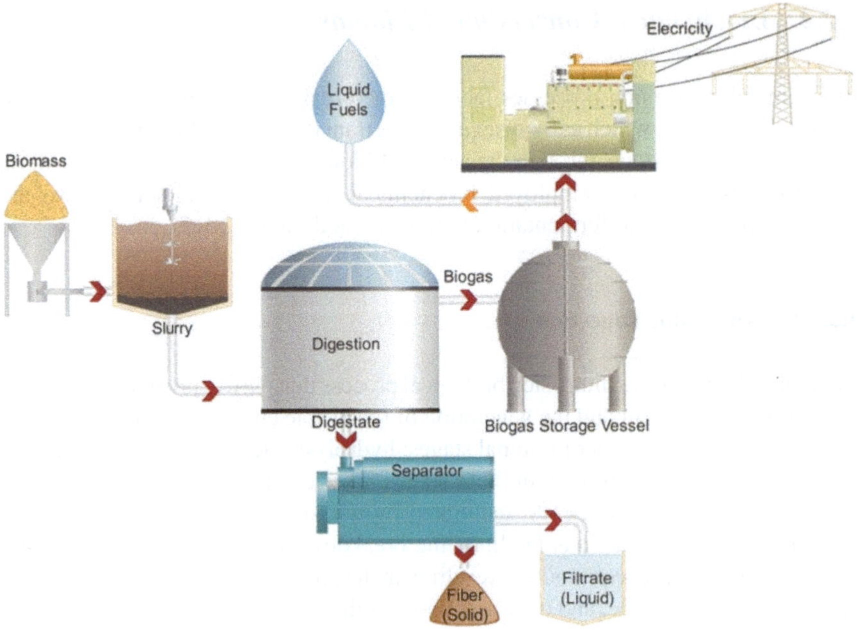

Fig. 4.6 Biomass conversion via anaerobic digestion process. Reproduced with permission from Ref. Tursi (2019)

factors, including the type of feedstock, temperature, pH, inoculum, and fermentation duration. The conversion of sugars into ethanol can occur via a number of different metabolic pathways, depending on the type of sugar that is being used. In particular, hexoses such as glucose are converted via glycolysis, which is also known as the Embden-Meyerhof pathway (Taherzadeh and Karimi 2008). In contrast, pentose sugars undergo conversion via the pentose phosphate pathway (PPP). In general, the conversion of hexoses is more rapid than that of pentoses. Once the conversion process is complete, ethanol is distilled and dehydrated, resulting in a concentrated alcohol product (Poornima et al. 2024; Ahmad et al. 2024). The solid residues that remain can be employed for a variety of purposes. They can be utilized as fuel in boilers for gas production or as livestock feed. A schematic representation of the biomass fermentation process is provided in Fig. 4.7.

4.2.3 Physico-Chemical Conversion of Biomass

The physicochemical conversion of biomass results in the generation of high-density biofuels. A range of vegetable oils and animal fats undergo a transformation into biodiesel through the processes of esterification and/or transesterification. The principal vegetable oils employed in the manufacture of first-generation biodiesel are

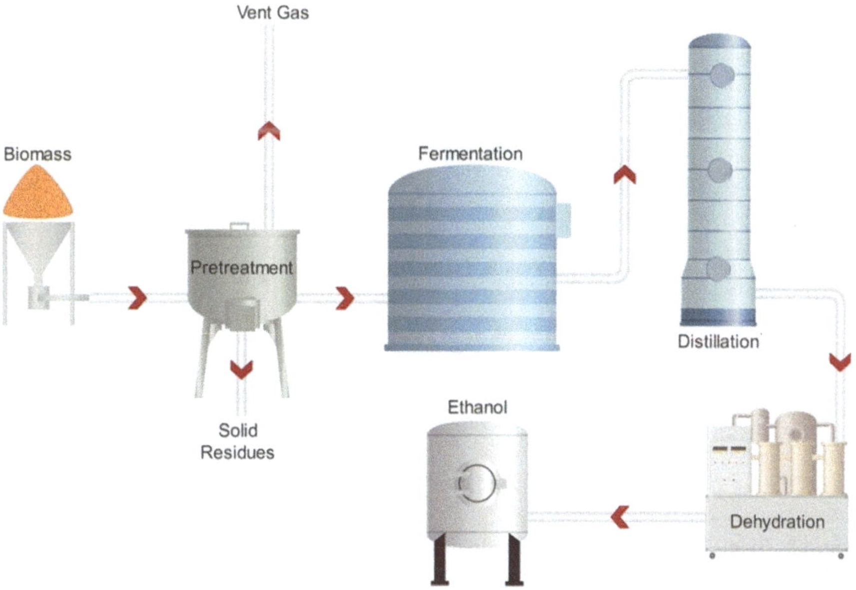

Fig. 4.7 Biomass conversion via fermentation process. Reproduced with permission from Ref. Tursi (2019)

rapeseed oil and sunflower oil, collectively accounting for approximately 80–85% and 10–15% of global biodiesel production, respectively (Fukuda et al. 2001). Furthermore, second- and third-generation biodiesel can be produced from waste oils, including waste cooking oil (WCO) and microbial oils such as algal oil. It is important to note that oils primarily consist of triglycerides, which are not usable as fuel in their raw form. The transformation of crude vegetable oil is essential; without this conversion, issues such as incomplete combustion and the buildup of residues in engines can occur. Consequently, the raw materials must be further processed-primarily through transesterification-to break down the triglyceride molecules into their individual components: fatty acids and glycerol. Glycerol is actually what contributes to the high viscosity of vegetable oils. During the transesterification reaction, triglycerides are converted into methyl or ethyl esters (biodiesel) by reacting with an excess of methyl or ethyl alcohol, usually in the presence of an alkaline catalyst, such as sodium hydroxide (NaOH) or potassium hydroxide (KOH). The transesterification of triglycerides occurs at low temperatures, typically between 50 and 70 °C, and under atmospheric pressure. Following this reaction, the resulting mixture undergoes neutralization and purification to remove glycerol (Leung et al. 2010). Due to its significantly higher density compared to biodiesel, glycerol can be effectively separated using a phase separator. After glycerol separation, any excess alcohol is usually eliminated through distillation or partial evaporation. The biodiesel then undergoes further purification via water washing, which helps to remove residual catalysts and soaps. Finally, the purified biodiesel is dried and stored for use (Fig. 4.8).

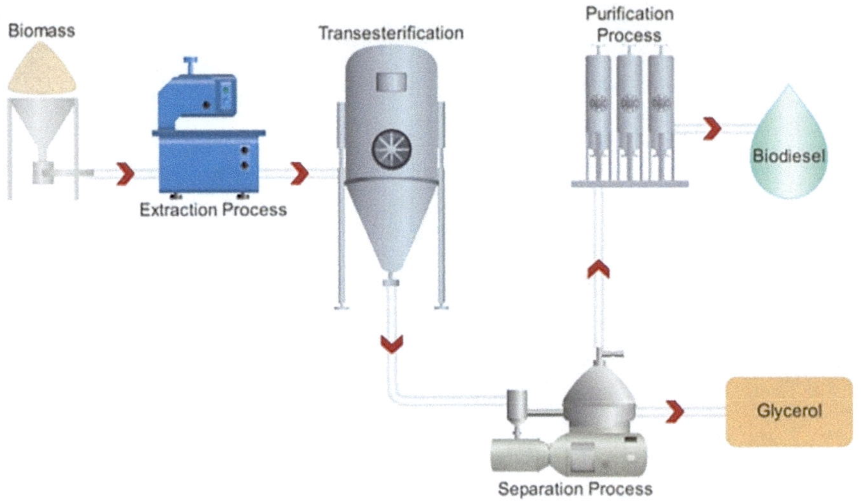

Fig. 4.8 Biomass conversion via transesterification process. Reproduced with permission from Ref. Tursi (2019)

In summary, the conversion of biomass plays a pivotal role in advancing sustainability for both the environment and the economy. Biomass conversion facilitates the transformation of organic materials into energy and valuable products, thereby contributing to renewable energy sources, waste management, and the reduction of greenhouse gas emissions. However, each biomass conversion technology presents a distinct set of benefits and limitations when implemented on an industrial scale. The table below provides a summary of these factors, highlighting the strengths and challenges associated with each technique. This comparison underscores the importance of selecting the appropriate biomass conversion method to maximize benefits while addressing specific limitations (Table 4.1).

4.3 Sustainability and Industrial Applications of Biomass

The conversion of biomass into energy represents a pivotal aspect of attaining the Sustainable Development Goals, offering a means to harness renewable energy sources, curtail waste generation, and mitigate environmental impacts. A variety of conversion technologies can be employed to transform biomass into biofuels, electricity, and other valuable products. This comprehensive strategy serves to reinforce energy security while simultaneously facilitating the evolution of a circular bioeconomy. The following sections present a detailed overview of the key aspects of biomass conversion in relation to sustainability and industrials applications of biomass.

Table 4.1 Advantages and limitations of each biomass conversion technology

Process	Technology	Product	Advantages	Limitations
Thermochemical conversion	Pyrolysis	Bio-oil, biochar, and syngas	High efficiency, adaptability, and superior fuel quality	High operational and capital expenses
	Liquefaction	Bio-oil, gases, liquid fuels and syngas	Versatile feedstock, improved yield and quality of bio-oil, the ability to process wet materials, and its positive environmental and economic benefits	The consumption of energy in high-pressure processes requires a longer residence time, which in turn results in higher costs
	Gasification	Syngas	Flexibility, improved emission control, and a range of applications for the produced syngas	The process is complex and comprises multiple stages, during which tars and char are formed
	Combustion	Thermal energy	High calorific value of biomass and the capacity for multiple fuel production	This process is only feasible for biomass with a moisture volume below 50%
Biochemical conversion	Fermentation	Ethanol (grain alcohol)	Reduced costs, high ethanol yield, and shorter processing times	Inability to utilize commonly used yeasts, instability of recombinant microorganisms, and techno-economic constraints
	Anaerobic digestion	Methane, CO_2, digestate	Economic expenses and the safe management of digestate	Complex products necessitate further processing to achieve refinement, leading to storage and processing challenges
Physico-chemical conversion	Transesterification	Liquid fuel (biodiesel)	Environmentally friendly process, with low processing temperatures and high ester yields	Extended processing times and high costs associated with product purification

4.3.1 *Role of Biomass in Reducing Greenhouse Gas Emissions*

The combustion and conversion of biomass for energy release has been demonstrated to result in a notable reduction in greenhouse gas emissions when compared to the use of fossil fuels. To illustrate, the utilization of biomass as a fuel in power generation or heating has been shown to result in a net reduction of CO_2 emissions, particularly when sourced in a sustainable manner. For instance, the production of ethanol from sugar beet requires a greater input of energy and provides a 50–60% reduction in emissions when compared with gasoline (Sadrul Islam and Ahiduzzaman 2012). Furthermore, biomass systems are capable of capturing methane from organic waste through anaerobic digestion, thereby reducing the greenhouse gas footprint even further. The substitution of conventional fuels with biomass can facilitate the attainment of climate objectives and bolster energy security, while simultaneously promoting rural development.

4.3.2 *Case Studies of Sustainable Biomass Energy Systems*

Bioenergy represents a pivotal element of the renewable energy sector, exhibiting considerable potential that varies across countries due to factors such as the availability of biomass resources, governmental policies, technological advancements, and economic conditions. The United States is distinguished by one of the most substantial bioenergy potentials globally, predominantly derived from agricultural residues, forestry by-products, and dedicated energy crops. Estimates indicate that the country possesses over 1 billion dry tons of biomass annually, which could potentially generate approximately 60 billion gallons of biofuel. Brazil is a global leader in bioenergy, particularly in the production of bioethanol from sugarcane. Its annual production capacity is approximately 37 billion litres of ethanol, which contributes approximately 18% of the country's energy mix. Germany has made considerable investments in bioenergy, particularly in the production of biogas from agricultural and organic waste. According to official reports, bioenergy accounted for approximately 8.5% of the total energy supply in 2021, with projections indicating a potential growth to 12% by 2030. China has experienced a rapid development of its bioenergy sector, with estimates indicating a biomass production capacity of over 120 million tons annually. The objective is for bioenergy to contribute 15% of the country's energy mix by 2025. In India, the potential for bioenergy is considerable, with estimates of approximately 30 million tons of biomass annually derived from agricultural residues. Additionally, the country has set a target to reach 10,000 MW of biomass-based power generation by 2022. In terms of renewable energy, Sweden is a frontrunner, with bioenergy representing approximately 35% of its total energy consumption. Furthermore, there is considerable scope for enhancing the utilization of biomass, particularly in the context of heating and electricity generation. Canada, with its abundant forest resources, is estimated to have the potential to produce approximately 23 billion liters of biofuels annually from forestry and agricultural residues, in addition to a growing biogas sector. The global bioenergy potential is considerable, with several countries, including the United States, Brazil, Germany, China, India, Sweden, and Canada, at the vanguard of its exploitation. These countries are playing a pivotal role in the advancement of renewable energy objectives and sustainability initiatives. It is of the utmost importance that continued investment in technology and infrastructure be made to fully realize the potential of bioenergy on a global scale (Strezov and Anawar n.d.).

4.3.3 Industrial Applications of Biomass

Biomass, derived from organic materials, has emerged as a versatile and sustainable resource with numerous industrial applications. The significance of biomass extends to a multitude of domains, including energy production, manufacturing, and waste management. The growing awareness of environmental issues and the promotion of renewable resources have been instrumental in driving the expansion of biomass utilisation across a range of sectors.

4.3.3.1 Energy Production

In recent years, biomass has made significant strides as a source of energy. According to the International Energy Agency (IEA), biomass accounted for approximately 5% of the total global primary energy supply in 2020, making it one of the largest sources of renewable energy after hydropower. The potential use of biomass for heating, electricity generation, and biofuels has garnered attention, with the global biomass energy market expected to reach $165 billion by 2027, growing at a CAGR of 5.8% from 2020, as per Fortune Business Insights.

4.3.3.2 Manufacturing and Raw Materials

Biomass also plays a pivotal role in the development of sustainable materials for a range of manufacturing industries. The recent advances in technology have facilitated the development of bioplastics derived from biomass sources, including corn starch, sugarcane, and even agricultural waste. As per the findings of a study conducted by Allied Market Research, the global bioplastics market is projected to reach approximately $44.09 billion by 2023. This growth is driven heavily by the demand for environmentally friendly alternatives to conventional plastics. Furthermore, biomass can serve as a source of valuable raw materials for a range of industrial sectors. For example, residual biomass from agricultural sources, such as rice husks and wheat straw, is being employed in the production of biochar, which is increasingly recognised for its capacity to enhance soil health and sequester carbon. It is anticipated that the market valuation of biochar will reach approximately $4.2 billion by 2026, reflecting an increasing recognition of its potential applications in agriculture and climate change mitigation strategies (He et al. 2024; Heo et al. 2021; Yeletsky et al. 2022).

4.3.3.3 Biomass in the Food Industry

The food industry is adopting biomass through the utilization of bio-packaging solutions that serve to minimize the environmental impact. The use of traditional plastic packaging has been identified as a significant contributor to environmental pollution. In contrast, bio-based packaging materials, including starch, cellulose, and proteins, offer biodegradable alternatives that can help reduce this impact. A study indicated that the substitution of conventional plastic with bioplastic packaging could result in a reduction of energy consumption in the food supply chain by up to 50% (Regadera-Macías et al. 2024; Zaini et al. 2023; Pérez-Torrado et al. 2015). In addition to their use in packaging, biomass-derived additives have been demonstrated to play a vital role in enhancing food quality and preservation. The use of natural antioxidants and emulsifiers derived from plants not only enhances food safety but also responds to the growing demand from consumers who prioritise health and wellbeing. The research indicates that the incorporation of biomass-derived additives can prolong the shelf life of a product and maintain its integrity without the need for synthetic chemicals. The aforementioned applications of biomass in the food industry contribute to sustainability while meeting consumer demands for environmentally friendly and health-oriented products.

4.3.3.4 Waste Management

The utilisation of biomass plays an integral role in the management of industrial waste. Techniques such as anaerobic digestion facilitate the conversion of organic waste into biogas, which can then be employed as a renewable energy source (Begum et al. 2024; Yang et al. 2024; Liu et al. 2024). The National Renewable Energy Laboratory (NREL) estimates that the United States has the potential to produce in excess of 8 billion gallons of biogas on an annual basis, thereby underscoring the resource's capability to address waste challenges while simultaneously generating energy. Furthermore, the repurposing of biomass waste is conducive to the development of a circular economy. An increasing number of industries are recognising the economic benefits of utilising agricultural residues, forestry by-products and food industry waste in the development of value-added products (Asaad et al. 2024; Roberts et al. 2015).

4.3.3.5 Challenges and Future Directions

Nevertheless, the expansion of biomass industries is confronted with a number of obstacles, most notably the competition with food production and the issue of land use. The allocation of land for biomass cultivation may conflict with the requirements of food production, thereby raising questions about the long-term sustainability of global food security. It is imperative to achieve a balance between biomass production and agricultural land use in order to address these concerns. Technological advancement is a crucial factor in enhancing the efficiency and cost-effectiveness of biomass conversion processes. Innovations in processing techniques, such as enzyme technology and fermentation methods, have the potential to enhance yield and reduce operational costs, thereby rendering biomass a more attractive option for industrial applications. Furthermore, the implementation of robust policy frameworks and incentives is essential for the promotion of sustainable biomass utilisation. It is incumbent upon governments to establish regulations and financial incentives that encourage investment in biomass technologies, research and development, and sustainable practices. By establishing a favorable environment for biomass industries, policymakers can facilitate the transition towards a more sustainable and circular economy.

4.4 Conclusion

This chapter book presents a compelling opportunity for sustainable energy production and resource utilization through biomass conversion. The transformation of organic materials into biofuels, bioproducts, and bioenergy can significantly reduce greenhouse gas emissions, decrease reliance on fossil fuels, and enhance energy security. Furthermore, the benefits of biomass conversion extend beyond environmental impacts, as it also supports rural economies by creating jobs in biomass collection, processing, and technology development. Nevertheless, to fully actualise the potential of biomass technologies, further research and development are imperative. Innovations in conversion processes, feedstock optimisation, and lifecycle analysis will facilitate improvements in efficiency and sustainability. It is therefore essential that collaborative efforts are undertaken between governments, research institutions, and industry stakeholders in order to overcome existing challenges and unlock the full potential of biomass.

References

Aghamiri AR, Lahijani P (2024) Catalytic conversion of biomass and plastic waste to alternative aviation fuels: a review. Biomass Bioenergy 183:107120. https://doi.org/10.1016/j.biombioe.2024.107120

Ahmad YH, Ibrahim MF, Banu JR, Al-Qaradawi SY (2024) Recent advances on the use of promoters in biochemical hydrogen production: a comprehensive review. Energy Convers Manag 317:118814. https://doi.org/10.1016/j.enconman.2024.118814

Akia M, Yazdani F, Motaee E, Han D, Arandiyan H (2014) A review on conversion of biomass to biofuel by nanocatalysts, vol 1

Amenaghawon AN, Anyalewechi CL, Okieimen CO, Kusuma HS (2021) Biomass pyrolysis technologies for value-added products: a state-of-the-art review. In: Environment, development and sustainability. Springer Science and Business Media B.V., pp 14324–14378. https://doi.org/10.1007/s10668-021-01276-5

Asaad SM, Inayat A, Ghenai C, Shanableh A (2024) Integration of waste heat recovery with biomass thermal conversion processes: a review. Process Saf Environ Protect 192:567–579. https://doi.org/10.1016/j.psep.2024.10.007

Ashokkumar V, Chandramughi VP, Kumar G, Ngamcharussrivichai C, Piechota G, Igliński B, Kothari R, Chen W-H (2024) Advancements in lignocellulosic biomass: a critical appraisal of fourth-generation biofuels and value-added bioproduct. Fuel 365:130751. https://doi.org/10.1016/j.fuel.2023.130751

Begum YA, Kumari S, Jain SK, Garg MC (2024) A review on waste biomass-to-energy: integrated thermochemical and biochemical conversion for resource recovery. Environ Sci Adv 3(9):1197–1216. https://doi.org/10.1039/d4va00109e

Bhushan S, Kalra A, Simsek H, Kumar G, Prajapati SK (2020) Current trends and prospects in microalgae-based bioenergy production. J Environ Chem Eng 8(5):104025. https://doi.org/10.1016/j.jece.2020.104025

Bisen D, Chouhan APS, Pant M, Chakma S (2025) Advancement of thermochemical conversion and the potential of biomasses for production of clean energy: a review. Renew Sustain Energy Rev 208:115016. https://doi.org/10.1016/j.rser.2024.115016

Brethauer S, Studer MH (2015) Biochemical conversion processes of lignocellulosic biomass to fuels and chemicals—A review. Chimia (Aarau) 69(10):572–581. https://doi.org/10.2533/chimia.2015.572

Chormare R, Moradeeya PG, Sahoo TP, Seenuvasan M, Baskar G, Saravaia HT, Kumar MA (2023) Conversion of solid wastes and natural biomass for deciphering the valorization of biochar in pollution abatement: a review on the thermo-chemical processes. Chemosphere 339:139760. https://doi.org/10.1016/j.chemosphere.2023.139760

Das N, Maheshwari DK, Pandey P (2024) Energy crop-based rhizoremediation and lignocellulosic biomass production as sustainable bioeconomy-driven solution for biofuel generation and waste mitigation. Prog Energy Combust Sci 103:101161. https://doi.org/10.1016/j.pecs.2024.101161

El Bast M, Allam N, Abou Msallem Y, Awad S, Loubar K (2023) A review on continuous biomass hydrothermal liquefaction systems: process design and operating parameters effects on biocrude. J Energy Inst 108:101260. https://doi.org/10.1016/j.joei.2023.101260

Fukuda H, Kondo A, Noda H (2001) Biodiesel fuel production by transesterification of oils. J Biosci Bioeng 92(5):405–416. https://doi.org/10.1016/S1389-1723(01)80288-7

Gollakota ARK, Kishore N, Gu S (2018) A review on hydrothermal liquefaction of biomass. Renew Sustain Energy Rev 81:1378–1392. https://doi.org/10.1016/j.rser.2017.05.178

He H, Wang Y, Sun Y, Sun W, Wu K (2024) From raw material powder to solid fuel pellet: a state-of-the-art review of biomass densification. Biomass Bioenergy 186:107271. https://doi.org/10.1016/j.biombioe.2024.107271

Heo JB, Lee Y-S, Chung C-H (2021) Conversion of inulin-rich raw plant biomass to 2,5-furandicarboxylic acid (FDCA): progress and challenge towards biorenewable plastics. Biotechnol Adv 53:107838. https://doi.org/10.1016/j.biotechadv.2021.107838

Jiang Z, Rahman Mahmud A, Maneengam A, Nassani AA, Haffar M, The Cong P (2022) Non linear effect of biomass, fossil fuels and renewable energy usage on the economic growth: managing sustainable development through energy sector. Fuel 326:124943. https://doi.org/10.1016/j.fuel.2022.124943

Kovalev AA, Kovalev DA, Zhuravleva EA, Laikova AA, Shekhurdina SV, Vivekanand V, Litti YV (2023) Biochemical hydrogen potential assay for predicting the patterns of the kinetics of semi-continuous dark fermentation. Bioresour Technol 376:128919. https://doi.org/10.1016/j.biortech.2023.128919

Lebaka VR (2013) Potential bioresources as future sources of biofuels production: an overview. In: Biofuel technologies: recent developments. Springer, Berlin, Heidelberg, pp 223–258. https://doi.org/10.1007/978-3-642-34519-7_9

Leung DYC, Wu X, Leung MKH (2010) A review on biodiesel production using catalyzed transesterification. Appl Energy 87(4):1083–1095. https://doi.org/10.1016/j.apenergy.2009.10.006

Liu Y, Xu X, Liu Y, Liu J, Hu W, Yang N, Jawad S, Wei Z (2024) Operational optimization of a rural multi-energy system supported by a joint biomass-solid-waste-energy conversion system and supply chain. Energy 312:133528. https://doi.org/10.1016/j.energy.2024.133528

Liu Y, Xu X, Xu L, Liu Y, Liu J, Hu W, Yang N, Jawad S, Luo Y (2025) Multi-timescale collaborative operation of renewable energy-based power system and agri-product supply chain considering dynamic energy consumption-based crop growth. Appl Energy 377. https://doi.org/10.1016/j.apenergy.2024.124359

Lombardelli G, Scaccabarozzi R, Conversano A, Gatti M (2024) Bio-methanol with negative CO_2 emissions from residual forestry biomass gasification: modelling and techno-economic assessment of different process configurations. Biomass Bioenergy 188:107315. https://doi.org/10.1016/j.biombioe.2024.107315

López Barreiro D, Prins W, Ronsse F, Brilman W (2013) Hydrothermal liquefaction (HTL) of microalgae for biofuel production: state of the art review and future prospects. Biomass Bioenergy 53:113–127. https://doi.org/10.1016/j.biombioe.2012.12.029

Matsumoto K (2024) An entrained flow biomass gasification technology with the fluidized bed concept for low-carbon fuel production. Appl Energy Combust Sci 20:100292. https://doi.org/10.1016/j.jaecs.2024.100292

Mondal P, Dang GS, Garg MO (2011) Syngas production through gasification and cleanup for downstream applications—Recent developments. Fuel Process Technol 92(8):1395–1410. https://doi.org/10.1016/j.fuproc.2011.03.021

Ni J, Qian L, Wang Y, Zhang B, Gu H, Hu Y, Wang Q (2022) A review on fast hydrothermal liquefaction of biomass. Fuel 327:125135. https://doi.org/10.1016/j.fuel.2022.125135

Paul S, Mazumder C, Mukherjee S (2024) Challenges faced in commercialization of biofuel from biomass energy resources. Biocatal Agric Biotechnol 60:103312. https://doi.org/10.1016/j.bcab.2024.103312

Peinemann JC, Pleissner D (2020) Continuous pretreatment, hydrolysis, and fermentation of organic residues for the production of biochemicals. Bioresour Technol 295:122256. https://doi.org/10.1016/j.biortech.2019.122256

Pérez-Torrado R, Gamero E, Gómez-Pastor R, Garre E, Aranda A, Matallana E (2015) Yeast biomass, an optimised product with myriad applications in the food industry. Trends Food Sci Technol 46(2, Part A):167–175. https://doi.org/10.1016/j.tifs.2015.10.008

Poornima S, Manikandan S, Prakash R, Deena SR, Subbaiya R, Karmegam N, Kim W, Govarthanan M (2024) Biofuel and biochemical production through biomass transformation using advanced thermochemical and biochemical processes—A review. Fuel 372:132204. https://doi.org/10.1016/j.fuel.2024.132204

Ravichandran SR, Venkatachalam CD, Sengottian M, Sekar S, Kandasamy S, Ramasamy Subramanian KP, Purushothaman K, Lavanya Chandrasekaran A, Narayanan M (2022) A review on hydrothermal liquefaction of algal biomass on process parameters, purification and applications. Fuel 313:122679. https://doi.org/10.1016/j.fuel.2021.122679

Regadera-Macías AM, Morales-Torres S, Pastrana-Martínez LM, Maldonado-Hódar FJ (2024) Optimizing filters of activated carbons obtained from biomass residues for ethylene removal in agro-food industry devices. Environ Res 248:118247. https://doi.org/10.1016/j.envres.2024.118247

Renewable Energy Statistics (2022) Statistiques D'énergie Renouvelable 2022 Estadísticas De Energía Renovable 2022 About Irena
Roberts DA, Paul NA, Cole AJ, de Nys R (2015) From waste water treatment to land management: conversion of aquatic biomass to biochar for soil amelioration and the fortification of crops with essential trace elements. J Environ Manag 157:60–68. https://doi.org/10.1016/j.jenvman.2015.04.016
Sadrul Islam AKM, Ahiduzzaman M (2012) Biomass energy: sustainable solution for greenhouse gas emission. AIP Conf Proc 1440:23–32. https://doi.org/10.1063/1.4704200
Saleem M (2022) Possibility of utilizing agriculture biomass as a renewable and sustainable future energy source. Heliyon 8(2):e08905. https://doi.org/10.1016/j.heliyon.2022.e08905
Saravanan A, Yaashikaa PR, Senthil Kumar P, Vickram AS, Karishma S, Kamalesh R, Rangasamy G (2023) Techno-economic and environmental sustainability prospects on biochemical conversion of agricultural and algal biomass to biofuels. J Clean Prod 414:137749. https://doi.org/10.1016/j.jclepro.2023.137749
Shahabuddin M, Alam MT, Krishna BB, Bhaskar T, Perkins G (2020) A review on the production of renewable aviation fuels from the gasification of biomass and residual wastes. Bioresour Technol 312:123596. https://doi.org/10.1016/j.biortech.2020.123596
Sikkema R, Proskurina S, Banja M, Vakkilainen E (2021) How can solid biomass contribute to the EU's renewable energy targets in 2020, 2030 and what are the GHG drivers and safeguards in energy- and forestry sectors? Renew Energy 165:758–772. https://doi.org/10.1016/j.renene.2020.11.047
Stolarski MJ, Dudziec P, Krzyżaniak M, Graban Ł, Lajszner W, Olba–Zięty E (2024) How do key for the bioenergy industry properties of baled biomass change over two years of storage? Renew Energy 224:120236. https://doi.org/10.1016/j.renene.2024.120236
Strezov V, Anawar HM (n.d.) Renewable energy systems from biomass efficiency, innovation, and sustainability edited by
Taherzadeh MJ, Karimi K (2008) Pretreatment of lignocellulosic wastes to improve ethanol and biogas production: a review. Int J Mol Sci 1621–1651. https://doi.org/10.3390/ijms9091621
Tursi A (2019) A review on biomass: importance, chemistry, classification, and conversion. Biofuel Res J (Green Wave Publishing of Canada) 962–979. https://doi.org/10.18331/BRJ2019.6.2.3
Wei Z, Cheng Z, Shen Y (2024) Recent development in production of pellet fuels from biomass and polyethylene (PE) wastes. Fuel 358:130222. https://doi.org/10.1016/j.fuel.2023.130222
Wu C, Yang H, He X, Hu C, Yang L, Li H (2022) Principle, development, application design and prospect of fluidized bed heat exchange technology: comprehensive review. Renew Sustain Energy Rev 157:112023. https://doi.org/10.1016/j.rser.2021.112023
Yang J, (Sophia) He Q, Yang L (2019) A review on hydrothermal co-liquefaction of biomass. Appl Energy 250:926–945. https://doi.org/10.1016/j.apenergy.2019.05.033
Yang B, Dong Z, Tan Z, Cai Y, Xie S (2024) Roles of carbon dioxide in the conversion of biomass or waste plastics. Sci Total Environ 955:176882. https://doi.org/10.1016/j.scitotenv.2024.176882
Yeletsky PM, Dubinin YV, Yazykov NA, Tabakaev RB, Okotrub KA, Yakovlev VA (2022) Conversion of natural feedstocks to porous carbons via carbonization in fluidized catalyst bed followed by leaching the feedstock mineral template phase: a comparison of biomass and sedimentary raw materials. Fuel Process Technol 226:107076. https://doi.org/10.1016/j.fuproc.2021.107076
Zaini HM, Saallah S, Roslan J, Sulaiman NS, Munsu E, Wahab NA, Pindi W (2023) Banana biomass waste: a prospective nanocellulose source and its potential application in food industry—A review. Heliyon 9(8):e18734. https://doi.org/10.1016/j.heliyon.2023.e18734
Zamani N, Ganjifar A, Karrabi M, Shahnavaz B, Dewil R (n.d.) Bioconversion of lignocellulosic biomass wastes to energy through anaerobic Co-2 digestion with sewage sludge (An Experimental and Kinetic-Based Study) 3:4. https://ssrn.com/abstract=4807776
Zhang H, Yang K, Tao Y, Yang Q, Xu L, Liu C, Ma L, Xiao R (2023) Biomass directional pyrolysis based on element economy to produce high-quality fuels, chemicals, carbon materials—A review. Biotechnol Adv 69:108262. https://doi.org/10.1016/j.biotechadv.2023.108262

Chapter 5
Development of New Sustainable Materials Based on Valorized Fish Biomass Waste

Sajjad Ahmad, Michelle Giltrap, and Gordon Chambers

Abstract Being an essential component of the world's economy and nourishment, fish aquaculture is at risk due to improper resource management, environmental damage, and climate change. Fish side streams are an intriguing resource and treasure trove of numerous healthful and physiologically active molecules, which have the potential to significantly contribute to the production of essential components for pharmaceutical, electrical, optical, biotechnological, functional, and nutritional compounds. The practical application of these bioactive compounds has encouraged research into sustainable extraction methods to optimize the recovery process. The present chapter highlights a thorough description of research performed considering extracting valuable compounds from fish waste. Integrating proteins, and bioactive substances from the muscles, tissues and fibers from the scales and pomace increased food items' nutritional content and applicability. Phenolic molecules, carotenoids, and other bioactive substances in fish demonstrated excellent antioxidant, antimicrobial, hypoglycaemic, and anticancer properties, suggesting good potential utilising in pharmaceutical products. Finally, the main possibilities and critical challenges, as well as recent developments, are described for practical usage of these compounds for reducing the adverse impacts of food waste and to improve the circular economy.

Keyword Sustainable development goals · Aquaculture · Bioactive compounds

---

S. Ahmad (✉)
Sustainability & Health Research Hub (SHRH), Technological University Dublin, City Campus, Grangegorman, D07 ADY7, Dublin, Ireland
e-mail: Sajjad.Ahmad@tudublin.ie

S. Ahmad · M. Giltrap
School of Food Science & Environmental Health, Technological University Dublin, City Campus, Central Quad, Grangegorman, D07 ADY7, Dublin, Ireland
e-mail: Michelle.Giltrap@tudublin.ie

S. Ahmad · G. Chambers
School of Physics, Clinical and Optometric Science, Technological University Dublin, City Campus, Central Quad, Grangegorman, D07 ADY7, Dublin, Ireland
e-mail: Gordon.Chambers@tudublin.ie

A. Hind (ed.), *Biomass Waste: Processing, Valorization, Industrial Applications*,
https://doi.org/10.1007/978-981-95-0991-1_5

5.1 Introduction

The consumption of fish has been rising annually during the past few decades due to the source of appropriate nutrition worth, particularly for its omega-3 fatty acids, docosahexaenoic acid (DHA) and eicosapentaenoic acid (EPA) throughout the world (Valcarcel et al. 2020). In the 1960s, global per capita consumption was over 9.0 kg (kg) and by 2013, this amount had almost doubled to reach more than 20.0 kg per person (Erasmus et al. 2021). In many developing and food-deficient nations, fish supplies are essential to maintaining the public's health because they serve as a vital source of the essential micronutrients required for early childhood growth and development (Suhartini et al. 2022). Fish waste production has increased significantly around the globe as a consequence of global population and consumption growth and the ensuing rapid urbanization and industrialization (Ahmad et al. 2021). Significant volumes of fish waste are produced in the major catch-fishing nations, including Canada, the United States, India, the Republic of Korea, China, Spain, and Norway (Ahuja et al. 2020). The world's fishing and aquaculture sectors produced 180 million tons of catch in 2018. However, fish's heads, skin, bones, and viscera contribute to about 60% of the biomass and are considered waste (Ozogul et al. 2021; Jafarpour et al. 2020). In a broad sense, food waste has constantly been receiving a greater degree of consideration on a global scale, primarily due to its connection to environmental contamination (Bhatt et al. 2022). The best way to lower disposal costs, protect the environment, and increase income is to recover marketable by-products from fishery refuse (Garofalo et al. 2023). Furthermore, the strain of industrial fishing on fish stocks may be lessened by the recovery of fish products from waste (Xu et al. 2019). This is accomplished through valorization techniques or nutrient recovery methods (Bhatt et al. 2023).

As part of its Sustainable Development Goals (SDG) strategy, the United Nations (UN) has set a goal of substantially reducing global food loss and waste by 2030 (Hemker et al. 2020). Therefore, research is stepping up efforts to create a circular economy, which seeks to increase the intrinsic worth of resources and accomplish sustainable consumption and production (Chaklader et al. 2023). It corresponds to a cutting-edge idea that attracts much interest in order to guarantee the standard of human existence through the effective and environmentally friendly utilization of resources, energy, and infrastructure (Ahmad et al. 2022a). Currently, fish waste is primarily tossed away and only partially used to make low-profitable seafood meal, fertilizers, and fish oil or used as a raw material for direct consumption in aquaculture by conventional methods (Sathish et al. 2023; Coppola et al. 2021). So, there is an enormous opportunity to make use of these materials that seem to be waste or low value by converting them into genuine value-added components, including lipids, proteins, carotenoids, chitin and chitosan, minerals, calcium, nucleic acids, pigments, and biologically active peptides by using clean and green technologies (Saranya et al. 2018).

Due to the lack of current knowledge, researchers throughout the world are motivated to keep looking into and increase their investigations in the field of bioprospecting through the development of new technological or extractive methods (Siddik et al. 2021). Additionally, transforming these raw materials using advanced valorization technologies offers a significant extra revenue opportunity to start businesses for producing delicate compounds, like pharmaceuticals and nanomaterials. However, to the support of more eco-friendly, socially, and economically sound business models, there is an urgent need for new approaches for the extraction and valorization of by-products from the aquaculture sector. In contrast to other reviews on the subject, our review emphasizes novel aspects of the valorization of biomass-based products, such as the capability of fish waste to produce energy from a biorefinery context within the broader framework of the circular economy (CE). In addition, this evaluation will assess possibilities for solutions for the obstacles that impede or restrict marine circularity. This study will generally catalyze researchers, entrepreneurs, and authorities to concentrate on essential technologies promoting a CE in aquaculture.

5.2 Strategies for Fish Waste Valorization

There are different kinds of strategies involved in utilizing fish waste. The suitability of these techniques could be analyzed by various parameters such as the selection of a method able to provide high-quality by-products at an acceptable level, the stability and shelf life of the method, and the last one which is more important it should be compatible with the environment and human resources. In the context of the circular economy, to achieve more sustainable methods for the development and utilization of by-products, process intensification (PI) could be a feasible option (Ahmad et al. 2022b). Smaller, less expensive, safer, and smoother have been generally used to describe the PI philosophy. According to one of the many definitions of PI, it includes any new machinery, techniques for processing, and methods for developing innovative procedures that significantly outperform existing ones in the production and processing of biochemicals (Monsiváis-Alonso et al. 2020). An additional PI technique recommends improving the fundamental functional principles of momentum, heat, and mass transfer in chemical, physical and biological reactions and designing operation units (Ahmad et al. 2022c). Nowadays, the valorization of fish wastes has been attempted by conventional methods such as sun drying, freezing, cooking, salting, and canning by applying toxic solvents such as HCl and NaOH. These methods must be revised to produce high and efficient yields of waste by-products (Saranya et al. 2020). Hence, the ultimate aim of this review literature is to provide new insights for valorizing fish waste and to understand the novel concepts and interactions with alternative approaches that are fundamentally smart, inexpensive, easy to handle, efficient, greener, environment friendly and facilitators to provide high-quality yield.

5.3 Highly Profitable By-Products of Fish Side-Stream Biomasses and Their Applications

5.3.1 Fish Oils

Oils from marine sources have a greater unsaturation content than terrestrial organisms. The degree of fat and composition of fatty acids in fish oils are affected by the fish's species, size and weight at growth, body components, the season and environment, and diet (Ideia et al. 2020). For instance, fish from arctic and boreal areas significantly differ in fat composition because of water temperature and viscous home adaptation (Välimaa et al. 2019). Fish oil is a significant source of long-chain polyunsaturated n-3 fatty acids (PUFAs), especially (EPA) and (DHA), which are particularly essential for human nutrition (Ahmed et al. 2022). Although the human body can also produce these fatty acids by converting alpha-linolenic acid, their yield is meagre due to poor enzymatic conversion efficiency. However, increasing their level consuming of fish food is the principal way (Zhang et al. 2020). Authorities and health organizations (such as FAO/WHO, the American Dietetic Association, or the American Heart Association) constantly update and revise the recommended daily consumption of PUFAs. Although, the current suggestions for daily intake of EPA, DHA and PUFAs range from 140 to 600 mg and 1.4 to 25 g, respectively (Marsol-Vall et al. 2022).

Fish (PUFAs) play a crucial role in the structural development of cell membranes and regulate their morphology and permeability by providing metabolic energy to the body (Zhao et al. 2019). They are utilized to synthesize eicosanoids, thromboxanes, leukotrienes, oxygenated metabolites prostaglandins, which are all engaged in a variety of cell-signalling pathways that control the regulation of the immunological, endocrine, cardiovascular, pulmonary, and reproductive functions (Monsiváis-Alonso et al. 2020). The promised health advantages of omega-6 fatty acids are linked to a variety of diseases, such as cardiovascular disease, infant neurodevelopment, cancer, Alzheimer's disease, cognitive function, rheumatoid arthritis, depression, attention deficit hyperactivity disorder, and certain eye conditions like age-related macular degeneration and dry eyes (Caruso et al. 2020; Patel et al. 2020). They could also delay or stop the synthesis of insulin resistance by enhancing the activities of the endothelial reticulum and mitochondrial bioenergetics (Charles et al. 2021). Recently, the extraction of fish oil was carried from European sea bass (*Dicentrarchus labrax*) and gilthead sea bream (*Sparus aurata*) and investigated their efficiency against cardiovascular diseases, antibacterial and antifungal impacts against food pathogens and cytotoxic effect against breast cancer cells. The findings of this study revealed that fish oils were rich in lipids and excellent agents for the protection of cardiovascular diseases. Fish oil's greatest cytotoxic effects t (GI50 = 38–46 μg/mL) were observed against the breast cancer cell lines (MCF-7). In addition, their contribution as an antioxidant and inflammatory against food pathogens was outstanding (de la Fuente et al. 2022). In another study, Pusceddu et al. (2016) found that fish oil is essential for the development and functions of the brain in children and adults.

Fish discards are rich in fat and are also efficient for the production of biodiesel. They are considered the most efficient renewable, sulphur-free, eco-friendly, oxygenated fuel and produce a smaller number of greenhouse gases (Viswanathan and Wang 2021). In a recent study, initially, the waste fish oil (WFO) was utilized to synthesize biodiesel and then by applying ultrasound irradiation, various formulations of (water/WFO) and (diesel/biodiesel) emulsion fuels were prepared. Following that, multiple emulsion fuel formulations were used at a uniform engine speed (1800 rpm) under a full load situation to examine a single-cylinder diesel engine's efficiency and exhaust emission levels. Results of this investigation indicated that the optimal emulsion fuel composition taking engine performance and emission considerations to be 3% WFO biodiesel, 6.13% water, and 1% surfactant. Regarding engine exhaust emissions, the optimum emulsion fuel showed a 42.86% reduction in carbon monoxide emission, a 34.02% reduction in unburned hydrocarbon emission, and a 25.53% reduction in nitrogen oxide emission compared to pure diesel fuel. Moreover, the engine torque and brake power decreased significantly by 6.33 and 7.86%, respectively, and the brake-specific fuel utilization increased by 9.52% (Khanjani and Sobati 2021).

5.3.2 Collagen and Gelatin

Collagen is a complex and abundant structural protein mainly found in animal tissues, which accounts for around 30% of all proteins. Conventionally, collagen has been mostly obtained from mammalian tissues; the richest sources include pig skin, bovine hide, pork and cattle bones, with a 46, 29.4, and 23% ratio, respectively (Felician et al. 2018). However, the consumption of collagen from terrestrial animals has been linked to potential hazards for various diseases, including foot-and-mouth disease, bovine spongiform encephalopathy, and other degenerative disorders of the central nervous system (Jongjareonrak et al. 2005). Additionally, halal and kosher meals cannot contain collagen that comes from pork (Sadowska et al. 2003). At least 28 varieties of collagen have been found to date, each with a distinctive amino acid sequence, structure, and function (Sionkowska et al. 2020). However, the category of collagen that makes up more than 90% of human collagen is fibrillar collagen, which comprises types I, II, III, V, and XI (Shori et al. 2021).

Fish tissue has been mainly demonstrated to have type I collagen, and it has been isolated and described from the by-products such as skin, bones, fins, and scales processing of seawater and freshwater fish (Ampitiya et al. 2023). Consequently, collagen type I is the most prevalent kind of collagen, with its primary functional role-playing in the skin and bone and, to a limited extent, in the extracellular matrix of various other tissues (Naomi et al. 2021). In terms of applications, the main difference between fish and mammalian collagen is related to cross-linkage development and their stability at high temperatures (Seixas et al. 2020). Furthermore, the species-specific environments affect the denaturation temperature of fish collagens. Amino acids (hydroxyproline and proline) make up the majority of collagen's primary structural components, and their presence directly positively correlates with both collagen's thermal stability and denaturation temperature (Amirrah et al. 2022). Nowadays, most studies investigate the characteristics of collagen type I, which has increased comprehension knowledge for various medicinal uses and benefited tissue engineering (Rezvani Ghomi et al. 2021). It is a naturally occurring biomaterial that is frequently employed in the creation of tissue replacements, controlled medication release (innovative scaffolds), wound care products, as a substrate for fragrance and cosmetic enhancing treatments, and the treatment of linked medical conditions like regeneration of bones (bin Mh Busra et al. 2016; Rico-Llanos et al. 2021). Due to its superior binding capacity, collagen contributed efficiently to drug delivery systems, growth factors and cell carriers (Bisht et al. 2021). In a recent study, distinctive tuna collagen hydrogels were developed as drug delivery systems with two structurally distinct antitumoral model's drugs, such as Doxorubicin and Crocin, to be injected locally in tissues with complicated human anatomies during surgical resection. The Doxorubicin and Crocin drug release was improved by the innovative gelatin hydrogels sustainably, highlighting the significance of structure–function in the material's chemical structure (Hermida-Merino et al. 2022). In another study, collagen extraction was efficiently obtained from the skin of *Scomberomorus lineolatus* fish. Results indicate that fish skin collagen is more functional and effectively employed as a biopolymer sheet for wound healing, food, drug delivery, tissue engineering, and pharmaceutical applications (Abinaya and Gayathri 2019).

Gelatin is created by partially hydrolysing collagen and heating it. In general, acid and alkaline hydrolysis produce types A and B of gelatin, respectively (Rana et al. 2023). The intermolecular and intramolecular connections in collagen fibres are broken down during the hydrolysis process that produces gelatin, and the structural feature of gelatin depends on the arrangement of structure, molecular weight, and biochemical profile of each subunit (Nurilmala et al. 2022). Two main factors affect the production of fish gelatin: the first is the lower percentage of intra- and extra-chain non-reducible crosslinks, and this will increase the susceptibility of fish skin collagen to deterioration; the second is that various aquatic species' skin will naturally have distinctive collagen molecules (Wasswa et al. 2007). Gelatin's chemical profile is similar to that of collagen but also comprises significant amounts of glycine, proline, and hydroxyproline molecules with a ratio of 21, 12, and 12%, respectively (Alipal et al. 2021). In contrast to collagen, gelatin is a multipurpose component that could be employed in various business sectors. The particular activity caused by gelatin's physical and chemical characteristics makes it useful in the food processing industry. Gelatin performs significant functions being a gelling agent, stabilizer, thickening up, water binder, coating, and an emulsifier in the food industry (Ghorbani et al. 2020). Gelatin is potentially helpful as colloidal protection in the photo and electroplating industries for covering metal. Gelatin also expands even in cold water, one of its other distinctive qualities. It may transform reversibly from sol to gel and vice versa, and these characteristics of gelatin make it superior to other substances like xanthan gum, carrageenan, and pectin (Tang et al. 2022). Edible gelatin is used for confectionary items, including fruit salads, ice cream, froth, and cottage cheese. For canned meat products like sausages, frankfurters, breads, and hams, gelatin is additionally employed to preserve lost liquids and offer an effective heat transmission medium during cooking (Ramos et al. 2016). Additionally, gelatin could be utilized as a gelling ingredient in cosmetic and health items such as face cream, body lotion, hair spray, bath salts, shampoo and conditioner, sunscreen, and shampoo (Lou and Chen 2023). Gelatin could be effectively utilized in air filters to analyze, preserve, and identify allergic-causing airborne microorganisms (Sun et al. 2019). However, fish-based gelatin proceeds to be one of the most viable substitutes for mammalian-based gelatin, or even its substitute equivalents, "veggie-gelatin," to meet the changing needs and uses of gelatin (Ranasinghe et al. 2022).

5.3.3 Chitin and Chitosan

Chitin, or poly (1-4) N-acetyl-D-glucosamine, is regarded as the second most abundant naturally occurring polycationic amino polysaccharide compound on Earth's surface after cellulose, whereas chitosan (1-4)-2-amino-2-deoxy-D-glucan is a polymer that is generated through the partial breakdown of chitin via chemical or biological techniques by the demineralization, deproteinization, decolourization, and deacetylation processes (Santos et al. 2020). This compound has a large molecular weight and is chemically made of N-acetyl-2-amino-2-deoxy-D-glucose units connected by glycosidic linkages (1-4), producing a linear chain with part of the deacetylated monomer units (Chen et al. 2019). In its natural state, chitin has no taste or smell and cannot be dissolved in water or organic solvents. After purification, it presents in either white or yellowish colour powder (Novikov et al. 2023). Chitin and its by-products are highly biocompatible and degrade quickly in the body's tissues. They also have high economic values and are demonstrated in a range of biological activities, including antibacterial, anti-cancer, anti-coagulant, antioxidants, antimutagenic, hypocholesterolemic, anti-proliferative, immunity-enhancing, as well as film-forming operations (Arrouze et al. 2019). Additionally, chitin is utilized in the food sector to produce biodegradable films and encapsulate flavouring agents (Rocha-Pimienta et al. 2023).

Chitin metabolites (chitosan) are employed in biomedical applications for the synthetic reconstruction of various tissues, including skin, bones, and cartilage, as well as in the pharmaceutical sector as a drug delivery candidate (Joseph et al. 2021). From the chemical point of view, the viscosity and molecular weight play a critical role in deciding their applications. For instance, a lower molecular weight of 2×10^6 Da and 7% weight-per-weight (w/w) nitrogen is appropriate for textile, confectionery items, and photographic, medicinal, and environmental implications (Boudouaia et al. 2019). Conversely, materials with more excellent crystallinity are stiffer and possess more muscular mechanical strength, recommending their implementation in tissue engineering (Joseph et al. 2022). Similarly, chitosan with a greater degree of acetylation depicted less degradation efficiency than lysozymes and other degradative enzymes and could be applied in various nutraceutical delivery systems (Liu et al. 2023). Chitosan-based membranes have demonstrated potential for adsorbing contaminants, including colours, phenolic chemicals, heavy metals, pesticides, and pharmaceuticals (Salehi et al. 2022). Additionally, the immobilization of chitosan with biological agents is considered an ideal tool for the environmental recovery and degradation of various pollutants in terrestrial and aquatic systems (Siddiqui et al. 2022). Irawan et al. (2018) concluded that the chitosan recovered from the fish scales of Papuyu fish enhanced the degradation of iron in the polluted water from 11.80 to 3.43 mg/L compared to commercial chitosan from shrimp cells. As can be discovered, the utilization of chitosan-based substances might provide a comprehensive and systemic strategy for revalorizing marine biomass waste, which would decrease or completely eradicate the contamination of the environment caused by the disposing of crustacean shell waste, thereby enhancing water quality and safeguarding aquatic ecosystems.

5.3.4 *Enzymes and Minerals*

An extensive number of enzymes with catalytic capabilities have been found through the bioprospecting of aquatic species and could prove a valuable asset in hundreds of biotechnological applications (Sridhar et al. 2021). For instance, proteases are contributed to around 60% of the enzyme market's commercial use, and fish proteinases have excellent activity across various pH and temperature conditions, making them a good choice for several industrial implications (Raveendran et al. 2018). The internal organs such as the stomach, pancreas and intestines are the most prevalent fish organs for purifying several enzymes, which are cold-active, exhibit higher catalytic activities and have greater stability in extreme environments (Caglayan et al. 2020). Proteases from fish stomachs are naturally salt-activated and are apt for preparing fish sauces or fermented products where certain other enzymes from varied sources get inhibited in sodium chloride (Osuna-Ruiz et al. 2019). In addition to proteases, fish also contain large amounts of the lipase group of enzymes. These enzymes are glycerol-ester hydrolases which are responsible for catalyzing the breakdown of ester bonds into various substrates compounds such as triglycerides, phospholipids, cholesteryl esters, and vitamin esters (Liang et al. 2022). Lipases are frequently employed in various sectors, such as food, cosmetics, medicine, detergents and the biodegradation of pollutants (Vivek et al. 2022). Further significant uses involve the manufacture of biosensors, agrochemicals, waxes, and biopolymers, as well as the production of tea and lubricants, which are ideal substitutes for traditional petroleum lubricants (Bolina et al. 2021). The efficacy of enzymes derived from fish is frequently verified using enzyme tests based on spectrophotometry. For instance, the efficacy of lysozyme enzymes derived from fish mucus tissues was evaluated at 450 nm by including freeze-dried lysosome-sensitive bacteria (Wang et al. 2019). While the alkaline phosphatase activation was measured at 405 nm via incubation of fish skin samples with p-nitrophenol (Khailova et al. 2020). The quantity of enzyme produced to transform the fish waste material into the desired product was measured and compared with the reference substances in both experiments.

Fish wastes are an excellent source of minerals, particularly bones contain 60–70% of fish-based minerals with an excellent quantity of hydroxyapatite, phosphorus, calcium, phosphate, zinc, selenium, and iron, which are excellent sources of food supplements (Caruso et al. 2020). Fishbone contains hydroxyapatite, which may be taken out and utilized for quick bone healing following significant surgery or damage. It can also be used in dental applications as a bone transplant (Maschmeyer et al. 2020). Bones from fish waste are often treated with phosphorus-solubilizing microorganisms and processed to create biofertilizers to increase agricultural output (Alfio et al. 2021). Additionally, selenium derived from fish is more digestible than other sources, such as yeast (Alhasan et al. 2019). Iron is a component of the haemoglobin molecule, which carries oxygen throughout the body's tissues, whereas zinc is primarily found as a cofactor in enzyme processes. Sodium and potassium interact to regulate electrolyte balance and aid in transmitting nerve impulses (Ashraf et al. 2020).

5.3.5 Fish Proteins and Amino Acids

Fish is an excellent source of proteins, with harmonious necessary amino acids, including lysine, valine, and phenylalanine (Wang et al. 2021). The frames and muscles of fish contain large amounts of proteins and amino acids that are highly bioactive and simple to digest, which control essential metabolic pathways and function as building blocks for the creation of physiologically relevant compounds (Alagawany et al. 2021). They are also necessary for the body's optimal development and growth, as well as for maintaining and mending tissues that have been worn out (Zhu et al. 2020). Most significantly, they contribute significantly to reducing protein-calorie malnutrition. Immunoglobulins, a protein, can aid in maintaining the ideal equilibrium of water and electrolytes in the human body and serve as a significant line of protection against bacterial and viral infections. Additionally, fish protein helps the body's various regulatory systems maintain balance (Balami et al. 2019). Fish proteins are very effective and have a variety of uses in the food industry due to their outstanding qualities, including their ability to emulsify and absorb oil, foaming and gelling capabilities, and water-holding capacities (Venugopal 2021). Fish protein can be beneficial for therapeutic purposes such as immune-modulating, antithrombotic, and antioxidant properties. The kidney, blood, fluids, and mucus tissues of fishing bycatch, including sea bass, were collected for their proteins, demonstrating outstanding agglutinating and antibacterial activity and a rich bioactive source (Khan et al. 2022).

On the other hand, amino acids are vital macromolecules that act as the building blocks of proteins and intermediaries throughout several metabolic processes. They are primarily derived through a protein-rich diet, and the nutritional value of dietary protein is determined by the proportion of essential to non-essential amino acids (Zhang et al. 2022). An inadequate or unbalanced amino acid ratio may impact body metabolism and homeostasis. As a result, consuming enough amino acids in people's diets can prevent such deficits or metabolic diseases (Viji et al. 2022). Glycine was extracted from the Senegalese sole (*Solea senegalensis*) and found that it contributed to controlling the expression of genes, the structure and activity of proteins, and a number of biological processes, including the production of glutathione. Lower plasma glycine contents are also frequently linked to type 2 diabetes and obesity (Alves Martins et al. 2013). In another study, phenylalanine amino acid was recovered from three different fish samples (*Catla catla*, *Labeo rohita* and *Cirrhinus mrigala*) and observed that it was involved in the precursor for tyrosine (Hussain et al. 2018). However, by the implications of sustainable extraction approaches, many fish trace and essential amino acids could be recovered and marketed as functional food supplements and other biomedical uses in the market worldwide.

5.3.6 Fish Protein Hydrolysates and Bioactive Peptides

Fish leftovers, including the head, frame, viscera, liver, kidneys, eggs, and skin, contain significant quantities of highly nutritious proteins (10–23%) that can be utilized as a base for the generation of hydrolysates (Idowu et al. 2021). Protein hydrolysates are the degradation compounds created when the protein is hydrolysed chemically (by an acid or an alkali) or biologically. They are composed of smaller fragments of peptides comprising (2–20) amino acid residues (Sharkey et al. 2020). It has been demonstrated that fish protein hydrolysate exhibits a range of biological activities, such as anti-inflammatory, anti-diabetic, ACE inhibitory, immunomodulatory, anti-hyaluronidase, and anti-tyrosinase (Espinales et al. 2023). Additionally, it has been found that fish protein hydrolysate has flavouring, emulsifying, foaming, Ca-binding, fat-binding, and iron-binding capabilities. Fish protein hydrolysates are also utilized in culinary, cosmetic, and medicinal products because of their wide range of biological activity and good digestibility (Korkmaz and Tokur 2022). Recent studies on radical scavenging discovered that fish hydrolysates containing polypeptides derived from the skin, cartilage, and bone had remarkable antioxidant properties. The peptides also showed a defence against kidney damage in mice under oxidative stress conditions (Lees and Carson 2020). Moreover, extremely high levels of β-secretase inhibitory properties are present in the neuroprotective actions of peptide sequences such as QGYRPLRGPEFL obtained from skate fish skin (Ryu et al. 2021). Fish waste-based proteins are primarily used in the healthcare industry, although polypeptides and hydrolysates are also frequently utilized in the food and nutraceutical sectors. In addition, protein hydrolysates of fish waste are progressively employed in formulations related to skin elasticity and wrinkle reduction in the cosmetics industry (Pezeshk et al. 2019).

Bioactive peptides are food-derived substances that added their nutritional significance and impose physiological impacts on the body. These physiological processes are significantly controlled by specific peptides encoded in the native protein sequences (Cipolari et al. 2020). The peptides which are found in the inactive form within the protein chains are turned into active forms after being hydrolyzed by various enzymes such as trypsin, proteinases, chymotrypsin, alcalde or pepsin (Venkatesan et al. 2017). Recently, there has been much interest in using bioactive peptides as functional foods and nutritional supplements. Bioactive peptides, which are generated through the hydrolysates of proteins from fish, have demonstrated a number of medical benefits, including antihypertensive, antioxidant, antibacterial, and antiproliferative activities (Chakrabarti et al. 2018). Additionally, bioactive peptides have been found to possess additional biological properties, including antithrombotic, opioid, cholesterol-lowering, immunomodulating, and antidiabetic properties (Bechaux et al. 2019). However, these bioactive peptides can exhibit favourable health effects and are thus regarded as essential for producing nutrients or functional food products based on several findings showing their potential therapeutic utilization.

5.3.7 Carotenoids or Natural Pigments

Fish and other seafood contain the majority of the fat-soluble carotenoids. Astaxanthin (3,3-dihydroxy-β, β-carotene-4,4-dione) is the most prevalent carotenoid in seafood and makes up 74–98% of all the pigments in the shells (Ashokkumar et al. 2022). The other crucial carotenoids which are frequently found in fish are lutein (greenish yellow), beta carotene (orange), alpha, beta-doradexanthins (yellow), zeaxanthin (yellow orange), canthaxanthin (orange red), and astaxanthin (yellow) (Singh et al. 2022). Seafood contains astaxanthin in its natural state, and esters form. In nature, it is non-toxic, extremely versatile, and easy to dissolve in both polar and non-polar solvents (Elbahnaswy and Elshopakey 2023). The structure of this substance consists of a chain of conjugated double bonds and both ketonic and hydroxylic functional groups, which are responsible for its strong antioxidant qualities. The antioxidant astaxanthin is essential in avoiding age-related muscle deterioration and cancers caused by chemical compounds (Huang et al. 2023). Additionally, it can strengthen the immune system and provide shelter to cells from harmful radiation. It is used extensively as a colourant, antioxidant, and source of vitamins (A and E) in the food and medical sectors (Yadollahi et al. 2021). Carotenoids also have been employed in sunscreen creams, lotions, and other antioxidants and algae bioactive compounds as preservatives in cosmetic products (Catanzaro et al. 2020). Furthermore, it has been demonstrated that the carotenoids lutein and zeaxanthin shield the eye's macula from blue light damage (Ashraf et al. 2020). The most widely used β-carotene is an orange-yellow pigment derived from marine sources. The pigment provides a distinctive shade and has the greatest vitamin-A activity due to the two-ionone rings, which play a crucial role in maintaining human health (Manzoor et al. 2021).

5.4 Current Bottlenecks and Future Perspective

Fish waste has been recognized as a nutritional, functional, and nutraceutical by-product with a broad range of benefits in different sectors, making it a viable alternative for social, economic, and environmental issues. Because of their unique qualities and durability, novel green technologies, including UAE, PEF, MAE, HAE, EE and SCE, are viewed as prospective technologies. Nevertheless, most of the developments are still in the lab or pilot projects, with a comparatively low stage of technological development. It is also essential to consider eliminating additional risk factors, such as hazardous chemicals, antinutritional elements, energy consumption, and instrument costs. The taste and texture of the food materials and customer acceptance are significant additional barriers to using fish waste as functional food components for producing and designing new food products. Furthermore, it is important to develop easy, low-cost processing methods and discover strategies to increase the shelf-life of these items. Meanwhile, a significant amount of research must be conducted to increase their industrial scalability, and a thorough analysis is required to determine the impact of these approaches on the characteristics of the bioactive recovered from fish waste. Additional studies may be carried out using multiple transducers for ultrasonic extractors, multiple orifices, or multiple ventilation devices in the pressurized hydraulic extraction system. However, it is essential to remember that from the farm to the consumption, novel insights and each person's responsibilities are crucial to ensuring and maintaining aquaculture food safety.

5.5 Conclusion

The notion of a biorefinery is further emphasized by the recent development of a health trend and environmental awareness among millennials, while countries throughout the world create legislation and encourage technical advancement in the valorization of fish waste to attain SDG targets of "zero discard." The recovery of bioactive materials from fish waste is attracted the consumer market and significantly contributes to the economy and environmental restoration. Extraction technologies are the primary domain of concern for sustainable valorization and utilization of bioactive substances. The current extraction methods are non-stable, time-consuming, expensive, and hazardous to delicate bioactive molecules. In order to accomplish the goals for sustainable development in attaining food safety and food security for the expanding global human population, novel green extraction methods are required for minimal processing, maximum quality and output, and product safety (non-toxic organic solvents). In this context, UAE, PEF, MAE, HAE, EE and SCE are considered cutting-edge, environment-friendly and sustainable techniques that increase the recovery of thermostable bioactive materials and maintain their quality and oxidative stability due to the fact these compounds have a variety of identified bioactivities, technological, and pharmaceutical characteristics and could be utilized as food antioxidants as well as active ingredients in nutraceutical and cosmetic products, with significant health benefits. Additionally, these green extraction techniques have the potential to be easily scaled up to the industrial level. In order to understand perspectives and research needs in this domain, researchers should evaluate each approach based on selection, optimization, effectiveness, required materials and energy consumption, the possible dangers, and novel products developed throughout the extraction process. It is also important to develop new understandings of the various approaches, such as fish cultivation methods, genetic varieties with desired traits, and breeding techniques to enhance the by-product's value.

References

Abinaya M, Gayathri M (2019) Biodegradable collagen from *Scomberomorus lineolatus* skin for wound healing dressings and its application on antibiofilm properties. J Cell Biochem 120(9):15572–15584. https://doi.org/10.1002/jcb.28824

Ahmad N, Qamar M, Yuan Y, Nazir Y, Wilairatana P, Mubarak MS (2022a) Dietary polyphenols: Extraction, identification, bioavailability, and role for prevention and treatment of colorectal and prostate cancers. Molecules 27(9):2831. https://doi.org/10.3390/molecules27092831

Ahmad S, Ahmad HW, Bhatt P (2022b) Microbial adaptation and impact into the pesticide's degradation. Arch Microbiol 204(5):288. https://doi.org/10.1007/s00203-022-02899-6

Ahmad S, Bhatt P, Ahmad HW, Cui D, Guo J, Zhong G (2022c) Enzymes involved in the bioremediation of pesticides, In: Bhatt P (ed) Industrial applications of microbial enzymes. CRC Press, Boca Raton, FL, pp 133–168. https://doi.org/10.1201/9781003202998-7

Ahmad S, Cui D, Zhong G, Liu J (2021) Microbial technologies employed for biodegradation of neonicotinoids in the agroecosystem. Front Microbiol 3510. https://doi.org/10.3389/fmicb.2021.759439

Ahmed I, Jan K, Fatma S, Dawood MA (2022) Muscle proximate composition of various food fish species and their nutritional significance: a review. J Anim Physiol Anim Nutr 106(3):690–719. https://doi.org/10.1111/jpn.13711

Ahuja I, Dauksas E, Remme JF, Richardsen R, Løes AK (2020) Fish and fish waste-based fertilizers in organic farming—with status in Norway: a review. Waste Manage 115:95–112. https://doi.org/10.1016/j.wasman.2020.07.025

Alagawany M, Taha AE, Noreldin A, El-Tarabily KA, Abd El-Hack ME (2021) Nutritional applications of species of *Spirulina* and *Chlorella* in farmed fish: a review. Aquaculture 542:736841. https://doi.org/10.1016/j.aquaculture.2021.736841

Alfio VG, Manzo C, Micillo R (2021) From fish waste to value: an overview of the sustainable recovery of omega-3 for food supplements. Molecules 26(4):1002. https://doi.org/10.3390/molecules26041002

Alhasan R, Nasim MJ, Jacob C, Gaucher C (2019) Selenoneine: a unique reactive selenium species from the blood of tuna with implications for human diseases. Curr Pharmacol Rep 5:163–173. https://doi.org/10.1007/s40495-019-00175-8

Alipal J, Pu'Ad NM, Lee TC, Nayan NHM, Sahari N, Basri H, Idiris MI, Abdullah HZ (2021) A review of gelatin: properties, sources, process, applications, and commercialisation. Mater Today: Proc 42:240–250. https://doi.org/10.1016/j.matpr.2020.12.922

Alves Martins D, Rocha F, Castanheira F, Mendes APousão-Ferreira P, Bandarra N, Coutinho J, Morais S, Yúfera M, Conceicao LEC, Martínez-Rodríguez G (2013) Effects of dietary arachidonic acid on cortisol production and gene expression in stress response in Senegalese sole (*Solea senegalensis*) post-larvae. Fish Physiol Biochem 39:1223–1238. https://doi.org/10.1007/s10695-013-9778-6

Amirrah IN, Lokanathan Y, Zulkiflee I, Wee MMR, Motta A, Fauzi MB (2022) A comprehensive review on collagen type I development of biomaterials for tissue engineering: from biosynthesis to bioscaffold. Biomedicines 10(9):2307. https://doi.org/10.3390/biomedicines10092307

Ampitiya AGDM, Gonapinuwala ST, Fernando CAN, De Croos MDST (2023) Extraction and characterisation of type I collagen from the skin offcuts generated at the commercial fish processing centres. J Food Sci Technol 60(2):484–493. https://doi.org/10.1007/s13197-022-05630-x

Arrouze F, Desbrieres J, Rhazi M, Essahli M, Tolaimate A (2019) Valorization of chitins extracted from North Morocco shrimps: comparison of chitin reactivity and characteristics. J Appl Polym Sci 136(30):47804. https://doi.org/10.1002/app.47804

Ashokkumar V, Flora G, Sevanan M, Sripriya R, Chen WH, Park JH, Kumar G (2022) Technological advances in the production of carotenoids and their applications—a critical review. Biores Technol 128215. https://doi.org/10.1016/j.biortech.2022.128215

Ashraf SA, Adnan M, Patel M, Siddiqui AJ, Sachidanandan M, Snoussi M, Hadi S (2020) Fish-based bioactives as potent nutraceuticals: exploring the therapeutic perspective of sustainable food from the sea. Mar Drugs 18(5):265. https://doi.org/10.3390/md18050265

Balami S, Sharma A, Karn R (2019) Significance of nutritional value of fish for human health. Malays J Halal Res 2(2):32–34. https://doi.org/10.2478/mjhr-2019-0012

Bechaux J, Gatellier P, Le Page JF, Drillet Y, Sante-Lhoutellier V (2019) A comprehensive review of bioactive peptides obtained from animal byproducts and their applications. Food Funct 10(10):6244–6266. https://doi.org/10.1039/C9FO01546A

Bhatt P, Ahmad S, Joshi S, Bhatt K (2022) Recent advancement in microbial enzymes and their industrial applications. In: Bhatt P (ed) Industrial applications of microbial enzymes. CRC Press, Boca Raton, FL, pp 1–17. https://doi.org/10.1201/9781003202998-1

Bhatt P, Chaudhary P, Ahmad S, Bhatt K, Chandra D, Chen S (2023) Recent advances in the application of microbial inoculants in the phytoremediation of xenobiotic compounds. In: Chandra D, Bhatt P (eds) Unravelling plant-microbe synergy. Elsevier, pp 37–48. https://doi.org/10.1016/B978-0-323-99896-3.00013-8

bin Mh Busra MF, Chowdhury SR, bin Ismail F, bin Saim A, Idrus RH (2016) Tissue-engineered skin substitute enhances wound healing after radiation therapy. Adv Skin Wound Care 29(3):120–129. https://doi.org/10.1097/01.ASW.0000480556.78111.e4

Bisht M, Martins M, Dias AC, Ventura SP, Coutinho JA (2021) Uncovering the potential of aqueous solutions of deep eutectic solvents on the extraction and purification of collagen type I from Atlantic codfish (*Gadus morhua*). Green Chem 23(22):8940–8948. https://doi.org/10.1039/D1GC01432C

Bolina IC, Gomes RA, Mendes AA (2021) Biolubricant production from several oleaginous feedstocks using lipases as catalysts: current scenario and future perspectives. Bioenergy Res 1–19. https://doi.org/10.1007/s12155-020-10242-4

Boudouaia N, Bengharez Z, Jellali S (2019) Preparation and characterization of chitosan extracted from shrimp shells waste and chitosan film: application for Eriochrome black T removal from aqueous solutions. Appl Water Sci 9(4):91. https://doi.org/10.1007/s13201-019-0967-z

Caglayan C, Taslimi P, Türk C, Gulcin İ, Kandemir FM, Demir Y, Beydemir Ş (2020) Inhibition effects of some pesticides and heavy metals on carbonic anhydrase enzyme activity purified from horse mackerel (*Trachurus trachurus*) gill tissues. Environ Sci Pollut Res 27:10607–10616. https://doi.org/10.1007/s11356-020-07611-z

Caruso G, Floris R, Serangeli C, Di Paola L (2020) Fishery wastes as a yet undiscovered treasure from the sea: Biomolecules sources, extraction methods and valorization. Mar Drugs 18(12):622. https://doi.org/10.3390/md18120622

Chaklader MR, Howieson J, Foysal MJ, Hanif MA, Abdel-Latif HM, Fotedar R (2023) Fish waste to sustainable additives: Fish protein hydrolysates alleviate intestinal dysbiosis and muscle atrophy induced by poultry by-product meal in *Lates calcarifer* juvenile. Front Nutr 10:1145068. https://doi.org/10.3389/fnut.2023.1145068

Chakrabarti S, Guha S, Majumder K (2018) Food-derived bioactive peptides in human health: challenges and opportunities. Nutrients 10(11):1738. https://doi.org/10.3390/nu10111738

Charles AL, Abdillah AA, Saraswati YR, Sridhar K, Balderamos C, Masithah ED, Alamsjah MA (2021) Characterization of freeze-dried microencapsulation tuna fish oil with arrowroot starch and maltodextrin. Food Hydrocoll 112:106281. https://doi.org/10.1016/j.foodhyd.2020.106281
Chen C, Li D, Yano H, Abe K (2019) Insect cuticle-mimetic hydrogels with high mechanical properties achieved via the combination of chitin nanofiber and gelatin. J Agric Food Chem 67(19):5571–5578. https://doi.org/10.1021/acs.jafc.9b00984
Cipolari OC, de Oliveira Neto XA, Conceição K (2020) Fish bioactive peptides: a systematic review focused on sting and skin. Aquaculture 515:734598. https://doi.org/10.1016/j.aquaculture.2019.734598
Coppola D, Lauritano C, Palma Esposito F, Riccio G, Rizzo C, de Pascale D (2021) Fish waste: from problem to valuable resource. Mar Drugs 19(2):116. https://doi.org/10.3390/md19020116
de la Fuente B, Pinela J, Calhelha RC, Heleno SA, Ferreira IC, Barba FJ, Berrada H, Caleja C, Barros L (2022) Sea bass (*Dicentrarchus labrax*) and sea bream (*Sparus aurata*) head oils recovered by microwave-assisted extraction: nutritional quality and biological properties. Food Bioprod Process 136:97–105. https://doi.org/10.1016/j.fbp.2022.09.004
Elbahnaswy S, Elshopakey GE (2023) Recent progress in practical applications of a potential carotenoid astaxanthin in aquaculture industry: a review. Fish Physiol Biochem 1–30. https://doi.org/10.1007/s10695-022-01167-0
Erasmus VN, Kadhila T, Gabriel NN, Thyberg KL, Ilungu S, Machado T (2021) Assessment and quantification of Namibian seafood waste production. Ocean Coast Manag 199:105402. https://doi.org/10.1016/j.ocecoaman.2020.105402
Espinales C, Romero-Peña M, Calderón G, Vergara K, Cáceres PCPJ (2023) Collagen, protein hydrolysates and chitin from by-products of fish and shellfish: an overview. Heliyon e14937. https://doi.org/10.1016/j.heliyon.2023.e14937
Felician FF, Xia C, Qi W, Xu H (2018) Collagen from marine biological sources and medical applications. Chem Biodivers 15(5):e1700557. https://doi.org/10.1002/cbdv.201700557
Garofalo SF, Cavallini N, Demichelis F, Savorani F, Mancini G, Fino D, Tommasi T (2023) From tuna viscera to added-value products: a circular approach for fish-waste recovery by green enzymatic hydrolysis. Food Bioprod Process 137:155–167. https://doi.org/10.1016/j.fbp.2022.11.006
Ghorbani K, Kouchaki S, Sadeghi N, Eslamifarsani E, Rabbani F, Beyramysoltan S (2020) Distinguishing tissue origin of bovine gelatin in processed products using LC/MS technique in combination with chemometrics tools. Food Chem 319:126302. https://doi.org/10.1016/j.foodchem.2020.126302
Hemker AK, Nguyen LT, Karwe M, Salvi D (2020) Effects of pressure-assisted enzymatic hydrolysis on functional and bioactive properties of tilapia (*Oreochromis niloticus*) by-product protein hydrolysates. Lwt 122:109003. https://doi.org/10.1016/j.lwt.2019.109003
Hermida-Merino C, Cabaleiro D, Lugo L, Valcarcel J, Vázquez JA, Bravo I, Long A, Salloum-Abou-Jaoude G, Solano E, Gracia-Fernández C, Piñeiro MM, Hermida-Merino D (2022) Characterization of tuna gelatin-based hydrogels as a matrix for drug delivery. Gels 8(4):237. https://doi.org/10.3390/gels8040237
Huang JN, Wen B, Li XX, Xu L, Gao JZ, Chen ZZ (2023) Astaxanthin mitigates oxidative stress caused by microplastics at the expense of reduced skin pigmentation in discus fish. Sci Total Environ 874:162494. https://doi.org/10.1016/j.scitotenv.2023.162494
Hussain B, Sultana T, Sultana S, Ahmed Z, Mahboob S (2018) Study on impact of habitat degradation on proximate composition and amino acid profile of Indian major carps from different habitats. Saudi J Biol Sci 25(4):755–759
Ideia P, Pinto J, Ferreira R, Figueiredo L, Spínola V, Castilho PC (2020) Fish processing industry residues: a review of valuable products extraction and characterization methods. Waste and Biomass Valorization 11:3223–3246. https://doi.org/10.1007/s12649-019-00739-1
Idowu AT, Igiehon OO, Idowu S, Olatunde OO, Benjakul S (2021) Bioactivity potentials and general applications of fish protein hydrolysates. Int J Pept Res Ther 27:109–118. https://doi.org/10.1007/s10989-020-10071-1

Irawan C, Nata IF, Putra MD, Marisa R, Asnia M, Arifin YF (2018) Biopolymer of chitosan from fish scales as natural coagulant for iron–contaminated groundwater treatment. J Rekayasa Kimia Lingkungan 13(2):93–99. https://doi.org/10.23955/rkl.v13i2.10601

Jafarpour A, Gomes RM, Gregersen S, Sloth JJ, Jacobsen C, Sørensen ADM (2020) Characterization of cod (*Gadus morhua*) frame composition and its valorization by enzymatic hydrolysis. J Food Compos Anal 89:103469. https://doi.org/10.1016/j.jfca.2020.103469

Jongjareonrak A, Benjakul S, Visessanguan W, Nagai T, Tanaka M (2005) Isolation and characterisation of acid and pepsin-solubilised collagens from the skin of Brownstripe red snapper (*Lutjanus vitta*). Food Chem 93(3):475–484. https://doi.org/10.1016/j.foodchem.2004.10.026

Joseph J, Manigundan K, Shamya Arokia Rajan M, Radhakrishnan M, Gopikrishnan V, Kumaran S, Thirunavukkarasu R, Aruni W, Shanmugam V (2022) Conversion of aquaculture waste into biomedical wealth: chitin and chitosan journey. Adv Mater Sci Eng. https://doi.org/10.1155/2022/2897179

Joseph SM, Krishnamoorthy S, Paranthaman R, Moses JA, Anandharamakrishnan C (2021) A review on source-specific chemistry, functionality, and applications of chitin and chitosan. Carbohydr Polym Technol Appl 2:100036. https://doi.org/10.1016/j.carpta.2021.100036

Khailova L, Robison J, Jaggers J, Lawson S, Treece A, Soranno D, Lujan SO, Davidson JA (2020) Tissue alkaline phosphatase activity and expression in an experimental infant swine model of cardiopulmonary bypass with deep hypothermic circulatory arrest. J Inflamm 17(1):1–13. https://doi.org/10.1186/s12950-020-00256-2

Khan S, Rehman A, Shah H, Aadil RM, Ali A, Shehzad Q, Ashraf W, Yang F, Karin A, Khaliq A, Xia W (2022) Fish protein and its derivatives: the novel applications, bioactivities, and their functional significance in food products. Food Rev Intl 38(8):1607–1634. https://doi.org/10.1080/87559129.2020.1828452

Khanjani A, Sobati MA (2021) Performance and emission of a diesel engine using different water/waste fish oil (WFO) biodiesel/diesel emulsion fuels: optimization of fuel formulation via response surface methodology (RSM). Fuel 288:119662. https://doi.org/10.1016/j.fuel.2020.119662

Korkmaz K, Tokur B (2022) Optimization of hydrolysis conditions for the production of protein hydrolysates from fish wastes using response surface methodology. Food Biosci 45:101312. https://doi.org/10.1016/j.fbio.2021.101312

Lees MJ, Carson BP (2020) The potential role of fish-derived protein hydrolysates on metabolic health, skeletal muscle mass and function in ageing. Nutrients 12(8):2434. https://doi.org/10.3390/nu12082434

Liang M, Ma L, Li X, Feng D, Zhang J, Sun Y (2022) Identification and characterization of two types of triacylglycerol lipase genes from *Neocaridina denticulata sinensis*. Fish Shellfish Immunol 131:980–990. https://doi.org/10.1016/j.fsi.2022.11.019

Liu X, Liao W, Xia W (2023) Recent advances in chitosan based bioactive materials for food preservation. Food Hydrocoll 108612. https://doi.org/10.1016/j.foodhyd.2023.108612

Lou L, Chen H (2023) Functional modification of gelatin-based biodegradable composite films: a review. Food Addit Contam Part A 1–22. https://doi.org/10.1080/19440049.2023.2222844

Manzoor M, Singh J, Gani A, Noor N (2021) Valorization of natural colors as health-promoting bioactive compounds: phytochemical profile, extraction techniques, and pharmacological perspectives. Food Chem 362:130141. https://doi.org/10.1016/j.foodchem.2021.130141

Marsol-Vall A, Aitta E, Guo Z, Yang B (2022) Green technologies for production of oils rich in n-3 polyunsaturated fatty acids from aquatic sources. Crit Rev Food Sci Nutr 62(11):2942–2962. https://doi.org/10.1080/10408398.2020.1861426

Maschmeyer T, Luque R, Selva M (2020) Upgrading of marine (fish and crustaceans) biowaste for high added-value molecules and bio (nano)-materials. Chem Soc Rev 49(13):4527–4563. https://doi.org/10.1039/C9CS00653B

Monsiváis-Alonso R, Mansouri SS, Román-Martínez A (2020) Life cycle assessment of intensified processes towards circular economy: omega-3 production from waste fish oil. Chem Eng Process Process Intensif 158:108171. https://doi.org/10.1016/j.cep.2020.108171

Naomi R, Ridzuan PM, Bahari H (2021) Current insights into collagen type I. Polymers 13(16):2642. https://doi.org/10.3390/polym13162642
Novikov VY, Rysakova KS, Shumskaya NV, Mukhortova AM, Kesarev KA (2023) King crab gills as a new source of chitin/chitosan and protein hydrolysates. Int J Biol Macromol 232:123346. https://doi.org/10.1016/j.ijbiomac.2023.123346
Nurilmala M, Suryamarevita H, Hizbullah HH, Jacoeb AM, Ochiai Y (2022) Fish skin as a biomaterial for halal collagen and gelatin. Saudi J Biol Sci 29(2):1100–1110. https://doi.org/10.1016/j.sjbs.2021.09.056
Osuna-Ruiz I, Espinoza-Marroquin MF, Salazar-Leyva JA, Pena E, Álvarez-González CA, Banuelos-Vargas I, Martínez-Montaño E (2019) Biochemical characterization of a semi-purified aspartic protease from sea catfish *Bagre panamensis* with milk-clotting activity. Food Sci Biotechnol 28(6):1785–1793. https://doi.org/10.1007/s10068-019-00614-8
Ozogul F, Cagalj M, Šimat V, Ozogul Y, Tkaczewska J, Hassoun A, Kaddour AA, Kuley E, Rathod NB, Phadke GG (2021) Recent developments in valorisation of bioactive ingredients in discard/seafood processing by-products. Trends Food Sci Technol 116:559–582. https://doi.org/10.1016/j.tifs.2021.08.007
Patel A, Karageorgou D, Rova E, Katapodis P, Rova U, Christakopoulos P, Matsakas L (2020) An overview of potential oleaginous microorganisms and their role in biodiesel and omega-3 fatty acid-based industries. Microorganisms 8(3):434. https://doi.org/10.3390/microorganisms8030434
Pezeshk S, Ojagh SM, Rezaei M, Shabanpour B (2019) Fractionation of protein hydrolysates of fish waste using membrane ultrafiltration: investigation of antibacterial and antioxidant activities. Probiotics Antimicrob Proteins 11:1015–1022. https://doi.org/10.1007/s12602-018-9483-y
Pusceddu MM, Kelly P, Stanton C, Cryan JF, Dinan TG (2016) N-3 polyunsaturated fatty acids through the lifespan: implication for psychopathology. Int J Neuropsychopharmacol 19(12). https://doi.org/10.1093/ijnp/pyw078
Ramos M, Valdés A, Beltran A, Garrigós MC (2016) Gelatin-based films and coatings for food packaging applications. Coatings 6(4):41. https://doi.org/10.3390/coatings6040041
Rana S, Singh A, Surasani VKR, Kapoor S, Desai A, Kumar S (2023) Fish processing waste: a novel source of non-conventional functional proteins. Int J Food Sci Technol 58(5):2637–2644. https://doi.org/10.1111/ijfs.16104
Ranasinghe RASN, Wijesekara WLI, Perera PRD, Senanayake SA, Pathmalal MM, Marapana RAUJ (2022) Functional and bioactive properties of gelatin extracted from aquatic bioresources—a review. Food Rev Intl 38(4):812–855. https://doi.org/10.1080/87559129.2020.1747486
Raveendran S, Parameswaran B, Ummalyma SB, Abraham A, Mathew AK, Madhavan A, Rebello S, Pandey A (2018) Applications of microbial enzymes in food industry. Food Technol Biotechnol 56(1):16. https://doi.org/10.17113/ftb.56.01.18.5491
Rezvani Ghomi E, Nourbakhsh N, Akbari Kenari M, Zare M, Ramakrishna S (2021) Collagen-based biomaterials for biomedical applications. J Biomed Mater Res B Appl Biomater 109(12):1986–1999. https://doi.org/10.1002/jbm.b.34881
Rico-Llanos GA, Borrego-González S, Moncayo-Donoso M, Becerra J, Visser R (2021) Collagen type I biomaterials as scaffolds for bone tissue engineering. Polymers 13(4):599. https://doi.org/10.3390/polym13040599
Rocha-Pimienta J, Navajas-Preciado B, Barraso-Gil C, Martillanes S, Delgado-Adámez J (2023) Optimization of the extraction of chitosan and fish gelatin from fishery waste and their antimicrobial potential as active biopolymers. Gels 9(3):254. https://doi.org/10.3390/gels9030254
Ryu B, Shin KH, Kim SK (2021) Muscle protein hydrolysates and amino acid composition in fish. Mar Drugs 19(7):377. https://doi.org/10.3390/md19070377
Sadowska M, Kołodziejska I, Niecikowska C (2003) Isolation of collagen from the skins of Baltic cod (*Gadus morhua*). Food Chem 81(2):257–262. https://doi.org/10.1016/S0308-8146(02)00420-X

Salehi E, Khajavian M, Sahebjamee N, Mahmoudi M, Drioli E, Matsuura T (2022) Advances in nanocomposite and nanostructured chitosan membrane adsorbents for environmental remediation: a review. Desalination 527:115565. https://doi.org/10.1016/j.desal.2022.115565

Santos VP, Marques NS, Maia PC, Lima MABD, Franco LDO, Campos-Takaki GMD (2020) Seafood waste as attractive source of chitin and chitosan production and their applications. Int J Mol Sci 21(12):4290. https://doi.org/10.3390/ijms21124290

Saranya R, Jayapriya JTSA (2018) Purification, characterization, molecular modeling and docking study of fish waste protease. Int J Biol Macromol 118:569–583. https://doi.org/10.1016/j.ijbiomac.2018.06.119

Saranya R, Tamil Selvi A, Jayapriya J, Aravindhan R (2020) Synthesis of fat liquor through fish waste valorization, characterization and applications in tannery industry. Waste Biomass Valoriz 11:6637–6647. https://doi.org/10.1007/s12649-020-00944-3

Sathish T, Ağbulut Ü, George SM, Ramesh K, Saravanan R, Roberts KL, Sharma P, Asif M, Hoang AT (2023) Waste to fuel: Synergetic effect of hybrid nanoparticle usage for the improvement of CI engine characteristics fuelled with waste fish oils. Energy 275:127397. https://doi.org/10.1016/j.energy.2023.127397

Seixas MJ, Martins E, Reis RL, Silva TH (2020) Extraction and characterization of collagen from elasmobranch byproducts for potential biomaterial use. Mar Drugs 18(12):617. https://doi.org/10.3390/md18120617

Sharkey SJ, Harnedy-Rothwell PA, Allsopp PJ, Hollywood LE, FitzGerald RJ, O'Harte FP (2020) A narrative review of the anti-hyperglycemic and satiating effects of fish protein hydrolysates and their bioactive peptides. Mol Nutr Food Res 64(21):2000403. https://doi.org/10.1002/mnfr.202000403

Shori AB, Hong YC, Baba AS (2021) Proteolytic profile, angiotensin-I converting enzyme inhibitory activity and sensory evaluation of *Codonopsis pilosula* and fish collagen cheese. Food Res Int 143:110238. https://doi.org/10.1016/j.foodres.2021.110238

Siddik MA, Howieson J, Fotedar R, Partridge GJ (2021) Enzymatic fish protein hydrolysates in finfish aquaculture: a review. Rev Aquac 13(1):406–430. https://doi.org/10.1111/raq.12481

Siddiqui JA, Khan MM, Bamisile BS, Hafeez M, Qasim M, Rasheed MT, Rasheed MA, Ahmad S, Shahid MI, Xu Y (2022) Role of insect gut microbiota in pesticide degradation: a review. Front Microbiol 860. https://doi.org/10.3389/fmicb.2022.870462

Singh A, Mittal A, Benjakul S (2022) Undesirable discoloration in edible fish muscle: impact of indigenous pigments, chemical reactions, processing, and its prevention. Compr Rev Food Sci Food Saf 21(1):580–603. https://doi.org/10.1111/1541-4337.12866

Sionkowska A, Adamiak K, Musiał K, Gadomska M (2020) Collagen based materials in cosmetic applications: a review. Materials 13(19):4217. https://doi.org/10.3390/ma13194217

Sridhar A, Krishnasamy Sekar R, Manikandan DB, Arumugam M, Veeran S, Ramasamy T (2021) Activity profile of innate immune-related enzymes and bactericidal of freshwater fish epidermal mucus extract at different pH. Environ Sci Pollut Res 28:33914–33926. https://doi.org/10.1007/s11356-020-11173-5

Suhartini S, Rohma NA, Santoso I, Paul R, Listiningrum P, Melville L (2022) Food waste to bioenergy: current status and role in future circular economies in Indonesia. Energy, Ecol Environ 7(4):297–339. https://doi.org/10.1007/s40974-022-00248-3

Sun M, Ding Z, Wang H, Yu G, Feng Z, Li B, Li P (2019) Gelatin filter capture-based high-throughput sequencing analysis of microbial diversity in haze particulate matter. J Biomed Res 33(6):414. https://doi.org/10.7555/JBR.33.20180121

Tang C, Zhou K, Zhu Y, Zhang W, Xie Y, Wang Z, Zhou H, Yang T, Zhang Q, Xu B (2022) Collagen and its derivatives: from structure and properties to their applications in food industry. Food Hydrocoll 107748. https://doi.org/10.1016/j.foodhyd.2022.107748

Valcarcel J, Sanz N, Vázquez JA (2020) Optimization of the enzymatic protein hydrolysis of by-products from seabream (*Sparus aurata*) and seabass (*Dicentrarchus labrax*), chemical and functional characterization. Foods 9(10):1503. https://doi.org/10.3390/foods9101503

Välimaa AL, Mäkinen S, Mattila P, Marnila P, Pihlanto A, Mäki M, Hiidenhovi J (2019) Fish and fish side streams are valuable sources of high-value components. Food Qual Saf 3(4):209–226. https://doi.org/10.1093/fqsafe/fyz024

Venkatesan J, Anil S, Kim SK, Shim MS (2017) Marine fish proteins and peptides for cosmeceuticals: a review. Mar Drugs 15(5):143. https://doi.org/10.3390/md15050143

Venugopal V (2021) Valorization of seafood processing discards: bioconversion and bio-refinery approaches. Front Sustain Food Syst 5:611835. https://doi.org/10.3389/fsufs.2021.611835

Viswanathan K, Wang S (2021) Experimental investigation on the application of preheated fish oil ethyl ester as a fuel in diesel engine. Fuel 285:119244. https://doi.org/10.1016/j.fuel.2020.119244

Vivek K, Sandhia GS, Subramaniyan SJBA (2022) Extremophilic lipases for industrial applications: a general review. Biotechnol Adv 108002. https://doi.org/10.1016/j.biotechadv.2022.108002

Wang S, Yan Z, Hänfling B, Zheng X, Wang P, Fan J, Li J (2021) Methodology of fish eDNA and its applications in ecology and environment. Sci Total Environ 755:142622. https://doi.org/10.1016/j.scitotenv.2020.142622

Wang W, Ishikawa M, Koshio S, Yokoyama S, Dawood MA, Hossain MS, Moss AS (2019) Effects of dietary astaxanthin and vitamin E and their interactions on the growth performance, pigmentation, digestive enzyme activity of kuruma shrimp (*Marsupenaeus japonicus*). Aquac Res 50(4):1186–1197. https://doi.org/10.1111/are.13993

Wasswa J, Tang J, Gu X (2007) Utilization of fish processing by-products in the gelatin industry. Food Rev Intl 23(2):159–174. https://doi.org/10.1080/87559120701225029

Xu C, Nasrollahzadeh M, Selva M, Issaabadi Z, Luque R (2019) Waste-to-wealth: biowaste valorization into valuable bio (nano) materials. Chem Soc Rev 48(18):4791–4822. https://doi.org/10.1039/C8CS00543E

Yadollahi F, Soltani M, Modarresi MH, Akhondzadeh Basti A (2021) Efficacy of vitamin E with or without probiotic, astaxanthin or rosemary extract on growth performance, survival, haematological parameters, antioxidant activity and liver enzymes in rainbow trout (*Oncorhynchus mykiss*). Aquac Res 52(11):5606–5616. https://doi.org/10.1111/are.15436

Zhang T, Ding M, Tao L, Liu L, Tao N, Wang X, Zhong J (2020) Octenyl succinic anhydride modification of bovine bone and fish skin gelatins and their application for fish oil-loaded emulsions. Food Hydrocoll 108:106041. https://doi.org/10.1016/j.foodhyd.2020.106041

Zhang Y, Guo X, Peng Z, Jamali MA (2022) A review of recent progress in reducing NaCl content in meat and fish products using basic amino acids. Trends Food Sci Technol 119:215–226. https://doi.org/10.1016/j.tifs.2021.12.009

Zhao YM, de Alba M, Sun DW, Tiwari B (2019) Principles and recent applications of novel non-thermal processing technologies for the fish industry—a review. Crit Rev Food Sci Nutr 59(5):728–742. https://doi.org/10.1080/10408398.2018.1495613

Zhu Z, Zhang B, Cai Q, Ling J, Lee K, Chen B (2020) Fish waste based lipopeptide production and the potential application as a bio-dispersant for oil spill control. Front Bioeng Biotechnol 8:734. https://doi.org/10.3389/fbioe.2020.00734

Chapter 6
Chemical Conversion Pathways for Enhanced Value of Biomass Waste: Catalytic Processes and Chemical Transformations

Hajar Rastegari

Abstract The sustainable conversion of biomass waste into added-value chemicals is critical for addressing global environmental challenges and promoting the transition toward circular bioeconomy. This chapter provides an overview of chemical conversion pathways that enhance the value of biomass waste through both catalytic and non-catalytic processes. Key valorization reactions discussed include hydrolysis, transesterification, esterification, and dehydration, which enable the transformation of complex biopolymers such as cellulose, hemicellulose, and triglycerides into platform molecules, biofuels, and specific chemicals. Emphasis is placed on the role of catalytic systems including homogenous and heterogeneous catalysts in improving conversion efficiency, selectivity, and product yield.

6.1 Introduction

One of the critical environmental challenges faced globally is the disposal of approximately 100 billion metric tons of biomass waste generated annually (Cho et al. 2020). Biomass waste, comprising carbon–neutral, renewable materials derived from plants and animals, is often discarded or considered waste and includes lignocellulosic, animal, and food waste (Basu 2018). However, these materials hold significant potential for conversion into value-added products through various pathways.

Transformation of biomass waste into useful chemicals plays a crucial role in promoting a circular economy by reducing reliance on finite resources. This conversion process helps lower greenhouse gas emissions and minimizes the environmental

The original version of the chapter has been revised. A correction to this chapter can be found at https://doi.org/10.1007/978-981-95-0991-1_16

H. Rastegari (✉)
Institute of Tropical Aquaculture and Fisheries, Universiti Malaysia Terengganu, Kuala Nerus, Terengganu, Malaysia
e-mail: hajar.rastegari@gmail.com

A. Hind (ed.), *Biomass Waste: Processing, Valorization, Industrial Applications*, https://doi.org/10.1007/978-981-95-0991-1_6

impacts associated with waste disposal methods like landfilling and incineration. Furthermore, biomass valorization, through the production of biofuels, provides a renewable energy source, thereby decreasing dependence on fossil fuels (Basu 2018). Converting biomass waste into valuable products alleviates the burden on waste management systems and creates economic opportunities through the generation of marketable products, fostering innovation and job creation. Additionally, valorised biomass can promote resource efficiency by ensuring the complete utilization of biomass components (Demirbas 2011).

Biomass valorization relies on the chemical constituents of biomass, such as carbohydrate and non-carbohydrate fractions, to produce commercially viable products (Cho et al. 2020). Various methods for valorising biomass waste include biological processes like composting, anaerobic digestion, and fermentation, thermochemical processes like pyrolysis and gasification, and chemical processes, both catalytic and non-catalytic. Among these, chemical valorisation methods are particularly significant as they can convert biomass into a wide array of high-value specific compounds with desirable properties, thus enhancing the economic viability and sustainability of biomass waste utilization (Pandey et al. 2019). Hence, this chapter explores the chemical transformations involved in converting biomass waste into high-value products, focusing on processes including hydrolysis, transesterification, esterification, and dehydration.

6.2 Biomass Waste Composition

As shown in Fig. 6.1, understanding biomass waste composition is key to effective valorization and environmental management, as its chemicals vary by source. In lignocellulosic waste, most (57%) comes from agriculture and crops, followed by forestry (21%), with smaller amounts from municipal (14%), industrial (8%), and other sources. Animal waste is mainly from arthropods (42%) and fish (29%), while mollusks and cnidarians make up 12%, and smaller portions come from mammals and birds (3%) and other sources (14%). In food waste, grains and starch make up the largest portion (37%), followed by fruit (25%) and vegetables (23%), with smaller amounts from animal products (13%) and paper and other materials (2%) (Chakhtouna et al. 2022; Pelé et al. 2021; Spierling 2011). Given the heterogeneous nature of biomass waste, the development of efficient and selective catalytic processes is essential for its effective conversion (Huber et al. 2006). In the following sections, we will examine chemical catalytic methods used in biomass conversion, the mechanism involved and evaluate the effectiveness of different catalysts.

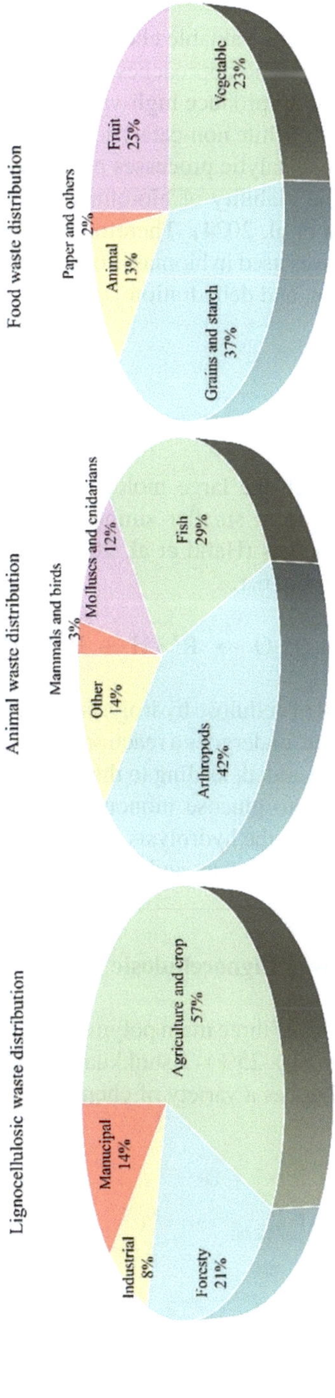

Fig. 6.1 Components of biomass waste regarding the source of it.

6.3 Catalytic Chemical Processes for Biomass Conversion

Biomass waste can be converted into valuable chemicals through both catalytic and non-catalytic processes, with catalytic approaches being highly preferred for their efficiency, selectivity, and ability to produce high-value products (Chen et al. 2024; Manyepedza and Makuve 2024). While non-catalytic methods are often simpler and more economically sustainable, catalytic processes play a critical role in improving the sustainability and economic viability of biorefineries, despite challenges such as catalyst expenses (Usanova et al. 2024). Therefore, the following sections will explore the key catalytic processes used in biomass conversion, including hydrolysis, transesterification, esterification, and dehydration.

6.3.1 Hydrolysis

Hydrolysis is a reaction in which the large molecule present in biomass, breaks down in presence of water to form smaller simpler molecules, which can then be processed into valuable products (Hafid et al. 2017). The general equation for hydrolysis reaction is shown in Eq. 6.1.

$$R^1{-}R^2 + H_2O \rightarrow R^1{-}H + R^2{-}OH \tag{6.1}$$

Figure 6.2 shows the process of cellulose hydrolysis. Cellulose, the most abundant macromolecule in biomass waste, undergoes a reaction where glycosidic linkages are broken, and water molecules are added, leading to the formation of stable oligomers, which eventually break down into glucose monomers (Hafid et al. 2017). In the following sections, we will discuss the hydrolysis of three major sources of biomass waste: lignocellulosic waste, animal waste, and food waste in more detail.

6.3.1.1 Catalytic Hydrolysis of Lignocellulosic Waste

Lignocellulosic biomass consists of three main polymers: cellulose (40–60%), hemicellulose (20–40%), and lignin (10–25%) (Ashokkumar et al. 2022). Hydrolysis of this type of biomass waste produces a variety of chemicals with different industrial

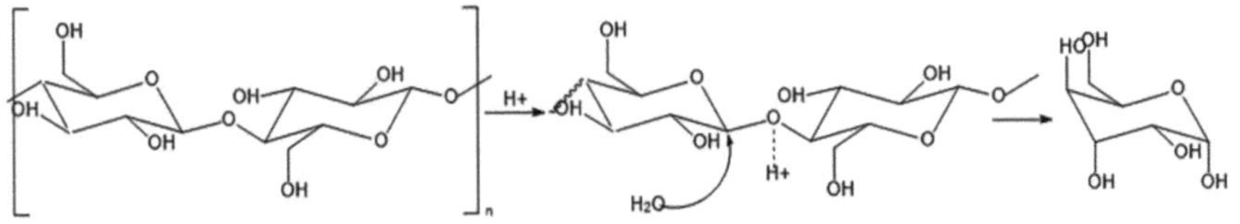

Fig. 6.2 Cellulose hydrolysis into oligomers/monomers under acidic conditions.

application. Table 6.1, summarizes the hydrolysates derived from lignocellulosic biomass, highlighting their chemical structure and their applications. Cellobiose and glucose, both of which serve important roles across various industries. Cellobiose is valued as a zero-calorie functional sweetener, soluble dietary fiber, prebiotic, and feed additive, while glucose is widely used as a food additive, biofuel precursor, and in pharmaceuticals (Basar et al. 2021; Zhou et al. 2016). Hemicellulose hydrolysis produces several sugars including glucose, mannose, galactose, rhamnose, xylose, and arabinose. These hydrolysates find applications in food, cosmetics, pharmaceuticals, animal breeding, and biotechnology. Mannose and galactose, for example, are essential in both the food and pharmaceutical industries, while xylose is pivotal in biofuel production. Arabinose also contributes to food products and biopolymer manufacturing (Basar et al. 2021; Carrasco-Reinado et al. 2019; Inokuma et al. 2024; Lu et al. 2020; Xing et al. 2016; Xu et al. 2018). These products underscore the versatility of lignocellulosic biomass in generating valuable bio-based compounds for sustainable industries.

Autohydrolysis is a process that hydrolyse cellulose and hemicellulose into oligomers and monomers using hot liquid water, without acidic or alkaline catalysts. Water autoionizes during feedstock autohydrolysis, generating hydronium ions that catalyse further hydrolysis. Typically conducted at 150–250 °C in pressurized reactors, this method minimizes capital and operating costs by avoiding corrosive substances. However, it is relatively slow and predominantly yields oligosaccharides, with higher temperatures often generate fermentation inhibitors, complicating downstream processes (Brodin et al. 2017). However, salt-assisted autohydrolysis can accelerate the autocatalytic depolymerization by compressing the electric double layer on cellulose surfaces and facilitating bond cleavage, enhance cellulose hydrolysis by disrupting hydrogen bonds and increasing surface acidity (Deng et al. 2023; Jiang et al. 2018).

In addition to autohydrolysis, catalytic hydrolysis plays a crucial role in breaking down biomass, particularly cellulose, into simpler, valuable chemicals (Ashokkumar et al. 2022). Homogeneous catalysts like mineral acids significantly improve polysaccharides hydrolysis. The complex crystalline structure of lignocellulose makes the type of lignocellulose and acid type and concentration crucial factors for efficient hydrolysis (Chen et al. 2018; Shreshtha et al. 2023).

The conventional acidic catalyst for the hydrolysis of lignocellulosic biomass is sulfuric acid (H_2SO_4) due to its effectiveness in hydrolysis of hemicellulose and cellulose into fermentable sugars (Mosier 2005). However, this acid is highly corrosive that require specialized, corrosion-resistant equipment, driving up capital costs. Additionally, it necessitates careful handling due to its hazardous nature, and post-hydrolysis neutralization produces large amounts of waste salts, raising environmental and waste treatment costs (Yang and Wyman 2008).

Hydrochloric acid (HCl) is another acidic catalyst which is used for the hydrolysis of cellulose and hemicellulose. This catalyst is effective in the hydrolysis process but less commonly used than H_2SO_4 due to handling and recovery challenges (Taherzadeh and Karimi 2007).

Table 6.1 Main hydrolysates derived from lignocellulosic biomass, their chemical structure and applications.

Hydrolysate	Structure	Application	References
Cellobiose	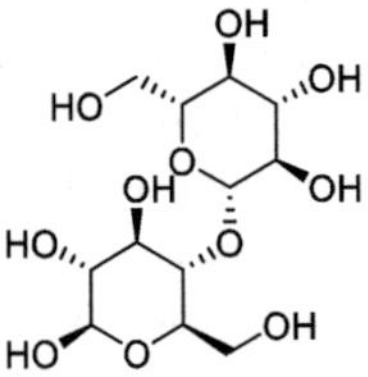	–Zero-calorie functional sweetener –Soluble dietary fiber –Prebiotic –Feed additive	Basar et al. (2021), Zhou et al. (2016)
Glucose	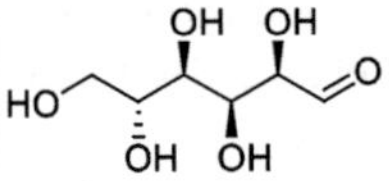	–Food additive –Biofuel –Pharmaceutical	Basar et al. (2021), Yang et al. (2020)
Mannose	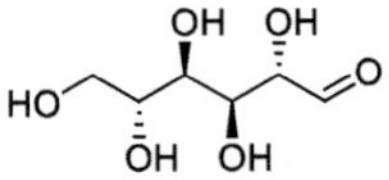	–Food –Cosmetic –Animal breeding –Pharmaceutical	Basar et al. (2021), Xing et al. (2016)
Galactose	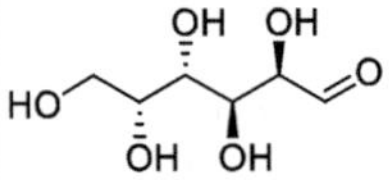	–Food –Pharmaceutical –Biotechnology	Basar et al. (2021), Lu et al. (2020)
Rhamnose	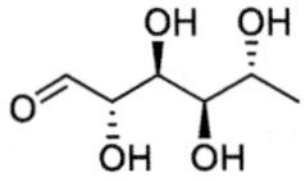	–Biotechnology	Basar et al. (2021), Chen et al. (2019)
Xylose	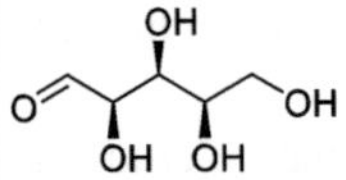	–Biofuel –Food	Basar et al. (2021), Nam (2024)
Arabinose	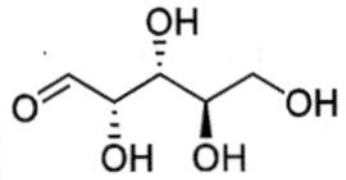	–Food –Biopolymers	Basar et al. (2021), Xu et al. (2018)

Nitric acid (HNO_3) is effective for lignin depolymerization in addition to cellulose hydrolysis, however, it's less common due to its oxidative properties (Zhang et al. 2011).

Phosphoric Acid (H_3PO_4) is able to hydrolyse hemicellulose and disrupt lignin (Vázquez et al. 2007). Less efficient than H_2SO_4, H_3PO_4 requires higher concentrations or longer reaction times, making it more costly. Additionally, its use can lead to phosphate runoff, which may contribute to algal blooms and eutrophication (Kim and Holtzapple 2005).

Organic acids offer a less corrosive alternative to mineral acids like H_2SO_4 and help mitigate unwanted side reactions during hydrolysis. Organic acids such as formic, acetic, oxalic, and maleic acid have demonstrated effectiveness in hemicellulose degradation. For example, acetic acid in a CO_2–H_2O system can efficiently convert wheat straw hemicellulose into xylo-oligosaccharides and xylose (Yang et al. 2020). While oxalic acid has shown high efficiency in breaking down bagasse hemicellulose into xylose and arabinose (Yan et al. 2018).

In contrast, the most common alkaline catalyst is sodium hydroxide (NaOH) which often used in combination with acids or enzymes to disrupt lignin and improve the accessibility of cellulose for hydrolysis. NaOH effectively solubilizes lignin, but this process can result in the loss of valuable biomass components. It also requires large amounts of water for washing and neutralization, adding complexity and cost, while its corrosiveness demands specialized materials for equipment construction (Sun et al. 2002).

On the other hand, heterogeneous solid acid catalysts offer significant advantages such as ease of separation, catalyst reuse, and reduced reactor corrosion, and lower wastewater production, but mass transfer limitations between the solid catalysts and the solid polysaccharides remain a challenge. To address this, solid acid catalysts are designed with strong adsorption sites, potent acid sites, and high specific surface areas. Heteropoly acids (HPAs), metal oxides, H-zeolites, acidic resins, acidic carbon catalysts, and ionic liquids are among the most widely studied ones for polysaccharide hydrolysis (Deng et al. 2023).

HPAs, consisting of hydrogen cations and polyoxometalate anions, are widely utilized in biomass conversion due to its strong Brønsted acidity, which aids in cellulose hydrolysis. Among these, $H_3PW_{12}O_{40}$ is the most effective in yielding glucose. However, HPAs are highly soluble in water and polar solvents, complicating their separation from reaction products and maintaining their recyclability (Shimizu et al. 2009). To address these challenges, two strategies have been employed: the substitution of protons with larger monovalent cations, such as Cs+, to form more stable solid catalysts, and the development of hybrid HPAs with lower solubility (Azlan et al. 2022). These advancements have resulted in more efficient and recyclable HPAs for biomass conversion.

Lewis acids, commonly metal salts, also facilitate polysaccharide hydrolysis by generating Brønsted acids through partial metal ions hydrolysis. Transition metal salts, such as $AlCl_3$, $CrCl_3$, $CuCl_2$, and $FeCl_3$, are particularly effective at hydrolysis of hemicellulose. Under mild conditions, these catalysts primarily yield oligosaccharides, while harsher conditions promote monosaccharide formation (Yu et al. 2011). For example, an $AlCl_3$–H_2O system can degrade over 90% of corncob hemicellulose, although this process often results in mixed degradation products (Jiang et al. 2021).

Metal oxides like ZrO_2 and TiO_2, often functionalized with H_2SO_4, show promise in catalysing in cellulose degradation due to their porous structures and abundant Lewis acid sites (Qiao et al. 2020). Complex metal oxides, such as ZrO_2–SiO_2 and $Zr(HPO_4)_2$–SiO_2, have shown notable selectivity for glucose production, with high glucose yields under specific conditions (Gliozzi et al. 2014).

H-zeolites, crystalline aluminosilicates, are another class of solid acid catalysts, characterized by their well-defined channels and cavities that facilitate hydrolysis. Although their acidity is lower than other solid acids, their catalytic performance is influenced by the Si/Al ratio and pore size. H-zeolites with lower Si/Al ratios poses stronger acidity, which enhances cellulose conversion and improve glucose yields (Chu et al. 2018). Additionally, the larger pore sizes in H-beta zeolites improve catalytic performance by enabling easier diffusion of cellulose-derived oligosaccharides (Zhou et al. 2016).

Acidic resins, particularly Amberlyst-15 and Nafion, have macro-reticular structures with sulfonic groups that enable polysaccharide hydrolysis. Amberlyst-15 can hydrolyze cellulose into cello-oligomers and glucose, though its thermal stability is limited (Karinen et al. 2011). Nafion resins, on the other hand, exhibit better thermal stability and maintain catalytic activity even at elevated temperatures (Hegner et al. 2010; Kim et al. 2010).

Acidic carbon catalysts, derived from natural organic materials and functionalized with various acidic groups, also show great potential for cellulose hydrolysis. These catalysts facilitate adsorption of cellulose through hydrogen bonding, leading to efficient hydrolysis, especially when combined with mix-milling pretreatment to overcome mass transfer limitations (Shrotri et al. 2018). Carbon-based catalysts functionalized with sulfonic acid groups, such as graphene oxide/activated carbon modified with sulfonic groups, have shown varied efficiency in converting sugars to levulinic acid. Modified graphene oxide demonstrated a high levulinic acid yield of 78%, attributed to its synergy between hydroxyl and Brønsted-Lowry acid sites. In contrast, modified activated carbon with lower hydrophobicity and excessive Lewis acid sites reduced its levulinic acid yield. Cellulase-mimetic catalysts and humin-derived catalysts offer alternative sustainable approaches, with humin-derived catalysts achieving a 65.9% levulinic acid yield. Additionally, combining these modified catalysts with $CrCl_3$ improved levulinic acid production due to the synergistic effects of Brønsted-Lowry and Lewis acids (Boonyakarn et al. 2019). Recently, a multifunctional carbon catalyst was designed by introducing boronic groups, weak acid sites, and high-density sulfonic acid onto mesophase pitch carbon, which successfully hydrolyzed cellulose into glucose with a 64% yield in water at 180 °C for 3 h. The catalyst demonstrated lower activation energy than homogeneous acids and remained effective for four cycles before significant loss of activity (Bedasso et al. 2024).

Brønsted-Lowry-Lewis acidic ionic liquids is another category of acid catalysts have been developed for biomass hydrolysis to levulinic acid (Jain et al. 2023). In general, incorporating Lewis acidic sites into ionic liquids (ILs) results in (1) stronger and more controllable acidity compared to using only Brønsted-Lowry acidic ILs; (2) uniformly distributed active sites that enhance the mass transfer rate of substrates and products; and (3) a positive, synergistic catalytic effect. Typically, ILs serve a dual role as both catalysts and solvents, following the common process of reaction-separation-regeneration in IL systems (Fang et al. 2020). Regenerating ILs is crucial for lowering material costs, as they are otherwise too expensive for widespread application. Among various ILs, imidazolium-based ILs are particularly favored for biomass conversion to levulinic acid, as they effectively target cellulose hydroxyl groups and promote cellulose depolymerization, yielding high levulinic acid outputs of 53–69% (Azlan et al. 2022).

Overall, while lignocellulosic biomass is structurally complex, advancements in pretreatment methods such as steam explosion and alkali treatment are crucial for increasing biomass porosity and reduce crystallinity, enhancing hydrolysis rates (Sun et al. 2002; Yan et al. 2018). However, the high costs and environmental impact of chemical hydrolysis, along with limited efficiency, remain significant challenges for commercialization (Usmani et al. 2021). Continued research is crucial for advancing more sustainable and cost-efficient approaches to produce value-added chemicals through the hydrolysis of lignocellulosic biomass.

6.3.1.2 Catalytic Hydrolysis of Animal Waste

Animal biomass waste, primarily consisting of proteinaceous materials (Fig. 6.1), includes sources such as human hair from personal care industries, avian feathers from poultry processing, animal fur from textile industry, and bovine hooves and horns from livestock operations (Karunarathne et al. 2024).

Converting to animal feed is a common approach for protein-rich waste biomass, although incineration is also used, especially when the biomass has low nutritional value or contains harmful substances. However, this process releases nitrogen into the environment, contributing to nitrate buildup in soil and surface water, and N_2O emissions, a potent greenhouse gas (De Schouwer et al. 2019). By transforming protein waste into nitrogen-based chemicals instead, the nitrogen cycle could be made more efficient, while also lowering energy use and reducing environmental harm. A key approach for valorising this type of biomass waste is hydrolysis, in which proteins are break down into oligopeptides and free amino acids (Fig. 6.3) (Kiewiet et al. 2018). These hydrolysates serve as the fundamental building blocks for further transformations into high-value products (Araujo et al. 2021).

Table 6.2 presents various hydrolysates derived from proteinaceous animal biomass. All these hydrolysates have applications in various industries including food, pharmaceuticals, nutraceuticals, and cosmetic (Czelej et al. 2022). The key products such as amino acids are the fundamental building blocks of proteins formed through complete hydrolysis. Partial hydrolysis yields polypeptides, which are short chains of amino acids that still possess functional properties. Additionally,

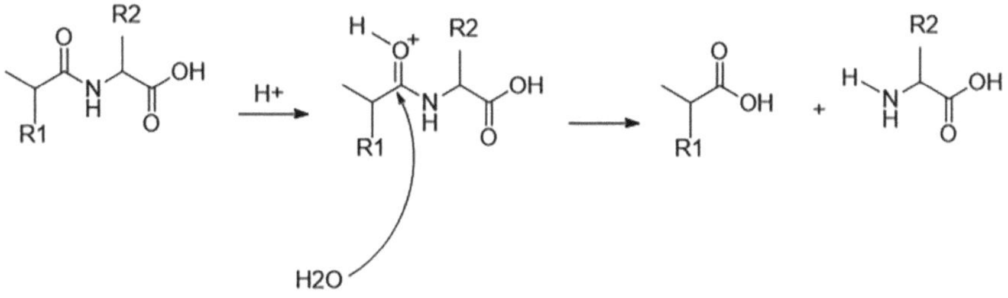

Fig. 6.3 Hydrolysis of proteins into oligopeptides and free amino acids.

collagen and gelatine hydrolysates consist of fragmented collagen chains, prized in cosmetics and pharmaceuticals for their skin-repairing and anti-aging benefits. Keratin hydrolysates, abundant in cysteine, are recognized for their strengthening effects, particularly in hair and nail care. Moreover, elastin hydrolysates, made up of bioactive peptides, improve tissue elasticity, finding applications in skincare and regenerative medicine (Czelej et al. 2022). Overall, these hydrolysates maintain significant bioactivity, enhancing their value in various applications.

In biorefineries of protein-rich biomass, the initial step involves separating proteins from other components, such as carbohydrates, lignin, and triglycerides. This separation is typically performed through extraction, either under acidic or alkaline conditions, with the latter being more commonly used. Alkaline conditions are favored because they more effectively break down cell wall components, saponify fats, and enhance protein solubility (Sari et al. 2015).

The second step involves hydrolysing proteins into oligopeptides and free amino acids in either acidic or alkaline environments (Khiari and Mason 2018). Highly

Table 6.2 Main hydrolysates derived from proteinaceous animal biomass and their chemical structure. Hydrolysate names are derived from Czelej et al. 2022); chemical structures were drawn by the author.

Hydrolysate	Structure
Amino acids	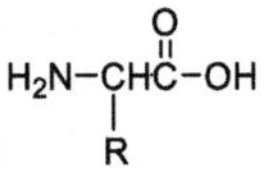
Polypeptides	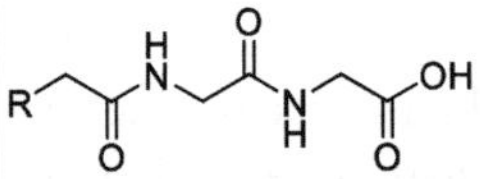
Collagen/gelatine hydrolysates	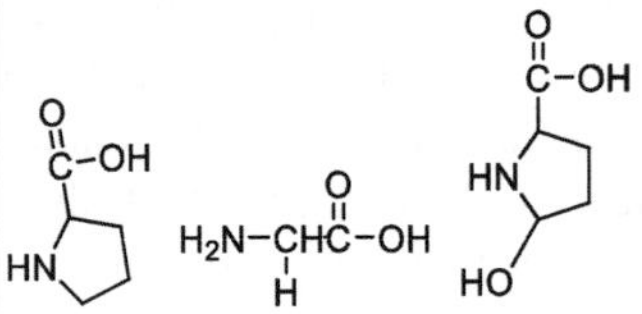
Keratin hydrolysates	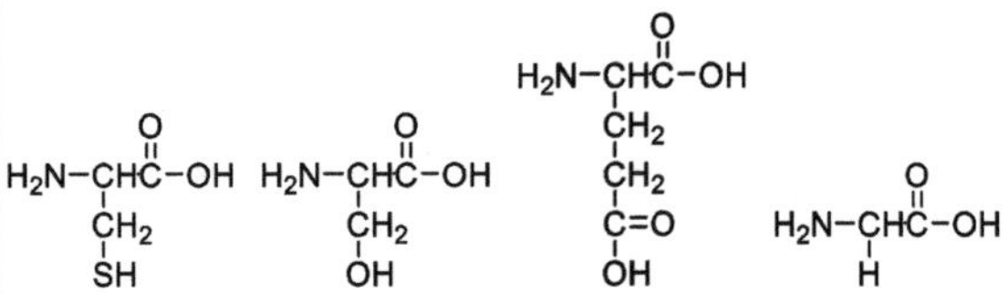
Elastin hydrolysates	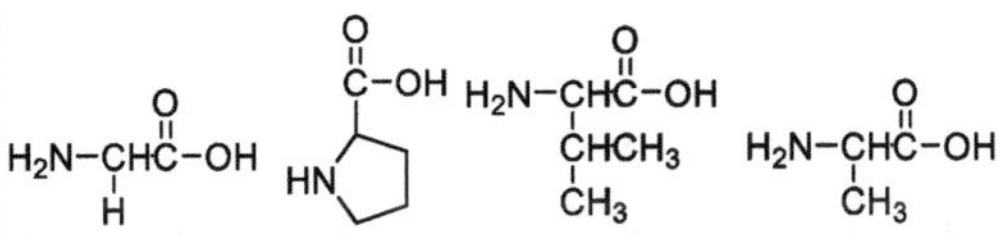

acidic conditions, such as the use of 6 M HCl at 110 °C for 24 h, typically achieve complete hydrolysis. However, this process can degrade amino acids like tryptophan, tyrosine, serine, and threonine and cause deamidation of glutamine and asparagine into glutamic and aspartic acids. For instance, industrial keratin hydrolysis from feathers uses 12 M HCl at 110 °C for 12 h. Lowering the concentration to 6 M HCl under the same conditions yields only 58.7% (Boonyakarn et al. 2019), though microwave-assisted acid-catalyzed hydrolysis has shown higher efficiency, achieving 64.12% conversion in shorter times (Brodin et al. 2017). Hydrolysis is also possible under highly alkaline environments. Heating such type of biomass waste in a 5 M NaOH at 100 °C for 24 h resulted in efficient hydrolysis; however, the primary challenge was degradation of lysine, cysteine, and threonine, along with an increased risk of racemization (Mulder 2010). Therefore, using both acidic and alkaline methods is limited in large-scale operations (Sari et al. 2014).

Subcritical water hydrolysis is a process that operates under conditions of 100–374 °C and 0.10–22 MPa, providing an alternative approach for converting proteins into oligopeptides and free amino acids (Powell et al. 2016). This method is considered environmentally friendly, as it does not require additional reagents. Under these conditions, the ionic product of water increases, leading to a greater degree of hydrolysis, which enhances the hydrolysis rate of complex organic materials. However, high energy consumption and the requirement for pressure-resistant equipment pose challenges for large-scale applications. Furthermore, prolonged exposure of free amino acids to elevated temperatures can result in their decomposition, resulting in the loss of nutritional value (Powell et al. 2016). These factors must be considered when evaluating the feasibility of the process.

ILs, such as methylimidazolium chloride and bromide have demonstrated the ability to dissolve proteins at mild temperatures. Studies have shown these liquids to be effective in dissolving wool and human hair keratin (Carrasco-Reinado et al. 2019). However, they need heating at 130 °C for 8–10 h (Chakhtouna et al. 2022).

Application of ultrasonic treatment for hydrolysing highly insoluble structural proteins like keratin allowed water and acids to penetrate more disordered regions, thereby enhancing solubility and promoting hydrolysis (Chingakham et al. 2019). Ultrasound has been shown to significantly improve the solubility of stable structural proteins and enhance the emulsification of protein suspensions (Chen et al. 2018). A novel approach combining ultrasonic treatment with ILs has yielded promising results for feather keratin hydrolysis. In this method, a solution of 8.4 M HCl was used with methylimidazolium chloride as a co-solvent. This technique achieved a keratin hydrolysis yield of 83.1%, reducing HCl consumption, shortening reaction times, and producing higher yields of amino acid hydrolysis products compared to previous methods (Ding et al. 2019).

The hydrolysis of animal biomass presents several challenges, including potential oxidative, degradation, and racemization reactions that can compromise the safety and bioactivity of the end products. These reactions are influenced by the freshness and handling of the by-products, as well as the hydrolysis conditions used. The complexity and variability of biomass composition further necessitate careful standardization of procedures to ensure consistent product quality (Nikoo et al. 2023).

Overall, optimizing the hydrolysis of animal biomass involves complex chemical reactions that can be approached through enzymatic, or chemical methods, each with its own advantages and limitations (Araujo et al. 2021; Gupta et al. 2022; Khiari and Mason 2018; Nikoo et al. 2023).

6.3.1.3 Catalytic Hydrolysis of Food Waste

Food waste refers to edible food that is discarded at various stages of the food supply chain, including primary production, logistics, storage operations, and post-consumer phases. In the primary sector, often referred to as the pre-consumer stage, food that does not reach consumers is either thrown away or recycled. Commonly discarded items at this stage include non-edible food scraps, products that do not meet organoleptic standards, and food that results from processing errors or contamination. The final stage, known as the post-consumer phase, involves waste from households and the food service industry (such as restaurants and hospitals), which typically includes items like peels, bones, and leftover food portions (Carmona-Cabello et al. 2018).

These food waste is rich source of diverse nutrients and bioactive compounds, most importantly carbohydrates, cellulose, proteins, chitin, pectin, and lipids (Pleissner et al. 2014), which can be effectively valorized through hydrolysis. By breaking down these complex compounds into simpler sugars, amino acids, and other valuable constituents, hydrolysis not only enhances the nutritional value of food waste but also opens up opportunities for their use in various applications, such as biofuels, bioplastics, and functional food ingredients (Sarker et al. 2024).

The pie chart in Fig. 6.1 highlights grains and starch, fruits, and vegetables as the most significant contributors to food waste. This is particularly important because their major chemical constituent, polysaccharides, can be broken down into simpler fermentable sugars (Favaretto et al. 2023; Pocan et al. 2018), as discussed in earlier. Hence, this section focuses specifically on the hydrolysis products of chitin and the main hydrolysates are listed in Table 6.3. Both hydrolysis products show significant versatility across various sectors. Chitosan is used in agriculture as a natural pesticide and growth enhancer, in the food industry for preservation, in water treatment for contaminant removal, in aquaculture to promote healthy fish growth, and in cosmetics for its skin benefits (Abidin et al. 2017). In contrast, glucosamine, often found in supplements, is mainly used in nutraceuticals to support joint health, in pharmaceuticals for its potential therapeutic effects, in cosmetics for improving skin hydration and elasticity, and as a food additive for texture enhancement (Abidin et al. 2017).

The second most abundant polysaccharide found in nature is chitin, which can be found in the chitinous biomass, such as shellfish waste. Chitin holds great potential for valorization in various applications. Chitin hydrolysis involves breaking down the chitin polymer into its monomeric components, primarily N-acetylglucosamine and chito-oligosaccharides (Fig. 6.4) (Roucan et al. 2018).

Table 6.3 Main hydrolysates of chitin, their chemical structure and applications. Hydrolysate names and applications are derived from Abidin et al. 2017); chemical structures were designed by the author.

Hydrolysate	Structure	Application
Chitosan	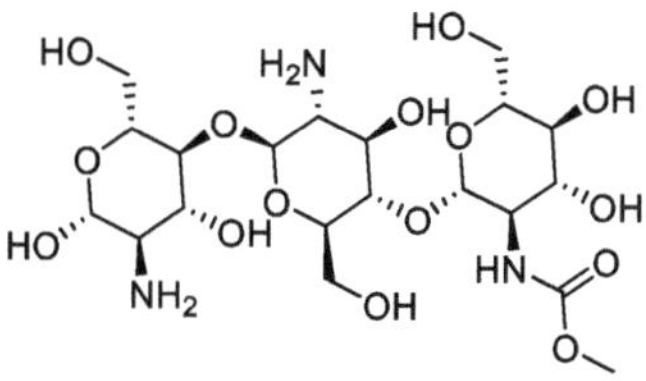	–Agriculture –Food –Water treatment –Aquaculture –Cosmetics
Glucosamine	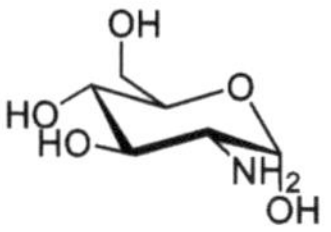	–Nutraceutical –Pharmaceutical –Cosmetics –Food

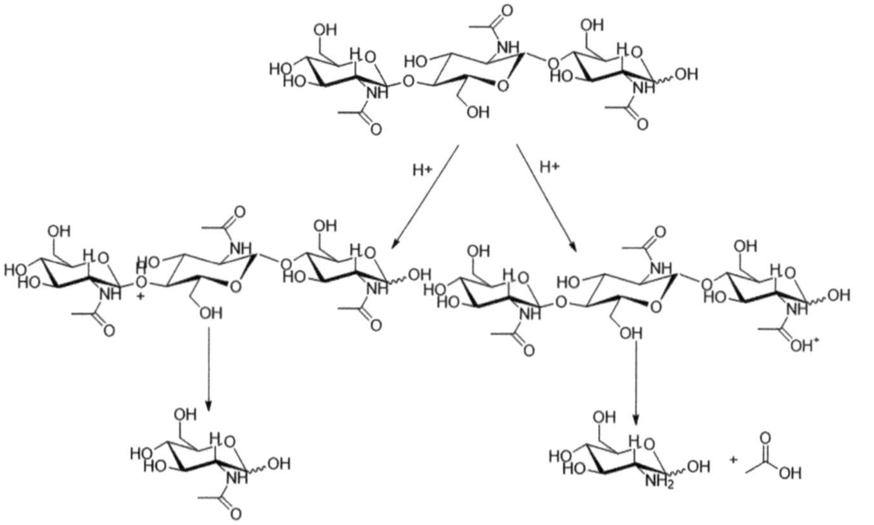

Fig. 6.4 Chitin hydrolysis into N-acetylglucosamine, and chitosan.

Chitin hydrolysis can be achieved through chemical or enzymatic methods, each with distinct mechanisms and reactions. Traditionally, it is processed using strong acids, such as HCl, to produce chito-oligosaccharides, N-acetylglucosamine, and chitosan depending on the process conditions. This method, while effective, poses environmental hazards due to the release of harmful effluents (Mathew et al. 2021). This process can be accelerated using microwave energy, which reduces the time and concentration of reactants needed, thus improving efficiency and reducing environmental impact (Zaeni et al. 2017).

Overall, chemical hydrolysis typically involves the use of strong acids, or bases. The choice of food waste hydrolysis method depends on the specific characteristics of the biomass waste and the desired end products. However, chemical methods can be more effective for certain types of fruit waste or when specific chemical products are desired (Gupta et al. 2022). The integration of these methods with other methods like enzymatic allows for efficient conversion of waste into various valuable products, contributing to sustainable waste management practices (Favaretto et al. 2023; Ilalova et al. 2020; Phitsuwan et al. 2013).

6.3.2 Transesterification

Biomass waste that is rich in lipids is a valuable feedstock for transesterification processes, in which oleaginous materials such as vegetable oils, animal fats, and algae are transformed into biodiesel (Sharmila et al. 2024). Transesterification involves the reaction of triglycerides, the main components of oils and fats, with an alcohol. This reaction takes place in the presence of a catalyst and produces fatty acid alkyl esters, commonly referred to as biodiesel. Glycerol is the main by-product of this process (Sharma et al. 2010). The general equation for transesterification is shown in Eq. 6.2 (Hamza et al. 2021).

$$\begin{aligned} &R^1C(O)OCH_2{-}CH\big(OC(O)R^2\big){-}CH_2C(O)OR^3 + 3R^4OH \\ &\rightarrow 3R^{(1-3)}C(O)OR^4 + CH_2OH{-}CHOH{-}CH_2OH \end{aligned} \tag{6.2}$$

The mechanism of transesterification involves three consecutive reversible reactions: the conversion of triglycerides to diglycerides, diglycerides to monoglycerides, and finally monoglycerides to glycerol, with each step releasing one molecule of fatty acid alkyl ester (Sharma et al. 2010). Initially, the triglyceride molecules react with the alcohol to form diglycerides and monoalkyl esters. This is followed by further reactions where diglycerides are converted to monoglycerides and additional monoalkyl esters, and finally, monoglycerides are converted to glycerol and more monoalkyl esters (Jon and Knothe 2005).

The type and concentration of the catalyst are key factors affecting both the yield and reaction rate of biodiesel production. Homogeneous catalysts are the most used due to their ease of use and shorter reaction times. These catalysts are typically alkaline, such as NaOH or KOH (Jain et al. 2023).

Alternatively, biomass-derived catalysts have attracted considerable attention for their environmental benefits, including biodegradability and the plentiful supply of raw materials like food waste, fruit peels, and plant or animal residues. By converting these materials into catalysts, they provide a more economical and sustainable alternative to traditional waste management practices (Farouk et al. 2024; Hamza et al. 2021; Rizwanul Fattah et al. 2020). Current research is focused on exploring their potential for eco-friendly biodiesel production, highlighting their role in advancing greener fuel technologies (Ao et al. 2024).

CaO derived from natural waste materials, particularly waste eggshells, has emerged as an effective and cost-efficient heterogeneous catalyst for biodiesel production. These waste shells, which are rich in calcium carbonate ($CaCO_3$), are calcined at high temperatures to produce CaO. The incorporation of microwave irradiation has significantly reduced reaction times, enhancing the efficiency and reusability of CaO catalysts, with biodiesel yields exceeding 90% (Pandit and Fulekar 2017). Further improvements in catalytic properties, such as increased basicity, surface area, and stability, have been achieved through enhancements like doping with alkali metals and combining CaO with silica (Pandit and Fulekar 2017).

The enhancement of the calcination-hydration-dehydration method increased biodiesel yields to over 93%, establishing CaO as a promising catalyst for sustainable biodiesel production (Pandit and Fulekar 2017), which has led to the utilization of various seashells for producing CaO catalysts.

These natural materials contribute to the synthesis of biodiesel effectively. For example, river snail shells calcined at 800 °C for 4 h demonstrated strong catalytic activity, converting 92.5% of used cooking oil into biodiesel (Roschat et al. 2016). CaO derived from snail shells has been successfully applied to various feedstocks, including palm oil, soybean oil, and waste cooking oil. The addition of co-solvents, such as tetrahydrofuran, has further improved biodiesel yields to 98.5% by lowering activation energy (Roschat et al. 2016). Similarly, mussel shell-derived CaO achieved a 95% biodiesel yield and maintained stability over ten cycles (Perea et al. 2016). To address the challenges of lower basicity and surface area, alkali metal doping has been employed to enhance catalytic performance; for instance, combining snail shell-derived CaO with kaolin and KBr resulted in a 98.5% yield (Liu et al. 2016), while mussel shell-derived CaO doped with activated carbon and NaOH achieved a yield of 95.1% (Hadiyanto et al. 2017). Despite these advancements, shell-derived CaO can deactivate rapidly due to hydration, carbonation, and byproduct adsorption, necessitating reactivation at high temperature (Ao et al. 2024).

Animal bones, rich in alkaline earth materials like calcium and phosphorus, can be calcined to produce highly basic calcium hydroxyapatite (Ca-HAP), which serves as an effective catalyst for biodiesel production. Studies have shown that Ca-HAP derived from ostrich bones achieved a 90.56% yield upon transesterification of waste cooking oil with a high methanol-to-oil ratio (MTOR) of 15:1, while cow bone-derived Ca-HAP resulted in a 96% yield when solar heated with a lower MTOR of 12:1, demonstrating its reusability across ten cycles with minimal activity loss (Corro et al. 2016). Further enhancements to catalyst performance have been explored, including soaking calcined bones in KOH, which yielded 96.1% biodiesel from Jatropha oil at a 9:1 ratio, showcasing better reusability due to increased basicity (Nisar et al. 2017). The low surface area of bone-derived catalysts presents a significant challenge. To address this, modifications such as hydrothermal treatment and impregnation onto supports like fly ash have been explored. This approach significantly improved surface area, allowing for biodiesel yields of 96% from hydrothermally treated catalysts and 90.4% from fly ash-supported Ca-HAP (Chingakham et al. 2019; Volli et al. 2019).

Magnetic catalysts are increasingly recognized for their significant potential to replace traditional catalysts in biodiesel production, primarily because of their easy and nearly complete recyclability, which can be readily monitored through performance retention. This makes them a cost-effective and environmentally friendly option. However, challenges persist in lowering the costs related to their more complex synthesis. Additionally, ensuring the consistent separation of magnetic and catalytic components over time remains a concern (Ao et al. 2024).

Future investigations should concentrate on improving biomass-derived catalysts, specifically regarding their durability and stability during repeated applications. Essential considerations should include enhancing the recycling of active sites through techniques such as magnetic recovery, as well as preventing the leaching of functional groups. Furthermore, priority should be given to developing high surface area supports and cost-effective methods for synthesizing multifunctional activated carbons.

6.3.3 Esterification

Esterification plays a vital role in the valorization of biomass waste, particularly in biodiesel production, where it converts free fatty acids (FFAs) into fatty acid methyl esters, the main components of biodiesel. Additionally, esterification is essential for transforming glycerol, the main by-product of biodiesel, into valuable fuel additives by enhancing its properties for efficient combustion. This process not only reduces waste but also supports biorefinery development, promoting circular economies and advancing sustainable energy solutions (Perez et al. 2024). The general equation for esterification reaction is shown in Eq. 6.3.

$$R^1COOH + R^2OH \rightarrow R^1COOR^2 + H_2O \tag{6.3}$$

Esterification of FFAs is a key industrial process before biodiesel production, enabling the use of cost-effective biomass waste feedstocks, which often include waste cooking oil, grease, animal fats, fish oil, oil-rich agricultural residues (such as rice bran and palm fatty acid distillate), and non-edible oils like Jatropha and Pongamia. To prevent soap formation during transesterification, FFAs must be converted into esters, making esterification essential for efficient biodiesel production (Zhang et al. 2023).

Methanol or ethanol is commonly used as the alcohol in this process to ensure the conversion of FFAs into biodiesel components. The esterification process can use either homogeneous or heterogeneous catalysts. Homogeneous acid catalysts, such as H_2SO_4 and HCl, are effective but can lead to equipment corrosion and toxic waste generation. Base catalysts like NaOH and KOH are suitable when FFA levels are low but may cause saponification if FFAs are high. Heterogeneous catalysts, on the other hand, are becoming more popular due to their reusability, easy separation, and lower environmental impact (Maheshwari et al. 2022).

Amberlyst-15 is a sulfonic ion exchange resin recognized for its significant catalytic activity in the esterification of FFAs derived from waste cooking oil (Ribeiro et al. 2024). Its efficacy is linked to favorable surface properties, including porosity and specific surface area, which are vital for liquid-phase reactions essential for biodiesel synthesis (Özbay et al. 2008). However, the major faced problem is its low thermal stability.

Transition metal oxides, when modified with sulphate ions, create highly acidic catalysts suitable for biodiesel feedstock esterification. The sulfonation process enhances surface acidity by introducing both Brønsted and Lewis acid sites, improving the adsorption of FFAs and reducing catalyst deactivation. For example, sulfated tin oxide $\left(SO_4^{2-}/SnO_2\right)$ exhibits heightened catalytic activity and thermal stability due to strong interactions between sulfate ions and the cassiterite phase of SnO_2 (Khder et al. 2008). Optimal conditions for these catalysts, such as sulphate content and calcination temperature, can maximize their acid site availability, enhancing their effectiveness in esterification reactions (Parangi and Mishra 2020).

More studies indicate that effective esterification catalysts should possess hydrophobic surfaces and adjustable pore structures. Carbon-based catalysts modified through sulfonation, characterized by a hydrophobic carbon framework combined with hydrophilic acidic groups, are particularly advantageous. These features enhance the selectivity and efficiency of the esterification process (Liu et al. 2013).

Solid carbon materials, including activated carbon, carbon nanotubes, fullerene, and graphene, serve as effective supports for catalytically active components, promoting the selective formation of desired products. Graphene oxide (GO) has been employed to develop acid catalysts, such as the GO-$PrSO_3H$ catalyst, which demonstrated significant efficacy in esterification reactions, yielding 60% methyl oleate from oleic acid and methanol (Masteri-Farahani et al. 2020). However, challenges remain, such as sulfur group leaching and limited adsorption due to low surface area. Addressing these issues may involve modifying raw materials to enhance porosity and improve the performance of biomass-derived carbon catalysts (Zhang et al. 2023).

Biomass primarily consists of lignin, hemicellulose, and cellulose. Each of these components has distinct chemical structures that influence their ability to support the attachment of active groups on catalysts (Zhang et al. 2023). A sulfonated carbon catalyst made from bamboo powder displayed outstanding performance, achieving a remarkable yield of 97.3% in its first use and maintaining an impressive yield of 82.9% after four times. These findings illustrate the catalyst's effectiveness in facilitating the desired reactions and highlight its versatility for various applications (Zhang et al. 2023). Likewise, sulfonated solid acid catalysts derived from palm seeds and bamboo bark reached a conversion of 95% for palm fatty acid distillate, further demonstrating their potential for biodiesel production (Farabi et al. 2019).

Crude glycerol, a major industrial waste byproduct of biodiesel production, contains several impurities. These include inorganic bases, sodium or potassium salts, methanol, and non-glycerol organic compounds such as FFAs and unreacted oils or fats. The presence of these impurities can significantly impact the production of glycerol esters. Therefore, it is crucial to evaluate catalysts that can process crude glycerol effectively without the need for expensive purification steps. Avoiding such costly processes is important to maintain profitability and prevent an increase in the final product price (Perez et al. 2024). In this context, activated peanut shells functionalized with H_2SO_4 have demonstrated high effectiveness as catalysts for esterifying crude glycerol with acetic acid, achieving full glycerol conversion, and selectivity of 55% to monoacetyl, 35% to diacetyl, and 10% to triacetyl glycerol. This performance is attributed to the catalyst's high surface area and strong acid sites from the sulfonic groups, although some surface area reduction occurred due to Na_2SO_4 adsorption (Perez et al. 2024).

Overall, by advancing catalyst technology, these developments seek to foster greater sustainability and efficiency in esterification, particularly in transforming waste biomass feedstocks into valuable products such as biodiesel and crude glycerol esters.

6.3.4 Dehydration

Dehydration reactions play a crucial role in the modification and valorization of biomass due to the high abundance of hydroxyl groups found in many natural compounds (Zhao et al. 2019). In these reactions, a water molecule is eliminated from a substrate, often an alcohol, leading to the formation of alkenes or other unsaturated products depending on the starting material (Hjerrild et al. 2020). The overall process of alcohol dehydration can be summarized by a general equation (Eq. 6.4).

$$R^1H_2C{-}CHR^2OH \rightarrow R^1HC{=}CHR^2 + H_2O \tag{6.4}$$

However, because the hydroxyl group (-OH) is not an effective leaving group on its own, the reaction typically requires catalysis by Brønsted or Lewis acids. When a Brønsted acid is used, it first protonates the hydroxyl group, transforming it into a better leaving group that can be eliminated as water. During this process, a C=C double bond is formed in the substrate's carbon skeleton, following Zaitsev's rule, which involves the release of a β-proton and the regeneration of the catalyst (Senger et al. 2012).

In contrast, Lewis acid catalysts initiate the reaction by bonding with the lone pair of electrons on the hydroxyl oxygen. This interaction reduces the electron density of the C–O bond in the alcohol, leading to the cleavage of this bond and the production of an alkene alongside the formation of Lewis acid hydroxide. Subsequently, the released β-proton reacts with the Lewis acid hydroxide, generating water and restoring the original catalyst species (Hjerrild et al. 2020).

Sulfonated ionic liquids facilitate the rapid conversion of disaccharides and polysaccharides with high selectivity for HMF in a renewable butanone-water system, achieving up to 65.1% HMF yield from disaccharides in just one minute. Notably, fructose-based saccharides, such as sucrose and inulin, exhibited the highest HMF selectivity, whereas glucose-based saccharides, like maltose and cellulose, generated a larger amount of undesirable by-products (Zhao et al. 2019).

Additionally, alcohol dehydration processes offer a route to convert C2–C4 alcohols into longer-chain hydrocarbons, with catalysts like SAPO-34 and HZSM-5 showing high activity in converting alcohols to olefins (Nash et al. 2016).

Dehydration reactions generally require acid catalysts to facilitate the removal of water from biomass-derived feedstocks. The choice of catalyst plays a crucial role in determining the process's efficiency, selectivity, and sustainability. SAPO-34, for instance, is highly effective for converting methanol into ethylene and propylene and has shown excellent performance in dehydrating alcohols like ethanol (Yu et al. 2022). Alcohol dehydration has been extensively explored using various solid acid catalysts, including zeolites like HZSM-5 (Manyepedza and Makuve 2024). Beyond zeolites, metal-incorporated mesoporous silica materials, such as Ni-MCM-41 and Zr-KIT-6, have also demonstrated effectiveness in converting alcohols into their corresponding olefins (Nash et al. 2016). The type and strength of acid sites on these catalysts significantly impact performance. Strong Brønsted acidic zeolites, while active, often show poor selectivity and are prone to coke deposition, reducing their operational lifespan. Studies suggest that catalysts with lower Brønsted to Lewis acid ratios tend to offer improved selectivity and stability, making them more suitable for alcohol dehydration reactions.

6.4 Conclusions

Biomass waste valorization through chemical processes, including hydrolysis, transesterification, esterification, and dehydration, offers significant potential for creating valuable biofuels and chemical intermediates. By employing advanced catalytic systems, such as bifunctional catalysts and solid acids, these processes transform low-cost biomass feedstocks like fruit waste and animal fats into fermentable sugars, biodiesel, and high-value compounds. Key innovations in catalyst design improving stability, reusability, and efficiency are driving the development of more sustainable and economically viable pathways for waste management and biofuel production.

Looking ahead, continued research into biomass-based catalysts and the exploration of underutilized waste streams will be critical for advancing sustainable technologies. The development of robust, multifunctional catalysts, particularly those derived from biomass, holds promise for reducing waste and environmental impact while promoting a circular economy. As these technologies evolve, they have the potential to reshape the field of renewable energy production and chemical synthesis, contributing to global efforts toward sustainability and environmental conservation.

References

Abidin MZ, Junqueira-Gonçalves MP, Khutoryanskiy VV, Niranjan K (2017) Intensifying chitin hydrolysis by adjunct treatments-an overview. J Chem Technol Biotechnol 92(11):2787–2798. https://doi.org/10.1002/jctb.5208

Ao S, Changmai B, Vanlalveni C, Chhandama MVL, Wheatley AEH, Rokhum SL (2024) Biomass waste-derived catalysts for biodiesel production: recent advances and key challenges. Renew Energy 223:120031. https://doi.org/10.1016/j.renene.2024.120031

Araujo J, Sica P, Costa C, Márquez MC (2021) Enzymatic hydrolysis of fish waste as an alternative to produce high value-added products. Waste Biomass Valoriz 12(2):847–855. https://doi.org/10.1007/s12649-020-01029-x

Ashokkumar V, Venkatkarthick R, Jayashree S, Chuetor S, Dharmaraj S, Kumar G, Chen W-H, Ngamcharussrivichai C (2022) Recent advances in lignocellulosic biomass for biofuels and value-added bioproducts—A critical review. Biores Technol 344:126195. https://doi.org/10.1016/j.biortech.2021.126195

Azlan NSM, Yap CL, Gan S, Rahman MBA (2022) Recent advances in the conversion of lignocellulosic biomass and its degraded products to levulinic acid: a synergy of Brønsted-Lowry acid and Lewis acid. Ind Crops Prod 181:114778. https://doi.org/10.1016/j.indcrop.2022.114778

Basar IA, Liu H, Carrere H, Trably E, Eskicioglu C (2021) A review on key design and operational parameters to optimize and develop hydrothermal liquefaction of biomass for biorefinery applications. Green Chem 23(4):1404–1446. https://doi.org/10.1039/D0GC04092D

Basu P (2018) Biomass gasification, pyrolysis and torrefaction: practical design and theory. Academic press

Bedasso GD, Lin C-Y, Wu C-H, Tzou D-LM, Chung P-W (2024) Hydrolysis of cellulose using boronic-acid multifunctional carbocatalysts. J Taiwan Inst Chem Eng 157:105435. https://doi.org/10.1016/j.jtice.2024.105435

Boonyakarn T, Wataniyakul P, Boonnoun P, Quitain AT, Kida T, Sasaki M, Laosiripojana N, Jongsomjit B, Shotipruk A (2019) Enhanced Levulinic acid production from cellulose by combined Brønsted hydrothermal carbon and lewis acid catalysts. Ind Eng Chem Res 58(8):2697–2703. https://doi.org/10.1021/acs.iecr.8b05332

Brodin M, Vallejos M, Opedal MT, Area MC, Chinga-Carrasco G (2017) Lignocellulosics as sustainable resources for production of bioplastics—A review. J Clean Prod 162:646–664. https://doi.org/10.1016/j.jclepro.2017.05.209

Carmona-Cabello M, Garcia IL, Leiva-Candia D, Dorado MP (2018) Valorization of food waste based on its composition through the concept of biorefinery. Curr Opin Green Sustain Chem 14:67–79. https://doi.org/10.1016/j.cogsc.2018.06.011

Carrasco-Reinado R, Escobar A, Carrera C, Guarnizo P, Vallejo RA, Fernández-Acero FJ (2019) Valorization of microalgae biomass as a potential source of high-value sugars and polyalcohols. Lwt-Food Sci Technol 114:108385. https://doi.org/10.1016/j.lwt.2019.108385

Chakhtouna H, Benzeid H, Zari N, Qaiss AE, Bouhfid R (2022) Recent advances in eco-friendly composites derived from lignocellulosic biomass for wastewater treatment. In: Biomass conversion and biorefinery. https://doi.org/10.1007/s13399-022-03159-9

Chen X, Cao X, Sun S, Yuan T, Shi Q, Zheng L, Sun R (2018) Evaluating the production of monosaccharides and xylooligosaccharides from the pre-hydrolysis liquor of kraft pulping process by acid and enzymatic hydrolysis. Ind Crops Prod 124:906–911. https://doi.org/10.1016/j.indcrop.2018.08.071

Chen X, Yue Z, Winberg PC, Dinoro JN, Hayes P, Beirne S, Wallace GG (2019) Development of rhamnose-rich hydrogels based on sulfated xylorhamno-uronic acid toward wound healing applications. Biomater Sci 7(8):3497–3509. https://doi.org/10.1039/C9BM00480G

Chen J-Y, Xiao Y, Guo F-S, Li K-M, Huang Y-B, Lu Q (2024) Single-atom metal catalysts for catalytic chemical conversion of biomass to chemicals and fuels. ACS Catal 14(7):5198–5226. https://doi.org/10.1021/acscatal.4c00208

Chingakham C, Tiwary C, Sajith V (2019) Waste animal bone as a novel layered heterogeneous catalyst for the transesterification of biodiesel. Catal Lett 149(4):1100–1110. https://doi.org/10.1007/s10562-019-02696-9

Cho EJ, Trinh LTP, Song Y, Lee YG, Bae H-J (2020) Bioconversion of biomass waste into high value chemicals. Biores Technol 298:122386. https://doi.org/10.1016/j.biortech.2019.122386

Chu S, Yang L, Guo X, Dong L, Chen X, Li Y, Mu X (2018) The influence of pore structure and Si/Al ratio of HZSM-5 zeolites on the product distributions of α-cellulose hydrolysis. Mol Catal 445:240–247. https://doi.org/10.1016/j.mcat.2017.11.032

Corro G, Sánchez N, Pal U, Bañuelos F (2016) Biodiesel production from waste frying oil using waste animal bone and solar heat. Waste Manag 47:105–113. https://doi.org/10.1016/j.wasman.2015.02.001

Czelej M, Garbacz K, Czernecki T, Wawrzykowski J, Waśko A (2022) Protein hydrolysates derived from animals and plants: a review of production methods and antioxidant activity. Foods 11(13):1953. https://doi.org/10.3390/foods11131953

Demirbas A (2011) Waste management, waste resource facilities and waste conversion processes. Energy Convers Manag 52(2):1280–1287. https://doi.org/10.1016/j.enconman.2010.09.025

Deng W, Feng Y, Fu J, Guo H, Guo Y, Han B, Jiang Z, Kong L, Li C, Liu H, Nguyen PTT, Ren P, Wang F, Wang S, Wang Y, Wang Y, Wong SS, Yan K, Yan N et al (2023) Catalytic conversion of lignocellulosic biomass into chemicals and fuels. Green Energy Environ 8(1):10–114. https://doi.org/10.1016/j.gee.2022.07.003

De Schouwer F, Claes L, Vandekerkhove A, Verduyckt J, De Vos DE (2019) Protein-rich biomass waste as a resource for future biorefineries: state of the art, challenges, and opportunities. Chemsuschem 12(7):1272–1303. https://doi.org/10.1002/cssc.201802418

Ding S, Sun Y, Chen H, Xu C, Hu Y (2019) An ultrasonic-ionic liquid process for the efficient acid catalyzed hydrolysis of feather keratin. Chin J Chem Eng 27(3):660–667. https://doi.org/10.1016/j.cjche.2018.05.008

Fang J, Zheng W, Liu K, Li H, Li C (2020) Molecular design and experimental study on the synergistic catalysis of cellulose into 5-hydroxymethylfurfural with Brønsted-Lewis acidic ionic liquids. Chem Eng J 385:123796. https://doi.org/10.1016/j.cej.2019.123796

Farabi MSA, Ibrahim ML, Rashid U, Taufiq-Yap YH (2019) Esterification of palm fatty acid distillate using sulfonated carbon-based catalyst derived from palm kernel shell and bamboo. Energy Convers Manag 181:562–570. https://doi.org/10.1016/j.enconman.2018.12.033

Farouk SM, Tayeb AM, Abdel-Hamid SMS, Osman RM (2024) Recent advances in transesterification for sustainable biodiesel production, challenges, and prospects: a comprehensive review. Environ Sci Pollut Res. https://doi.org/10.1007/s11356-024-32027-4

Favaretto DPC, Rempel A, Lanzini JR, Silva ACM, Lazzari T, Barbizan LD, Brião VB, Colla LM, Treichel H (2023) Fruit residues as biomass for bioethanol production using enzymatic hydrolysis as pretreatment. https://doi.org/10.21203/rs.3.rs-2465028/v1

Gliozzi G, Innorta A, Mancini A, Bortolo R, Perego C, Ricci M, Cavani F (2014) Zr/P/O catalyst for the direct acid chemo-hydrolysis of non-pretreated microcrystalline cellulose and softwood sawdust. Appl Catal B 145:24–33. https://doi.org/10.1016/j.apcatb.2012.12.035

Gupta K, Rather MA, Kumar P, Bardhan P, Mahnot NK, Mandal M, Kataki R (2022) Physical and chemical hydrolysis methods for breaking down the complex waste biomass to the fermentable sugars and value-added products. Thermochem Catal Convers Technol Future Biorefineries 2:59–75

Hadiyanto H, Afianti AH, Navi'a UI, Adetya NP, Widayat W, Sutanto H (2017) The development of heterogeneous catalyst C/CaO/NaOH from waste of green mussel shell (Perna varidis) for biodiesel synthesis. J Environ Chem Eng 5(5):4559–4563. https://doi.org/10.1016/j.jece.2017.08.049

Hafid HS, Rahman NAA, Shah UKM, Baharuddin AS, Ariff AB (2017) Feasibility of using kitchen waste as future substrate for bioethanol production: a review. Renew Sustain Energy Rev 74:671–686. https://doi.org/10.1016/j.rser.2017.02.071

Hamza M, Ayoub M, Shamsuddin RB, Mukhtar A, Saqib S, Zahid I, Ameen M, Ullah S, Al-Sehemi AG, Ibrahim M (2021) A review on the waste biomass derived catalysts for biodiesel production. Environ Technol Innov 21:101200. https://doi.org/10.1016/j.eti.2020.101200

Hegner J, Pereira KC, DeBoef B, Lucht BL (2010) Conversion of cellulose to glucose and levulinic acid via solid-supported acid catalysis. Tetrahedron Lett 51(17):2356–2358. https://doi.org/10.1016/j.tetlet.2010.02.148

Hjerrild P, Tørring T, Poulsen TB (2020) Dehydration reactions in polyfunctional natural products. Nat Prod Rep 37(8):1043–1064. https://doi.org/10.1039/D0NP00009D

Huber GW, Iborra S, Corma A (2006) Synthesis of transportation fuels from biomass: chemistry, catalysts, and engineering. Chem Rev 106(9):4044–4098. https://doi.org/10.1021/cr068360d

Ilalova G, Safin R, Mukhametzyanov S, Gazizullina A (2020) Hydrolysis as a basis for processing vegetable waste into bioplastics. In: E3S web of conferences 221:03009. https://doi.org/10.1051/e3sconf/202022103009

Inokuma K, Toyohara K, Hamada T, Kondo A, Hasunuma T (2024) One-pot synthesis of cellobiose from sucrose using sucrose phosphorylase and cellobiose phosphorylase co-displaying Pichia pastoris as a reusable whole-cell biocatalyst. Sci Rep 14(1):18540. https://doi.org/10.1038/s41598-024-69676-x

Jain S, Kumar N, Singh VP, Mishra S, Sharma NK, Bajaj M, Khan TMY (2023) Transesterification of algae oil and little amount of waste cooking oil blend at low temperature in the presence of NaOH. Energies 16(3):1293. https://doi.org/10.3390/en16031293

Jiang Z, Fan J, Budarin VL, Macquarrie DJ, Gao Y, Li T, Hu C, Clark JH (2018) Mechanistic understanding of salt-assisted autocatalytic hydrolysis of cellulose. Sustain Energy Fuels 2(5):936–940. https://doi.org/10.1039/C8SE00045J

Jiang Z, Gao M, Ding W, Huang C, Hu C, Shi B, Tsang DCW (2021) Selective degradation and oxidation of hemicellulose in corncob to oligosaccharides: from biomass into masking agent for sustainable leather tanning. J Hazard Mater 413:125425. https://doi.org/10.1016/j.jhazmat.2021.125425

Jon VG, Knothe G (2005) Basics of the transesterification reaction. In: The biodiesel handbook, pp 26–41

Karinen R, Vilonen K, Niemelä M (2011) Biorefining: heterogeneously catalyzed reactions of carbohydrates for the production of furfural and hydroxymethylfurfural. Chemsuschem 4(8):1002–1016. https://doi.org/10.1002/cssc.201000375

Karunarathne M, Kandegama WW, Faraj TK, Baig MB, Xiaoyu H, Gefei H (2024) Valorization of proteinaceous animal waste to amino acids reduces environmental pollution while catalyzing plant functions in sustainable agriculture. https://doi.org/10.20944/preprints202406.1689.v1

Khder AS, El-Sharkawy EA, El-Hakam SA, Ahmed AI (2008) Surface characterization and catalytic activity of sulfated tin oxide catalyst. Catal Commun 9(5):769–777. https://doi.org/10.1016/j.catcom.2007.08.022

Khiari Z, Mason B (2018) Comparative dynamics of fish by-catch hydrolysis through chemical and microbial methods. LWT 97:135–143. https://doi.org/10.1016/j.lwt.2018.06.032

Kiewiet MBG, Faas MM, De Vos P (2018) Immunomodulatory protein hydrolysates and their application. Nutrients 10(7):904. https://doi.org/10.3390/nu10070904

Kim S, Holtzapple MT (2005) Lime pretreatment and enzymatic hydrolysis of corn stover. Biores Technol 96(18):1994–2006. https://doi.org/10.1016/j.biortech.2005.01.014

Kim S-J, Dwiatmoko AA, Choi JW, Suh Y-W, Suh DJ, Oh M (2010) Cellulose pretreatment with 1-n-butyl-3-methylimidazolium chloride for solid acid-catalyzed hydrolysis. Biores Technol 101(21):8273–8279. https://doi.org/10.1016/j.biortech.2010.06.047

Liu T, Li Z, Li W, Shi C, Wang Y (2013) Preparation and characterization of biomass carbon-based solid acid catalyst for the esterification of oleic acid with methanol. Biores Technol 133:618–621. https://doi.org/10.1016/j.biortech.2013.01.163

Liu H, Guo HS, Wang XJ, Jiang JZ, Lin H, Han S, Pei SP (2016) Mixed and ground KBr-impregnated calcined snail shell and kaolin as solid base catalysts for biodiesel production. Renew Energy 93:648–657. https://doi.org/10.1016/j.renene.2016.03.017

Lu L, Guo L, Wang K, Liu Y, Xiao M (2020) β-Galactosidases: a great tool for synthesizing galactose-containing carbohydrates. Biotechnol Adv 39:107465. https://doi.org/10.1016/j.biotechadv.2019.107465

Maheshwari P, Haider MB, Yusuf M, Klemeš JJ, Bokhari A, Beg M, Al-Othman A, Kumar R, Jaiswal AK (2022) A review on latest trends in cleaner biodiesel production: role of feedstock, production methods, and catalysts. J Clean Prod 355:131588. https://doi.org/10.1016/j.jclepro.2022.131588

Manyepedza T, Makuve N (2024) Catalyst for biomass upgrading: transfomrming 5-hydroxymethyl furfural into 2,5-furan dicarboxylic acid for fine chemicals, featuring Mil-100(Fe) Mof as a catalyst. ChemsitrySelect 9(28):e202401412. https://doi.org/10.1002/slct.202401412

Masteri-Farahani M, Hosseini M-S, Forouzeshfar N (2020) Propyl-SO3H functionalized graphene oxide as multipurpose solid acid catalyst for biodiesel synthesis and acid-catalyzed esterification and acetalization reactions. Renew Energy 151:1092–1101. https://doi.org/10.1016/j.renene.2019.11.108

Mathew GM, Huang CC, Sindhu R, Binod P, Sirohi R, Awsathi MK, Pillai S, Pandey A (2021) Enzymatic approaches in the bioprocessing of shellfish wastes. 3 Biotech 11(8):367. https://doi.org/10.1007/s13205-021-02912-7

Mosier N (2005) Features of promising technologies for pretreatment of lignocellulosic biomass. Biores Technol 96(6):673–686. https://doi.org/10.1016/j.biortech.2004.06.025

Mulder WJ (2010) Proteins in biomass streams. In: Wageningen UR food biobased research, p 1134

Nam KH (2024) Engineering xylose isomerase for industrial applications. Catalysts 14(9):597. https://doi.org/10.3390/catal14090597

Nash CP, Ramanathan A, Ruddy DA, Behl M, Gjersing E, Griffin M, Zhu H, Subramaniam B, Schaidle JA, Hensley JE (2016) Mixed alcohol dehydration over Brønsted and Lewis acidic catalysts. Appl Catal A 510:110–124. https://doi.org/10.1016/j.apcata.2015.11.019

Nikoo M, Regenstein JM, Haghi Vayghan A, Walayat N (2023) Formation of oxidative compounds during enzymatic hydrolysis of byproducts of the seafood industry. Processes 11(2):543. https://doi.org/10.3390/pr11020543

Nisar J, Razaq R, Farooq M, Iqbal M, Khan RA, Sayed M, Shah A, Rahman IU (2017) Enhanced biodiesel production from Jatropha oil using calcined waste animal bones as catalyst. Renew Energy 101:111–119. https://doi.org/10.1016/j.renene.2016.08.048

Özbay N, Oktar N, Tapan NA (2008) Esterification of free fatty acids in waste cooking oils (WCO): role of ion-exchange resins. Fuel 87(10–11):1789–1798. https://doi.org/10.1016/j.fuel.2007.12.010

Pandey A, Larroche C, Gnansounou E, Khanal SK, Dussap C-G, Ricke S (2019) Biomass, biofuels, biochemicals: biofuels: alternative feedstocks and conversion processes for the production of liquid and gaseous biofuels. Academic press

Pandit PR, Fulekar MH (2017) Egg shell waste as heterogeneous nanocatalyst for biodiesel production: optimized by response surface methodology. J Environ Manag 198:319–329. https://doi.org/10.1016/j.jenvman.2017.04.100

Parangi T, Mishra MK (2020) Solid acid catalysts for biodiesel production. Comments Inorg Chem 40(4):176–216. https://doi.org/10.1080/02603594.2020.1755273

Pelé M, Georges J-Y, Matsuzawa T, Sueur C (2021) Editorial: perceptions of human-animal relationships and their impacts on animal ethics, law and research. Front Psychol 11:631238. https://doi.org/10.3389/fpsyg.2020.631238

Perea A, Kelly T, Hangun-Balkir Y (2016) Utilization of waste seashells and *Camelina sativa* oil for biodiesel synthesis. Green Chem Lett Rev 9(1):27–32. https://doi.org/10.1080/17518253.2016.1142004

Perez FM, Gatti MN, Fermanelli CS, Saux C, Renzini MS, Pompeo F (2024) Crude glycerol esterification using biomass-derived carbon acid catalysts. Next Mater 2:100125. https://doi.org/10.1016/j.nxmate.2024.100125

Phitsuwan P, Sakka K, Ratanakhanokchai K (2013) Improvement of lignocellulosic biomass in planta: a review of feedstocks, biomass recalcitrance, and strategic manipulation of ideal plants designed for ethanol production and processability. Biomass Bioenerg 58:390–405. https://doi.org/10.1016/j.biombioe.2013.08.027

Pleissner D, Kwan TH, Lin CSK (2014) Fungal hydrolysis in submerged fermentation for food waste treatment and fermentation feedstock preparation. Biores Technol 158:48–54. https://doi.org/10.1016/j.biortech.2014.01.139

Pocan P, Bahcegul E, Oztop MH, Hamamci H (2018) Enzymatic hydrolysis of fruit peels and other lignocellulosic biomass as a source of sugar. Waste Biomass Valoriz 9:929–937

Powell T, Bowra S, Cooper HJ (2016) Subcritical water processing of proteins: an alternative to enzymatic digestion? Anal Chem 88(12):6425–6432. https://doi.org/10.1021/acs.analchem.6b01013

Qiao Y, Zhai C, Liu F, Chen L, Na H, Chen J, Zhu J (2020) Highly efficient microwave driven assisted hydrolysis of cellulose to sugar with the utilization of ZrO2 to inhibit recrystallization of cellulose. Carbohyd Polym 228:115358. https://doi.org/10.1016/j.carbpol.2019.115358

Ribeiro FCP, Santos JL, Araujo RO, Santos VO, Chaar JS, Tenório JAS, De Souza LKC (2024) Sustainable catalysts for esterification: sulfonated carbon spheres from biomass waste using hydrothermal carbonization. Renew Energy 220:119653. https://doi.org/10.1016/j.renene.2023.119653

Rizwanul Fattah IM, Ong HC, Mahlia TMI, Mofijur M, Silitonga AS, Rahman SMA, Ahmad A (2020) State of the art of catalysts for biodiesel production. Front Energy Res 8:101. https://doi.org/10.3389/fenrg.2020.00101

Roschat W, Siritanon T, Kaewpuang T, Yoosuk B, Promarak V (2016) Economical and green biodiesel production process using river snail shells-derived heterogeneous catalyst and co-solvent method. Biores Technol 209:343–350. https://doi.org/10.1016/j.biortech.2016.03.038

Roucan M, Kielmann M, Connon SJ, Bernhard SSR, Senge MO (2018) Conformational control of nonplanar free base porphyrins: towards bifunctional catalysts of tunable basicity. Chem Commun 54(1):26–29. https://doi.org/10.1039/C7CC08099A

Sari YW, Alting AC, Floris R, Sanders JPM, Bruins ME (2014) Glutamic acid production from wheat by-products using enzymatic and acid hydrolysis. Biomass Bioenerg 67:451–459. https://doi.org/10.1016/j.biombioe.2014.05.018

Sari YW, Mulder WJ, Sanders JPM, Bruins ME (2015) Towards plant protein refinery: review on protein extraction using alkali and potential enzymatic assistance. Biotechnol J 10(8):1138–1157. https://doi.org/10.1002/biot.201400569

Sarker A, Ahmmed R, Ahsan SM, Rana J, Ghosh MK, Nandi R (2024) A comprehensive review of food waste valorization for the sustainable management of global food waste. Sustain Food Technol 2(1):48–69. https://doi.org/10.1039/D3FB00156C

Senger NA, Bo B, Cheng Q, Keeffe JR, Gronert S, Wu W (2012) The element effect revisited: factors determining leaving group ability in activated nucleophilic aromatic substitution reactions. J Org Chem 77(21):9535–9540. https://doi.org/10.1021/jo301134q

Sharma YC, Singh B, Korstad J (2010) Application of an efficient nonconventional heterogeneous catalyst for biodiesel synthesis from Pongamia pinnata Oil. Energy Fuels 24(5):3223–3231. https://doi.org/10.1021/ef901514a

Sharmila VG, Shanmugavel SP, Banu JR (2024) A review on emerging technologies and machine learning approaches for sustainable production of biofuel from biomass waste. Biomass Bioenerg 180:106997. https://doi.org/10.1016/j.biombioe.2023.106997

Shimizu K, Furukawa H, Kobayashi N, Itaya Y, Satsuma A (2009) Effects of Brønsted and Lewis acidities on activity and selectivity of heteropolyacid-based catalysts for hydrolysis of cellobiose and cellulose. Green Chem 11(10):1627. https://doi.org/10.1039/b913737h

Shreshtha S, Ingle AP, Moharil MP (2023) Nanobiocatalysts: the promising alternatives for conventional biocatalysts used in lignocellulosic biomass hydrolysis. Nanotechnol Biorefinery 143–162

Shrotri A, Kobayashi H, Fukuoka A (2018) Cellulose depolymerization over heterogeneous catalysts. Acc Chem Res 51(3):761–768. https://doi.org/10.1021/acs.accounts.7b00614

Spierling RE (2011) Anaerobic co-digestion of microalgae with food waste and wastewater sludge. California Polytechnic State University. https://doi.org/10.15368/theses.2011.142

Sun R, Sun XF, Liu GQ, Fowler P, Tomkinson J (2002) Structural and physicochemical characterization of hemicelluloses isolated by alkaline peroxide from barley straw. Polym Int 51(2):117–124

Taherzadeh MJ, Karimi K (2007) Acid-based hydrolysis processes for ethanol from lignocellulosic materials: a review

Usanova KI, Dhall H, Chandna M, Pandey N, Mouli KC, Vyas A (2024) Catalytic conversion of biomass to biofuels using green nanocatalysts. In: E3S web of conferences 537:07011. https://doi.org/10.1051/e3sconf/202453707011

Usmani Z, Sharma M, Awasthi AK, Lukk T, Tuohy MG, Gong L, Nguyen-Tri P, Goddard AD, Bill RM, Nayak SC, Gupta VK (2021) Lignocellulosic biorefineries: the current state of challenges and strategies for efficient commercialization. Renew Sustain Energy Rev 148:111258. https://doi.org/10.1016/j.rser.2021.111258

Vázquez M, Oliva M, Téllez-Luis SJ, Ramírez JA (2007) Hydrolysis of sorghum straw using phosphoric acid: evaluation of furfural production. Biores Technol 98(16):3053–3060. https://doi.org/10.1016/j.biortech.2006.10.017

Volli V, Purkait MK, Shu C-M (2019) Preparation and characterization of animal bone powder impregnated fly ash catalyst for transesterification. Sci Total Environ 669:314–321. https://doi.org/10.1016/j.scitotenv.2019.03.080

Xing H, Shi Y, Zhang P, Miao M, Zhang T, Jiang B (2016) D-Mannose: properties, production, and applications: an overview. Compr Rev Food Sci Food Saf 15(4):773–785. https://doi.org/10.1111/1541-4337.12211

Xu W, Zhang W, Zhang T, Jiang B, Mu W (2018) L-arabinose isomerases: Characteristics, modification, and application. Trends Food Sci Technol 78:25–33. https://doi.org/10.1016/j.tifs.2018.05.016

Yan Y, Zhang C, Lin Q, Wang X, Cheng B, Li H, Ren J (2018) Microwave-assisted oxalic acid pretreatment for the enhancing of enzyme hydrolysis in the production of xylose and arabinose from bagasse. Molecules 23(4):862. https://doi.org/10.3390/molecules23040862

Yang B, Wyman CE (2008) Pretreatment: the key to unlocking low-cost cellulosic ethanol. Biofuels Bioprod Biorefin 2(1):26–40. https://doi.org/10.1002/bbb.49

Yang X, Wang B, Lu T, Zhou L (2020) Depolymerization of wheat straw to produce glucose by self-catalyzed hydrolysis. Energy Fuels 34(5):5990–5996. https://doi.org/10.1021/acs.energyfuels.0c00525

Yu Q, Zhuang X, Yuan Z, Qi W, Wang Q, Tan X (2011) The effect of metal salts on the decomposition of sweet sorghum bagasse in flow-through liquid hot water. Biores Technol 102(3):3445–3450. https://doi.org/10.1016/j.biortech.2010.10.084

Yu W, Wu X, Cheng B, Tao T, Min X, Mi R, Huang Z, Fang M, Liu Y (2022) Synthesis and applications of SAPO-34 molecular sieves. Chem A Eur J 28(11):e202102787

Zaeni A, Safitri E, Fuadah B, Sudiana IN (2017) Microwave-assisted hydrolysis of chitosan from shrimp shell waste for glucosammine hydrochlorid production. J Phys: Conf Ser 846:012011. https://doi.org/10.1088/1742-6596/846/1/012011

Zhang B, Gao M, Tang W, Wang X, Wu C, Wang Q, Cheung SM, Chen X (2023) Esterification efficiency improvement of carbon-based solid acid catalysts induced by biomass pretreatments: intrinsic mechanism. Energy 263:125606. https://doi.org/10.1016/j.energy.2022.125606

Zhang R, Lu X, Sun Y, Wang X, Zhang S (2011) Modeling and optimization of dilute nitric acid hydrolysis on corn stover. J Chem Technol Biotechnol 86(2):306–314

Zhao Y, Lu K, Xu H, Qu Y, Zhu L, Wang S (2019) Comparative Study on the dehydration of biomass-derived disaccharides and polysaccharides to 5-hydroxymethylfurfural. Energy Fuels 33(10):9985–9995. https://doi.org/10.1021/acs.energyfuels.9b02863

Zhou L, Liu Z, Bai Y, Lu T, Yang X, Xu J (2016) Hydrolysis of cellobiose catalyzed by zeolites—The role of acidity and micropore structure. J Energy Chem 25(1):141–145. https://doi.org/10.1016/j.jechem.2015.11.010

Chapter 7
Biological Approaches to Fish Biomass Waste Processing: Microbial Fermentation and Enzymatic Conversion Strategies

Sajjad Ahmad, Michelle Giltrap, and Gordon Chambers

Abstract The demand for improved valorization is focusing attention on methods for extracting significant bioactive compounds from seafood byproducts, such as bones, heads, skin, frames, fins, shells, guts, and viscera. It is also important to use green extraction techniques to produce these bioactive compounds in a high-quality and efficient manner. The percentages of these viscera fluctuate based on the species, habitat, diet and taxonomy. Fish waste is an efficient source of various bioactive materials such as amino acids, peptides, enzymes, gelatine, collagen, chitin, vitamins, polyphenolic components, carotenoids and omega-3 long-chain polyunsaturated fatty acids. Fish wastes are additionally considered to be an effective and affordable source of material for making biodiesel. Depending on their functional and structural properties, these high-value compounds have application across a variety of industrial sectors, such as food, nutraceuticals, pharmacology, medicine, and agriculture. As compared to conventional methods, biobased methods considered environment friendly, and sustainable techniques including enzymatic hydrolysis and microbial fermentation extraction procedures offer additional benefits which promote sustainability and enable to achieve number of United Nation's Sustainable Development Goals (SDGs). In enzymatic hydrolysis process fish waste breakdown into proteins, peptides and amino acids using various enzymes rather than harsh chemicals. Similarly, microbial fermentation technologies minimize pollution and

S. Ahmad (✉)
Sustainability & Health Research Hub (SHRH), Technological University Dublin, City Campus, Grangegorman, D07 ADY7, Dublin, Ireland
e-mail: Sajjad.Ahmad@tudublin.ie

S. Ahmad · M. Giltrap
School of Food Science & Environmental Health, Technological University Dublin, City Campus, Central Quad, Grangegorman, D07 ADY7, Dublin, Ireland
e-mail: Michelle.Giltrap@tudublin.ie

G. Chambers
School of Physics, Clinical and Optometric Science, Technological University Dublin, City Campus, Central Quad, Grangegorman, D07 ADY7, Dublin, Ireland
e-mail: Gordon.Chambers@tudublin.ie

A. Hind (ed.), *Biomass Waste: Processing, Valorization, Industrial Applications*,
https://doi.org/10.1007/978-981-95-0991-1_7

negative effects on the environment by turning fish waste into beneficial chemicals, biofuels, fertilizers, and animal feed. The ultimate purpose of this study is to prevent environmental contamination from inappropriate waste disposal and boost the nutritional value of fish waste through microbial fermentation and enzymatic hydrolysis methods. Additionally, a brief discussion of the value-added goods and other practical applications of fish waste by-products is provided.

Keywords Fish waste · Microbial fermentation · Enzymatic hydrolysis · Bioactive compounds

7.1 Introduction

The importance of marine fisheries and aquaculture to the world's food security is growing, particularly with the context of an increase in the consumption and production of fish and the United Nations Sustainable Development Goals (SDG's) for 2030 (Meharoof et al. 2023). The world's fishing sector is a vital component of the food chain, providing substantial contributions to economic growth, nutrition, and food security. Annually, more than 20 million tons of fish tissue are discarded globally, accounting for approximately 70% of the total fish biomass (Shaw et al. 2022). In addition to seriously impeding the utilization of low-value fish, the common practice of burying or burning these by-products also has negative effects on the environment, human health, and the economy (Sampantamit et al. 2020). Consequently, there has been a significant increase in interest in the value-adding products produced from fish biomass waste. The advantages of fish wastes including high nutritional content, easy accessibility, low cost, and speedy processing, have greatly expedited the study of protein and its derivatives in several newly emerging sectors (Gatto et al. 2023).

The bioeconomy is a key driver for accomplishing sustainability objectives related to natural resources management and environmental preservation, and it is an essential part of the larger circular economy framework (Holden et al. 2023). The concepts of a biorefinery approach align with technological developments focused on using biomass as a substrate for the creation of bio-based goods (Abad-Segura et al. 2021). The ability of waste-focused biorefineries to convert industrial byproducts into products with additional value has drawn interest from researchers all around the world. Fish waste is processed by conventional techniques such as drying, salting, solvent extraction, wet pressing, acid digestion, and anaerobic fermentation. However, these methods frequently need strong solvents, high temperatures, and large energy input and increased the cost of the end-product (Del Castillo-Llamosas et al. 2021). As compared to these conventional methods, biological methods present a viable and effective means of transforming waste into compounds that have high economic worth and considered environment friendly (Freitas et al. 2021). Additionally, the utilization of bio-based technologies, including microbial fermentation and enzymatic conversion strategies, offers a sustainable substitute. These methods convert fish biomass waste into useful chemicals including proteins, peptides, lipids, biofuels, and other bioactive substances using the metabolic capacities of microbes and the specificity of enzymes (Gupta et al. 2022). The move toward biological processes is in line with the ideas of the circular economy and green chemistry, where waste as a resource that can be recovered and utilized again rather than as a problem that needs to be disposed of Kosamia et al. (2022).

The European Commission initiated a long-term strategy called "Blue Growth" in 2012 with the goal of optimizing the use of renewable resources. The plan specifically targeted fish stocks to prevent industrial pollution of the environment (Hoerterer et al. 2020). From the perspective of enhancing the integrated usage of fish resources, raising the added value of products, minimizing waste, and decreasing environmental pollution, a comprehensive use of fish processing by-products is required to accomplish the Sustainable Development Goals (SDGs) by 2030. However, in this chapter, we describe an efficient microbial fermentation and enzymatic hydrolysis-based valorisation approach to support sustainable agriculture and SDG 2 (Zero Hunger) by turning fish waste into high-protein and other bioactive compounds. Additionally, SDGs 12 and 14 are further supported by these procedures, which encourage sustainable waste management, the sustainable use of marine resources, and the establishment of sustainable business models and economic prospects, respectively. However. in this chapter, we assess the effectiveness of enzymatic hydrolysis and microbial fermentation in managing fish waste. Furthermore, a quick and preliminary economic analysis of this bioconversion process is suggested to demonstrate its viability and consequently support the development of sustainable byproducts.

7.2 Fish Biomass Waste: An Underutilized Asset

Fish waste is a valuable reservoir of minerals, lipids, proteins, and bioactive substances that can be converted into goods for the food, medicine, beauty, and biofuel sectors. The protein found in fish biomass waste is especially significant as it contains essential amino acids that can be used in human food dietary supplements, animal feed, or bioactive peptides that have anti-inflammatory, anti-hypertensive, antimicrobial, and antioxidant effects (Alfio et al. 2021). Furthermore, waste-derived fish oils have a high concentration of omega-3 polyunsaturated fatty acids (PUFAs), which are widely recognized for their positive effects on cognitive and cardiovascular health (Begum et al. 2021). The food processing, pharmaceutical, and cosmetic industries have considerable demand for collagen and gelatin derived from fish skin and bones. However, the entire potential of marine biomass is largely untapped by present fish waste management procedures, which results in negative environmental effects as well as financial losses (Tropea et al. 2021). The efficient exploitation of fish biomass waste can be achieved by the application of biological processes, particularly enzymatic hydrolysis and microbial fermentation. These methods concentrate on the biotransformation and biodegradation of organic matter, maximizing beneficial byproducts while releasing fish waste's biochemical potential (Maksimenko et al. 2024).

7.3 Microbial Fermentation

Biological processing is highly suggested due to the limitations of chemically valorizing fish wastes. The substantial amounts of acids and bases used in chemical processes make them dangerous since they harm the environment (Qian et al. 2022). In contrast to alkali-based chemicals, several studies have shown that recovering valuable compounds using microbial fermentation is a more affordable, sustainable, and efficient method for valorizing fish wastes (Mathew et al. 2021). Microbial fermentation is a metabolic procedure that comprises acquiring energy from the organic compounds without implementing any external oxidizing candidate (Marti-Quijal et al. 2020a). Fermentation processes can produce more valuable products by breaking down complex organic molecules into simpler chemicals through the metabolic activities of bacteria, yeasts, and fungus.

Lactic acid bacteria, a kind of aero-tolerant Gram-positive bacilli or cocci, are utilized for this concern. They are purely fermentative and organotrophic and produce lactic acid as the primary end product of fermentation (Rene et al. 2023). Since the beginning, fermentation has been employed to preserve various food items. Nevertheless, it is currently more important for nutrition and health due to increased extraction quantities, the qualities of beneficial chemicals and encouraging hydrolysis to reduce anti-nutrients (Nirmal et al. 2022). Recently, Suresh et al. (2023) investigated a study to extract chitin from fish scales by the successive fermentation of lactic acid bacteria (*Stenotrophomonas koreensis*) isolated from the soil. The determined proteolytic bacteria *S. koreensis* could generate protease to hydrolyse proteins found in fish scales to meet its carbon and nitrogen demands. Additionally, fermentation of the high sugar content in the media results in the synthesis of lactic acid, which causes the calcium carbonate in the fish scales to convert to calcium lactate, ultimately resulting in the extraction of chitin. Additionally, this study concluded that 28% of chitin was recovered, and their properties were evaluated through differential scanning calorimetry (DSC), (FTIR), (XRD), and thermo gravimetric analysis (TGA). Findings revealed the biologically produced chitosan's remarkable antioxidant capacity and high biodegradability readily available to develop chitosan films as active packaging materials for the food sector. In another study, the conversion of Nile tilapia (*Oreochromis niloticus*) waste into chitin by the fermentation of two novel lactic acid bacteria strains (*Lactobacillus futsaii* LAB06 and *L. plantarum* LAB14) was evaluated. Outcomes demonstrated that the maximum amount of lactic acid produced after 48 h was 100 g/L, which allowed for a 30% recovery of the chitin and protein-rich liquid (Ximenes et al. 2019).

Recently, Chen et al. (2022) compared the efficiency of microbial fermentation with a chemical-based method for the extraction of gelatin from the *Nile tilapia*

(*Oreochromis niloticus*). The fish skin was immersed in two chemical reagents (sodium hydroxide and butanol), and the fermentation process was carried by the bacterial strain *Bacillus velezensis* FEL-BM21. Results showed microbial fermentation process was more efficient and able to recover 49.99% of gelatin with high continuous filamentous structure, gelling point, melting point, and gel strength compared to the implementation of chemical solvents. In another study, the recovery of antioxidant compounds was studied by the fermentation of bacteria from sea bass discards. A bacterial strain, *Lactobacillus plantarum*, was cultivated from the sea bass stomach, intestine and colon, and several proteins, phenolic acids, and antioxidant properties were assessed after the fermentation process. Results indicated two primary phenolic acids (DL-3-phenyl-lactic acid and benzoic acid), with 466 and 314 ppb concentrations, respectively. Additionally, the highest antioxidant activity was observed in the extracts followed through the fermentation of fish by-products by consortium cultivated from the colon and stomach, respectively (Martí-Quijal et al. 2020b).

However, these studies advocated that microbial fermentation is a sustainable, economical, and environment-friendly technique that generates various biochemicals, including bioactive peptides and aromatic compounds (Khubber et al. 2021). Despite this, the adoption of microbial fermentation could offer several advantages, such as the efficiency of antioxidant peptides that operate with the combination of glutathione to protect against damage produced by oxidative stress (Siddiqui et al. 2023). Additionally, more digestible proteins are manufactured through the fermentation process because of their hydrolyzation and break down into smaller peptides and amino acids. So mainly, fermentation is an effective strategy for enhancing the nutritional value of fish products utilized as a protein source (Ucak et al. 2021).

7.4 Enzymatic Hydrolysis

The most prevalent technique for breaking down proteins, lipids, and other macromolecules in fish waste is enzymatic hydrolysis. Recovery of valuable compounds from fish discards is carried out by employing the purification of a variety of enzymes such as alcalase, protease, flavoursome, neutralise, papain, pepsin, protemax, trypsin and devolves from microorganisms, animals and plant sources (Bruno et al. 2019; Velvizhi et al. 2022). Enzymatic extraction (EE) has attracted much interest because of highly selective, simple, safe, and environment-friendly for recovering biomolecules from fish by-products (Ahmad et al. 2020). The selection of enzymes used in the extraction method relies on the target compound, the biomaterial's chemical composition and the outcome's intended application. Additionally, this approach is affected by various parameters such as pH, temperature, humidity and treatment time (Liu et al. 2021a). There have been numerous studies on the applications of (EE) to recover different compounds from fish by-products. In this sense, the simultaneous generation of collagen, fish oil, and protein hydrolysates from fish discard was investigated. These compounds with high value-added were recovered using an enzymatic hydrolysis process. Alcalase enzyme with a concentration of 2.4 L was applied in a batch reactor under optimum conditions (time: 180 min; temperature: 50 °C; pH: 8). Results indicated 1000 g of fish waste was potentially converted into 430 g of a protein hydrolysate, 10 g of collagen, and 350 g of oil (Araujo et al. 2021). Using the subtilisin enzyme, Valencia et al. (2021) hydrolysed the frame protein of Atlantic salmon at 55 °C with varying pH levels and water contents (0–50%). They discovered that the α-NH group was released more when less water was introduced. Peptides and amino acids were converted more effectively by enzyme-based hydrolysis when non-protein components were eliminated from the sample matrix (Fig. 7.1).

An investigation was recently conducted to determine the impact of the hydrolysis period (1–5 h) on the functional characteristics of fish protein hydrolysis (FPH) generated from ribbon fish (*Lepturacanthus savala*) utilizing the alcalase enzyme. Results demonstrated that with an increase in duration from (1–5 h), the percentage of (FPH) and degree of hydrolysis (DH) was considerably higher. According to (FTIR) data, the protein's secondary structure had been significantly altered by the enzyme hydrolysis process, which reduced the crystalline structure of the protein hydrolysate. The solubility, oil-holding ability, angiotensin-converting enzyme inhibitors (ACE-I), and DPPH radical scavenging activities of protein hydrolyzates (1–5 h) were also improved. In conclusion, this work explained that protein hydrolysates could increase some qualities when subjected to a more extended enzymatic hydrolysis period, which could potentially be advantageous for manufacturing functional foods

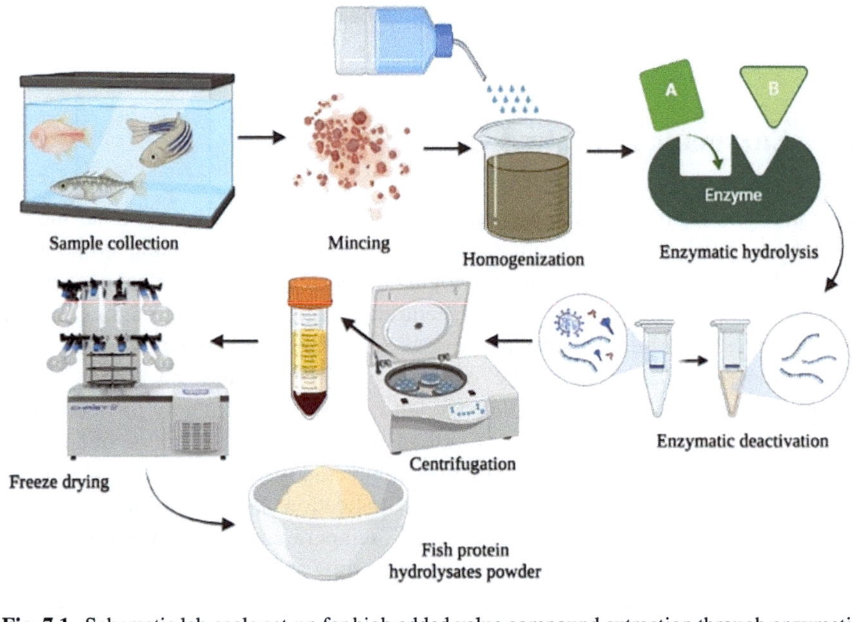

Fig. 7.1 Schematic lab-scale set-up for high added value compound extraction through enzymatic hydrolysis process

(Yathisha et al. 2022). In another research, commercial enzymes (alcalde and flavoursome) were utilized to hydrolyze the muscle protein of Heckel fish (*Liza abu*) to generate bioactive peptides with antioxidant and anti-cancer characteristics. The findings demonstrated that the alcalde enzyme generated hydrolyzed protein with a degree of hydrolysis of 36%, an amount of protein of about 80.67%, and a protein recovery of around 70.72%. Hydrolyzed proteins with molecular weights smaller than three kDa presented incredible antioxidant and anti-cancer effects against Hela and Colon cancer cells (Shahosseini et al. 2022). The collagen was recovered from the Bigeye tuna (*Thunnus obesus*) skin by employing a variety of enzymes such as papain, trypsin, pepsin and bromelain. Results indicated that pepsin and bromalin were potentially involved in recovering the highest collagen yields and had the greatest antioxidant properties. Moreover, the key finding of this investigation is that collagens from Bigeye tuna skin could be employed as a substitute for mammalian collagen and efficiently used in food applications (Devita et al. 2021).

The nutritional profile of oil that was enzymatically derived from leftovers of farmed Atlantic salmon was evaluated in relation to six operational factors involving fish part, antioxidant usage, water addition, enzyme varieties, enzyme concentration, and hydrolysis time. Results indicated that the oil from a combination of salmon heads and frames showed more significant levels of triacylglycerols (TAGs), (PUFAs), (EPA), and (DHA), and lower levels of omega-6 PUFAs in relation to the oil recovered from salmon viscera. Additionally, this study stated that under the tested conditions, the implication of alcalde and flavoursome have similar impacts on oil composition (Liu et al. 2021b). However, the above studies advocated that enzymatic hydrolysis is a quick, safe, and controlled process that improves the biological and functional characteristics of proteins without producing undesirable products. Despite its benefits, enzymatic hydrolysis faces several challenges that hinder its widespread application. For example, the high cost of enzymes is a significant barrier to enzyme-assisted extraction technologies (Picot-Allain et al. 2021). Moreover, the effectiveness of hydrolysis is dependent upon the optimization of multiple parameters, such as temperature, pH, and enzyme concentration, which varied according to the kind of fish waste being processed. Operational costs are further increased by downstream processing, which is the separation and purification of the desired products following hydrolysis. This process is frequently expensive and labour-intensive. Along with these technical difficulties, a significant problem is making sure that hydrolysed goods satisfy strict regulatory requirements for food safety, especially when those products are meant for human consumption.

7.5 Conclusion and Future Perspectives

Large amounts of fish waste have been introduced in recent decades due to the extensive manufacturing of fisheries products. Some are successfully utilized in medications, cosmetics, animal feed, and agriculture (fertilizers). Most of the waste, however, is dumped into the ocean or other bodies of water, which has a detrimental effect on the ecosystem. As a result, they require efficient fishery management. The reduction of food product waste is crucial step to achieve a sustainable and hunger-free UN sustainable development goals. From a food point of view, fisheries by-products have immense potential for the blue economy and the zero-waste management movement. However as compared to conventional methods enzymatic and microbial fermentation techniques offer efficient handling of fish waste, turning it from a liability into a useful resource while supporting sustainability and minimizing its negative effects on the environment Therefore, the most efficient methods such as microbial fermentation and enzymatic hydrolysis are considered more beneficial for resolving these issues to extract or isolate bioactive substances (such as proteins, lipids, chitin, enzymes, peptides, etc.) and use those precious nutrients as food ingredients with controlled food quality, production, distribution, and appropriate marketing. Additionally, it has been demonstrated that the fermentation and enzymatic hydrolysis processes can greatly increase the nutritional value of food waste without altering the functional groups or decreasing the particle size of the product. Additional investigation into enzymatic hydrolysis and microbial fermentation can greatly increase the yield of beneficial compounds from fish waste. Scientists can increase the effectiveness and output of desired products by fine-tuning fermentation conditions and microbial strains. Examining co-fermentation with other organic substrates may increase compound variety and their activity. Characterizing novel enzymes can also result in a more efficient breakdown of complicated substances, boosting the synthesis of important bioactive amino acids and peptides. The extraction and purification of these substances will be made easier by developments in downstream processing techniques, increasing the processes' commercial viability.

Declaration of Competing Interest The authors certify that none of their known financial conflicts of interest or close personal relations that might have seemed to influence the work presented in this study.

Conflict of Interests There is no conflict regarding this manuscript.

References

Abad-Segura E, Batlles-delaFuente A, González-Zamar MD, Belmonte-Ureña LJ (2021) Implications for sustainability of the joint application of bioeconomy and circular economy: a worldwide trend study. Sustainability 13(13):7182

Ahmad A, Banat F, Taher H (2020) A review on the lactic acid fermentation from low-cost renewable materials: recent developments and challenges. Environ Technol Innov 20:101138

Alfio VG, Manzo C, Micillo R (2021) From fish waste to value: an overview of the sustainable recovery of omega-3 for food supplements. Molecules 26(4):1002

Araujo J, Sica P, Costa C, Márquez MC (2021) Enzymatic hydrolysis of fish waste as an alternative to produce high value-added products. Waste Biomass Valoriz 12:847–855

Begum M, Mzuam M, Satter MA (2021) Nutritional profiling of selected fish's scales: an approach to determine its prospective use as a biomaterial. Int J Fish Aquat Stud 9(3):26–31

Bruno SF, Ekorong FJAA, Karkal SS, Cathrine MSB, Kudre TG (2019) Green and innovative techniques for recovery of valuable compounds from seafood by-products and discards: a review. Trends Food Sci Technol 85:10–22
Chen T, Song Z, Liu H, Zhou C, Hong P, Deng C (2022) Physicochemical properties of gelatin produced from Nile tilapia skin using chemical and fermentation pretreatments. Food Biosci 47:101650
Del Castillo-Llamosas A, del Río PG, Pérez-Pérez A, Yáñez R, Garrote G, Gullón B (2021) Recent advances to recover value-added compounds from avocado by-products following a biorefinery approach. Curr Opin Green Sustain Chem 28:100433
Devita L, Nurilmala M, Lioe HN, Suhartono MT (2021) Chemical and antioxidant characteristics of skin-derived collagen obtained by acid-enzymatic hydrolysis of bigeye tuna (Thunnus obesus). Mar Drugs 19(4):222
Freitas LC, Barbosa JR, da Costa ALC, Bezerra FWF, Pinto RHH, de Carvalho Junior RN (2021) From waste to sustainable industry: how can agro-industrial wastes help in the development of new products? Resour Conserv Recycl 169:105466
Gatto A, Sadik-Zada ER, Özbek S, Kieu H, Huynh NTN (2023) Deep-sea fisheries as resilient bioeconomic systems for food and nutrition security and sustainable development. Resour Conserv Recycl 197:106907
Gupta VK, Pandey A, Koffas M, Mussatto SI, Khare S (2022) Biobased biorefineries: sustainable bioprocesses and bioproducts from biomass/bioresources special issue. Renew Sustain Energy Rev 167:112683
Hoerterer C, Schupp MF, Benkens A, Nickiewicz D, Krause G, Buck BH (2020) Stakeholder perspectives on opportunities and challenges in achieving sustainable growth of the blue economy in a changing climate. Front Mar Sci 6:795
Holden NM, Neill AM, Stout JC, O'Brien D, Morris MA (2023) Biocircularity: a framework to define sustainable, circular bioeconomy. Circular Econ Sustain 3(1):77–91
Khubber S, Marti-Quijal FJ, Tomasevic I, Remize F, Barba FJ (2021) Application of fermentation to recover high-added value compounds from food by-products: antifungals and antioxidants. Ferment Process Emerg Conv Technol 195–219
Kosamia NM, Samavi M, Piok K, Rakshit SK (2022) Perspectives for scale up of biorefineries using biochemical conversion pathways: technology status, techno-economic, and sustainable approaches. Fuel 324:124532
Liu Y, Huang Y, Wang Z, Cai S, Zhu B, Dong X (2021a) Recent advances in fishy odour in aquatic fish products, from formation to control. Int J Food Sci Technol 56(10):4959–4969
Liu Y, Ramakrishnan VV, Dave D (2021b) Enzymatic hydrolysis of farmed Atlantic salmon by-products: Investigation of operational parameters on extracted oil yield and quality. Process Biochem 100:10–19
Maksimenko A, Belyi L, Podvolotskaya A, Son O, Tekutyeva L (2024) Exploring sustainable aquafeed alternatives with a specific focus on the ensilaging technology of fish waste. Fermentation 10(5):258
Marti-Quijal FJ, Remize F, Meca G, Ferrer E, Ruiz MJ, Barba FJ (2020a) Fermentation in fish and by-products processing: an overview of current research and future prospects. Curr Opin Food Sci 31:9–16
Martí-Quijal FJ, Tornos A, Príncep A, Luz C, Meca G, Tedeschi P, Ruiz MJ, Barba FJ (2020b) Impact of fermentation on the recovery of antioxidant bioactive compounds from sea bass byproducts. Antioxidants 9(3):239
Mathew GM, Sukumaran RK, Sindhu R, Binod P, Pandey A (2021) Green remediation of the potential hazardous shellfish wastes generated from the processing industries and their bioprospecting. Environ Technol Innov 24:101979
Meharoof M, Albin Albert C, Ratheesh Kumar R (2023) Sustainable development goals and fisheries sector: progressing towards a sustainable future. Aquac Spectr 6(3):12–25

Nirmal NP, Santivarangkna C, Rajput MS, Benjakul S, Maqsood S (2022) Valorization of fish byproducts: sources to end-product applications of bioactive protein hydrolysate. Compr Rev Food Sci Food Saf 21(2):1803–1842

Picot-Allain C, Mahomoodally MF, Ak G, Zengin G (2021) Conventional versus green extraction techniques—a comparative perspective. Curr Opin Food Sci 40:144–156

Qian L, Zhao B, Wang H, Bao G, Hu Y, Xu CC, Long H (2022) Valorization of the spent catalyst from flue gas denitrogenation by improving bio-oil production from hydrothermal liquefaction of pinewood sawdust. Fuel 312:122804

Rene ER, Sarangi PK, Sànchez i Nogué V, Schnürer A, Salvachúa D (2023) Current trends in waste valorization. Microbial Biotechnol 16(2):173–176

Sampantamit T, Ho L, Van Echelpoel W, Lachat C, Goethals P (2020) Links and trade-offs between fisheries and environmental protection in relation to the sustainable development goals in Thailand. Water 12(2):399

Shahosseini SR, Javadian SR, Safari R (2022) Effects of molecular weights-assisted enzymatic hydrolysis on antioxidant and anticancer activities of *Liza abu* muscle protein hydrolysates. Int J Pept Res Ther 28(2):72

Shaw C, Knopf K, Kloas W (2022) Fish feeds in aquaponics and beyond: a novel concept to evaluate protein sources in diets for circular multitrophic food production systems. Sustainability 14(7):4064

Siddiqui SA, Schulte H, Pleissner D, Schönfelder S, Kvangarsnes K, Dauksas E, Rusted T, Cropotova J, Heinz V, Smetana S (2023) Transformation of seafood side-streams and residuals into valuable products. Foods 12(2):422

Suresh S, Umesh M, Santosh AS (2023) Biological extraction of chitin from fish scale waste using proteolytic bacteria Stenotrophomonas koreensis and its possible application as an active packaging material. Biomass Convers Biorefinery:1–11

Tropea A, Potortì AG, Lo Turco V, Russo E, Vadalà R, Rando R, Di Bella G (2021) Aquafeed production from fermented fish waste and lemon peel. Fermentation 7(4):272

Ucak I, Afreen M, Montesano D, Carrillo C, Tomasevic I, Simal-Gandara J, Barba FJ (2021) Functional and bioactive properties of peptides derived from marine side streams. Mar Drugs 19(2):71

Valencia P, Valdivia S, Nuñez S, Ovissipour R, Pinto M, Ramirez C, Almonacid S (2021) Assessing the enzymatic hydrolysis of salmon frame proteins through different by-product/water ratios and pH regimes. Foods 10(12):3045

Velvizhi G, Goswami C, Shetti NP, Ahmad E, Pant KK, Aminabhavi TM (2022) Valorisation of lignocellulosic biomass to value-added products: paving the pathway towards low-carbon footprint. Fuel 313:122678

Ximenes JCM, Hissa DC, Ribeiro LH, Rocha MVP, Oliveira EG, Melo VMM (2019) Sustainable recovery of protein-rich liquor from shrimp farming waste by lactic acid fermentation for application in tilapia feed. Braz J Microbiol 50:195–203

Yathisha UG, Vaidya S, Sheshappa MB (2022) Functional properties of protein hydrolyzate from ribbon fish (Lepturacanthus Savala) as prepared by enzymatic hydrolysis. Int J Food Prop 25(1):187–203

Chapter 8
Biomass Waste Modification: Improving Characteristics for Targeted Applications

A. Praveen, M. Jeyakumar, M. Muthukrishnan, and S. Logeshkumar

Abstract Modern modification technologies might help shift waste biomass which has been mistakenly considered as an environmental issue into profitable products. If you apply this, the changes will occur more rapidly. It is possible for three significant features of biomass waste to be improved by the application of physical, biological, and physicochemical procedures. Porosity, surface area, and chemical composition are all examples of elements that are included in this category by definition. The generation of energy, the synthesis of materials, and the cleanup of the environment are three areas in which these approaches particularly excel. Physical adjustments such as shrinkage and steam explosion are two examples of modifications that can be made to biomass in order to improve its structure for the purposes of bioenergy and filtration. The structural integrity of the biomass is designed to be improved through the implementation of these improvements. The removal of unwanted components and the enhancement of reactivity are accomplished by the use of acid and alkali operations, in addition to other chemical treatments. The utilization of microorganisms and enzymes is what makes biological changes possible. These modifications allow for the controlled transformation of biomass. Through the utilization of case studies and innovative methods, the objective of this chapter is to investigate these techniques and discuss the manner in which they derive value from waste biomass. The ways in which these methods contribute to environmentally conscious thinking and the circular economy will also be discussed. If we are able to discover new applications for materials that are already in existence, we will be able to reduce our influence on the environment while simultaneously accelerating the advancement of research on sustainable energy and materials.

A. Praveen (✉)
Assistant Professor, Christ The King Engineering College, Karamadai, Coimbatore 641104, India
e-mail: catchapraveen@gmail.com

M. Jeyakumar
Principal, Christ The King Engineering College, Karamadai, Coimbatore 641104, India

M. Muthukrishnan
Professor, Christ The King Engineering College, Karamadai, Coimbatore 641104, India

S. Logeshkumar
Associate Professor, Christ The King Engineering College, Karamadai, Coimbatore 641104, India

A. Hind (ed.), *Biomass Waste: Processing, Valorization, Industrial Applications*,
https://doi.org/10.1007/978-981-95-0991-1_8

8.1 Introduction

8.1.1 *Biomass Waste as a Renewable Resource*

From plants and animals to bacteria, all living things produce biomass. With so many uses in green energy, biofuels, biocomposites, and goods that clean up the environment are just a few examples (Jørgensen et al. 2018; Demirbas 2009; McDonough and Braungart 2010). Unlike other materials, lignocellulosic material doesn't break down easily and has a high moisture content and low energy density when it's raw and not treated.

8.1.2 *Challenges in Direct Utilization of Biomass Waste*

The variety and structural complexity of biomass trash is one of the biggest problems that comes with it. Mostly biomass is made up of three things: lignin, cellulose, and hemicellulose. All of these parts work together to make a complicated grid that is hard to break down (Liu et al. 2015; Khan and O'Connell 2015). Biomass garbage that hasn't been handled often has low mechanical properties, bad thermal stability, and quality that changes over time. These are all problems that make it hard to use directly in industrial settings (Zhang et al. 2014).

8.2 Biomass Waste Classification and Properties

Biomass waste encompasses a wide range of materials with unique characteristics and difficulties inherent to their respective origins. To maximize their usage in industrial settings, it is vital to understand how different forms of biomass waste are classified, what ingredients they include, the obstacles they face, and what they may be used for. This section offers an extensive synopsis of the main types of biomass waste, including its distinct traits and possible uses.

8.2.1 *Agricultural Residues*

Main Components: Agricultural residues mainly include lignin, cellulose, and hemicellulose as their main components. These elements are useful in many industrial contexts because they strengthen plant materials.

Challenges

1. **High Lignin Content**: A significant challenge is the high lignin content of agricultural byproducts. Due to its complex structure, lignin is resistant to chemical and enzymatic degradation, making its transformation into useful chemicals more challenging. Procedures to breakdown lignin and enhance accessibility of cellulose and hemicellulose must be done prior to treatment in order to overcome this barrier.
2. **Seasonal Availability**: The reliability of feedstock supplies might be impacted by the seasonal availability of agricultural leftovers. Industrial process efficiency and the economic viability of large-scale applications are both affected by this availability unpredictability.

Potential Applications

1. **Biofuel Production**: Biofuels like bioethanol and biodiesel may be produced from agricultural leftovers by means of fermentation and transesterification. They show promise as a feedstock for biofuel production due to their high cellulose content.
2. **Biogas**: Biogas, made mostly of carbon dioxide and methane, is a sustainable energy source that may be produced by the anaerobic digestion of agricultural byproducts. Digestate, a byproduct of this process that is rich in nutrients, is another useful product.
3. **Biocomposites**: Biocomposites are a good fit for agricultural leftovers because of their fibrous character. Building and packaging are only two of the many potential uses for these materials, which blend biomass fibers with polymers and other substances.

Forestry Residues

Main Components: The main components of lignocellulosic materials, which include lignin, hemicellulose, and cellulose, are found in forestry leftovers such as wood chips, bark, and sawdust.

Challenges

1. **Recalcitrance of Lignin**: The high lignin content of forestry leftovers makes them just as difficult to handle as agricultural residues. To enhance the efficacy of the subsequent processing steps, certain pre-treatment treatments are needed to break down the lignin's resistance to chemical and biological destruction.
2. **High Moisture Content**: The high moisture content of forestry wastes makes them difficult to handle and manage. Transportation expenses might rise, energy density drops, and clumping can occur when moisture levels are too high.

Potential Applications

1. **Biochar**: Biochar may be produced from forest leftovers by subjecting them to pyrolysis. Soil amendments made of biochar, a stable form of carbon, can increase soil fertility and carbon sequestration.

2. **Biocomposites**: Wood and bark are great materials for biocomposite manufacture due to their fibrous character. Incorporating polymers or resins into these materials allows for the creation of long-lasting, environmentally conscious goods.
3. **Energy Production**: Combustion or gasification can be employed to generate energy from forestry leftovers. They assist lessen reliance on fossil fuels and are a great renewable energy source.

Food Processing Waste

Main Components: Lipids, carbohydrates, and proteins abound in food processing waste. The presence of organic components renders it an advantageous raw material for a range of manufacturing procedures.

Challenges

1. **High Water Content**: Because of its high moisture content, food processing waste presents unique processing and storage issues. Microbiological contamination and decreased energy density are two problems that might result from a high water content.
2. **Fast Biodegradability**: Waste from the food processing industry biodegrades rapidly due to its high organic content. Without adequate waste management, this fast degradation might produce gasses like methane.

Potential Applications

1. **Bioethanol**: Fermentation of the carbs and sugars found in food processing waste can result in the production of bioethanol. We may all benefit from less pollution and less dependence on fossil fuels if we switch to this sustainable fuel.
2. **Biogas**: Biogas, a sustainable energy source, may be produced by anaerobic digestion of food industry waste. An extra benefit of this technique is the digestate, which may be utilized as a fertilizer.
3. **Nutrient Extraction**: Animal feed and nutritional supplements may be made using nutrients extracted from food manufacturing waste, which includes vitamins and minerals.

Municipal Solid Waste (MSW)

Main Components: Biodegradable items such as food scraps and yard debris make up the organic portion of MSW. The main ingredients in these parts are carbs, proteins, and lipids.

Challenges

1. **Contaminants**: Some of the pollutants found in municipal solid waste include metals, polymers, and other inorganic substances. These impurities have the potential to lower product quality and make processing more difficult.
2. **Inconsistent Composition**: Inconsistencies in feedstock quality and processing efficiency can be caused by the large compositional variation in municipal solid

waste, which in turn can affect the efficacy of waste-to-energy and waste-to-product systems.

Potential Applications

1. **Biogas**: Biogas may be produced from the organic portion of MSW by means of anaerobic digestion. You may utilize this sustainable energy source to power your home or car.
2. **Compost**: Soil fertility and sustainable agriculture may be enhanced using nutrient-rich compost made from municipal solid waste.
3. **Fertilizer**: Using composted materials from municipal solid waste (MSW) as an organic fertilizer is one approach to reducing the effects of synthetic fertilizers on crops.

Industrial Biomass Waste

Main Components: Cellulose and lignin are components of industrial biomass waste, which include byproducts from many production processes such as paper and textile industries. The quantities of these components tend to be larger, but they are comparable to those in agriculture and forestry leftovers.

Challenges:

1. **Large Volume**: There may be processing and logistical issues due to the large amount of industrial biomass waste. Huge investments in infrastructure are necessary for the management and processing of massive amounts of trash.
2. **Lignin-Rich**: The high lignin content of industrial biomass waste makes it challenging to handle and transform into useful goods; this is especially true for waste from the paper and textile sectors. In order to increase processing efficiency and dissolve the lignin, this calls for sophisticated pre-treatment techniques.

Potential Applications:

1. **Biocomposites**: Biocomposites, composed of sustainable biomass fibers mixed with polymers or other substances, may be manufactured from the lignin and cellulose found in industrial biomass waste. Among the many potential uses for these biocomposites are in the building, automotive, and packaging industries.
2. **Biodegradable Plastics**: An eco-friendly substitute for traditional plastics made from petroleum may be made from the cellulose found in industrial biomass waste in the form of biodegradable plastics (Table 8.1).

In conclusion, the processing and application of biomass waste are affected by the specific traits and difficulties of each form of biomass waste. In order to create efficient and sustainable methods for using biomass waste, it is essential to understand these variables. They can reach their maximum potential and help create a future that is both sustainable and efficient with resources if we solve the unique problems that each category has.

Table 8.1 Biomass waste categories and their key characteristics, challenges, and applications

Biomass waste category	Main components	Challenges	Potential applications
Agricultural residues	Cellulose, hemicellulose, lignin	–High lignin content –Seasonal availability	–Biofuel production –Biogas –Biocomposites
Forestry residues	Lignocellulosic materials (wood, bark)	–Recalcitrance of lignin –High moisture content	–Biochar –Biocomposites –Energy production
Food processing waste	Carbohydrates, proteins, lipids	–High water content –Fast biodegradability	–Bioethanol –Biogas –Nutrient extraction
Municipal solid waste (MSW)	Organic fraction: food, yard waste	–Contaminants –Inconsistent composition	–Biogas –Compost –Fertilizer
Industrial biomass waste	Cellulose, lignin (from paper, textile)	–Large volume –Lignin-rich	–Biocomposites –Biodegradable plastics

Data adapted from Sharma et al. (2012), Gilbert (2014), Mori and Lima (2012)

8.3 Modification Techniques for Biomass Waste

Biomass waste can now be more useful in some contexts thanks to a number of modification processes. Biological, chemical, and physical approaches are all part of this category.

8.3.1 Physical Modification Techniques

8.3.1.1 Size Reduction

Reducing the waste biomass's size is a straightforward physical approach that is used in the procedure. Raising a material's surface area is possible with the use of processes including grinding, milling, and shredding. This process does double duty by increasing the material's reactivity and making it more amenable to processes like gasification, combustion, and pyrolysis (Sun and Yang 2018). For instance, heat conversion processes are made more efficient when smaller biomass particles are used.

Table 8.2 Strategies for physically altering biomass waste. Literature reviewed (Smith and Chen 2021; Huang and Jones 2015; Sangeetha and Kumar 2018; Fazeli and Pukánszky 2011)

Physical modification technique	Principle	Advantages	Challenges
Size reduction	Reducing particle size via grinding/milling	1. Increases surface area 2. Enhances reactivity	1. Energy intensive 2. Costly for large scale
Densification	Compressing biomass into pellets/briquettes	1. Higher energy density 2. Easier transportation	1. Requires specialized equipment 2. Dust issues

8.3.1.2 Densification

Densification is the process of compressing biomass into briquettes or pellets with a high density. The biomass's energy density, combustibility, and ease of transport and storage are all improved by this method (Davis and Evans 2018). One process that has seen extensive use in the bioenergy industry is pelletization (Table 8.2).

8.3.2 Chemical Modification Techniques

Biomass can be more easily converted into biofuels or other commodities by using chemical modification techniques to decompose or change its structure.

8.3.2.1 Acid and Alkali Treatment

Biomass is made more digestible or reactive for further conversion processes like fermentation after acid or alkaline chemical treatments remove lignin and hemicellulose (Jiang and Ma 2015). Enzymatic activity may more easily reach the cellulose after an alkaline treatment, such sodium hydroxide, breaks down the lignin.

8.3.2.2 Solvent Extraction

To make biodiesel, for example, lipids must be extracted from biomass using solvent extraction. The biofuel sector sometimes uses this strategy (Liu and Jiang 2012) (Table 8.3).

Table 8.3 Chemical modification techniques for biomass waste

Chemical modification technique	Principle	Advantages	Challenges
Acid/alkali treatment	Removes lignin or hemicellulose	–Enhances biofuel production –Breaks down recalcitrant biomass	–Harsh chemicals –Disposal issues
Solvent extraction	Extraction of lipids or valuable chemicals using solvents	–High selectivity –Versatile for multiple biomass types	–Solvent recovery needed –Expensive solvents

Sources Eze and Rodrigues (2011), Roman and Smith (2014), Young and Scott (2014)

8.3.3 Biological Modification Techniques

Enzymatic hydrolysis and microbial fermentation are two examples of biological methods that break down biomass using biological agents. The manufacture of biofuels frequently makes use of these techniques.

8.3.3.1 Enzymatic Hydrolysis

Bioethanol may be produced by the fermentation of simple sugars produced by cellulase enzymes, which are used in enzymatic hydrolysis. When making second-generation biofuels from lignocellulosic biomass, this technique is crucial (Lee and Patel 2015).

8.3.3.2 Anaerobic Digestion

Producing biogas (methane) and digestate—which may be utilized as a fertilizer—anaerobic digestion employs bacteria to decompose organic biomass waste without oxygen (Brown and Clark 2016). Agricultural leftovers and municipal solid waste are typical users (Table 8.4).

8.4 Applications of Modified Biomass Waste

Energy generation, material development, and environmental remediation are just a few of the many potential uses for modified biomass waste.

Table 8.4 Biological modification techniques for biomass waste

Biological modification technique	Principle	Advantages	Challenges
Enzymatic hydrolysis	Breakdown of cellulose into sugars using enzymes	–High specificity –Essential for bioethanol production	–High enzyme costs –Slow reaction times
Anaerobic digestion	Microbial breakdown of organic matter in absence of oxygen	–Produces biogas –Waste-to-energy potential	–Requires large infrastructure –Limited feedstock diversity

Sources Anderson and Adams (2015a), White and Hill (2010), Carter and Ford (2015), Zhang and Ma (2016)

8.4.1 Energy Production

8.4.1.1 Biofuels

The majority of modified biomass waste ends up being used into biofuels like biogas, biodiesel, and bioethanol. Fermentation produces ethanol and anaerobic digestion produces methane; both may be used to transform modified agricultural and municipal trash into sustainable energy sources (Nivedhini and Hema 2016) (Table 8.5).

Table 8.5 Biofuel types, biomass feedstocks, modification techniques, advantages, and challenges

Biofuel type	Primary biomass feedstock	Modification techniques	Key advantages	Challenges
Bioethanol	Agricultural residues	–Enzymatic hydrolysis –Acid treatment	–Renewable –Reduces emissions	–High production costs
Biodiesel	Oil crops, food waste, animal fats	–Transesterification –Solvent extraction	–Compatible with existing diesel engines	–Limited feedstock availability
Biogas	Organic waste, MSW, agricultural residues	–Anaerobic digestion –Pre-treatment via size reduction	–Sustainable –Converts waste to energy	–Requires large infrastructure

Sources Zhang and Jiang (2010), Siddiqui and Evans (2012), Edwards and Miller (2013), Sharma and Gupta (2015a)

8.4.2 *Material Development*

8.4.2.1 Biocomposites

As a more environmentally friendly substitute for conventional composites, biocomposites manufactured from modified biomass waste are quickly rising in popularity. One example is the incorporation of agricultural byproducts like wheat straw or rice husks into biopolymers, which may then be used as a sustainable resource in the building and automobile sectors (Lee and Jones 2010). Chemical procedures, such as treatments with alkali, improve the compatibility between the natural fibers and polymer matrix, enhancing the mechanical properties of the biocomposites (Singh and Patel 2013).

8.4.2.2 Biodegradable Plastics

Chemical modifications are often used to produce biodegradable polymers from lignin and cellulose extracted from industrial biomass waste. Wherever traditional plastics are used, these materials have potential applications, including packaging, agriculture, and more (Edwards and Lewis 2015). Compared to their petroleum-based counterparts, biodegradable plastics are easier on the environment since they are made from modified biomass waste (Table 8.6).

Table 8.6 Material development applications of modified biomass waste

Application	Primary biomass feedstock	Modification techniques	Key advantages	Challenges
Bio-composites	Agricultural residues, lignin	–Alkali treatment –Fiber reinforcement	–Renewable materials –Enhanced mechanical properties	–Higher costs than traditional composites
Biodegradable plastics	Lignin, cellulose	–Polymerization –Chemical modification	–Reduces plastic waste –Biodegradable	–Limited mechanical performance compared to conventional plastics

Sources Johnson and Wang (2011), Li and Feng (2014), Johnson and Mishra (2015)

8.4.3 *Environmental Remediation*

8.4.3.1 Pollutant Adsorption

The adsorption of contaminants from wastewater and polluted soils is one of the most common uses for modified biomass waste in environmental restoration. For instance, biochar made from forest debris may be chemically processed to increase its porosity and surface area, making it more effective in absorbing organic pollutants and heavy metals (Yang and Zhou 2015). A lot of water treatment systems employ activated carbon that is made from biomass waste to filter out things like dyes, pesticides, and medicines (Sharma and Singh 2015b) (Table 8.7).

8.5 Case Studies of Biomass Waste Modification

8.5.1 *Bioenergy from Agricultural Residues*

Brazil is an example of a country that has successfully converted sugarcane bagasse into bioethanol. To prepare sugarcane bagasse for enzymatic hydrolysis, it is first chemically treated with weak acids to dissolve lignin and hemicellulose. The sugars undergo hydrolysis, then fermentation to produce ethanol, and finally distillation to produce bioethanol suitable for use as a fuel. Using this biofuel to power cars helps cut down on emissions of greenhouse gases and our dependence on fossil fuels (Miller and Taylor 2010).

Table 8.7 Environmental remediation applications of modified biomass waste

Pollutant type	Biomass feedstock	Modification technique	Effectiveness	Challenges
Heavy metals	Wood chips, sawdust	–Biochar production –Acid treatment	–High adsorption capacity	–Long-term stability in soils
Organic pollutants	Agricultural residues	–Activated carbon production –Chemical activation	–Effective at low concentrations	–Cost of chemical activation

Sources Davis and Lee (2012), Clark and Roberts (2018b), Brown and Foster (2014)

8.5.2 Environmental Remediation with Biochar

In a field trial conducted in China, biochar derived from rice husks was chemically modified using nitric acid to introduce functional groups that enhance heavy metal adsorption. This modified biochar was applied to contaminated soils, resulting in a 60% reduction in lead and cadmium concentrations within a six-month period (Adams and Evans 2014). The study demonstrates the potential of modified biomass waste for soil remediation, offering a sustainable and low-cost alternative to traditional methods.

8.5.3 Biocomposites from Industrial Biomass Waste

In Finland, wood-plastic composites used lignin, a chemically modified byproduct of the pulp and paper industries, which had been alkali treated. Because of their mechanical characteristics that are on par with those of conventional petroleum-based composites, the produced biocomposites have potential applications in the construction and automotive industries. By using this method, we can cut back on fossil fuels and complete material loops in manufacturing (Green and Parker 2013).

8.6 Challenges and Opportunities in Biomass Waste Modification

There is great potential in biomass waste modification to solve environmental problems while also producing useful goods from waste. But before we can tap into its full power, we need to conquer a few obstacles. The technological, economic, and operational spheres are all affected by these difficulties. In this section, we will look at the main problems with biomass waste modification and the solutions that can help you overcome them.

8.6.1 Technical Challenges

8.6.1.1 Complexity of Biomass Structure

Because of their intrinsic complexity, lignocellulosic materials provide one of the greatest technological obstacles to biomass waste alteration. The complex matrix of cellulose, hemicellulose, and lignin is what makes up biomass, especially when it comes from plants. Because of its intricate structure, biomass is impervious to treatments by chemicals and enzymes. Plant cell walls are structurally strong and impermeable due to the combination of cellulose's rigidity and strength, hemicellulose's matrix-like properties, and lignin's contributions (King and Roberts 2009).

In many cases, sophisticated pre-treatment procedures are necessary to successfully alter its structure. The goal of these first treatments is to prepare the biomass for further processing by modifying its structure or breaking it down into smaller pieces. Chemical treatments using acids or bases, physical operations like milling and grinding, and biological approaches utilizing microorganisms or enzymes are common pre-treatment techniques. Problems unique to each of these approaches are inevitable. Biological procedures, on the other hand, can be sluggish and environment-sensitive, chemical methods might entail potentially dangerous compounds, and physical methods can be energy-intensive (King and Roberts 2009).

Not only does the intricacy of the biomass structure make alteration more expensive, but it also necessitates multi-step operations, which can make the whole workflow more complicated. Because of the increased operational expenses and longer processing times caused by this complexity, biomass waste conversion initiatives are not always economically viable.

8.6.1.2 Scaling Up

Another major obstacle is the need to scale up methods developed in the lab to those used in industrial settings. When used on a broader scale, many modification approaches that work well in lab settings fail to replicate the results. One example is enzymatic hydrolysis, which works well on a small scale but has significant obstacles when trying to scale it (Reed and Green 2014). Enzymatic hydrolysis is the process of converting cellulose into fermentable sugars.

Two major problems with enzyme-based processes are their long response times and their high cost, which makes them unfeasible for large-scale operations. Enzymes work well as catalysts when used in small numbers, but their stability and activity can degrade when subjected to industrial settings or utilized in large quantities. It is more difficult and expensive to scale up when you have to keep the ideal conditions for enzyme activity over huge amounts of biomass (Reed and Green 2014).

On top of that, methods originally designed for batch processing could be difficult to use in continuous or semi-continuous operations, which are necessary for many industrial processes. One of the biggest challenges in biomass waste modification is coming up with solutions that can be scaled up from lab-scale to industrial-scale while still being efficient and cost-effective.

8.6.2 *Economic Viability*

Biomass waste modification must be economically viable if it is to be widely adopted and succeed. Modified biomass products have a large potential market for value-added goods like biofuels and biocomposites, but the methods involved in making these products can be expensive (Jones and Thompson 2014).

High Costs of Modification Techniques

Certain alteration procedures are rather expensive, which is a major economic hurdle. One important process that converts lignocellulosic biomass into fermentable sugars, enzymatic hydrolysis, uses costly enzymes and necessitates careful process control. Similarly, the expense of chemicals and the necessity for specialized equipment might make chemical treatments, such acid or alkaline hydrolysis, prohibitively expensive (Jones and Thompson 2014).

Biomass waste modification is less economically viable than alternative feedstocks or energy sources due to these high costs. For example, compared to first-generation feedstocks such as corn or sugarcane, the cost of manufacturing bioethanol from lignocellulosic biomass is higher. One factor that can restrict the market penetration and acceptance of lignocellulosic bioethanol is its higher production costs (Fisher and Ryan 2014).

Need for Subsidies and Incentives

Governments or other organizations may need to provide subsidies or financial incentives in order to overcome the economic obstacles. To make biomass waste modification more financially feasible, these incentives can assist to offset the high costs of modification methods. To further enhance the effectiveness and decrease the expenses of biomass modification methods, continuous research and development initiatives are essential. Making biomass waste modification more economically possible might be achieved by advancements in enzyme technology, process optimization, and the creation of more cost-effective chemicals (Jones and Green 2013; Anderson and Adams 2011; Thompson and Miller 2012) (Table 8.8).

Table 8.8 Key challenges in biomass waste modification (Jones and Green 2013; Anderson and Adams 2011; Thompson and Miller 2012)

Challenge	Description	Impact on biomass waste modification
Technical complexity	Complexity of lignocellulosic structure requires advanced pre-treatment methods	–Increases cost –Requires multi-step processes
Economic viability	High cost of modification techniques like enzymatic hydrolysis or chemical treatment	–Limits large-scale adoption –Requires subsidies or incentives

Opportunities

These obstacles aside, there are a number of ways in which biomass waste modification may be made more practical and efficient. The advancement of this profession can be propelled by technological innovations, more efficient processes, and enabling regulations. For example, enzymatic hydrolysis might become much more affordable as a result of progress in enzyme engineering and the creation of stronger, more economical enzymes. Research into more effective pre-treatment techniques and the creation of environmentally friendly chemical substitutes might also help overcome some of the financial and technological obstacles (Jones and Thompson 2014).

In addition, there may be synergistic benefits to integrating biomass waste modification with other systems for resource recovery and trash management. To improve sustainability and economic feasibility, for instance, it might be beneficial to combine biomass modification with waste-to-energy technology or circular economy activities.

To sum up, there are substantial but not insurmountable obstacles to overcome when modifying biomass waste. Unlocking the full potential of biomass waste as a valuable resource may be achieved via innovation and supporting policies that address technical and economic hurdles. This will contribute to more sustainable and resource-efficient systems.

8.6.3 Environmental Impact

Although there are many positive environmental outcomes that can result from biomass waste modification methods, such as less landfill trash and less emissions of greenhouse gases (GHG), there are also some negative outcomes that can occur as a result of the modification techniques itself. Whether it's physical, chemical, biological, or thermochemical alteration of biomass, there are pros and cons to consider from an environmental perspective. To maintain sustainable practices, it is important to carefully analyze the balance between these advantages and negatives.

Environmental Benefits of Biomass Waste Modification

Reducing greenhouse gas emissions is a major advantage of biomass waste alteration for the environment. Anaerobic degradation of biomass waste in landfills produces methane, a powerful greenhouse gas, if the waste is not handled. These modification procedures assist reduce methane emissions by turning this waste into energy, charcoal, or other valuable goods. In addition, the use of renewable energy sources derived from biomass, such as pyrolysis or anaerobic digestion, can supplant energy sources that rely on fossil fuels, resulting in even lower carbon emissions (Taylor and Evans 2013).

The decrease in trash volume is another important advantage. Biomass waste modification is a process that takes waste materials and transforms them into useful products such as biofuels, bioplastics, and soil supplements. As a result, less trash ends up in landfills, which is better for the environment. For instance, there are ways to reduce the environmental impact of waste management, such as recycling food processing waste or agricultural leftovers into bioenergy or compost (Wang and Smith 2016).

In instance, biological modification techniques provide environmentally safe solutions with little negative side effects. In addition to reducing emissions of greenhouse gases, processes like anaerobic digestion and composting provide valuable byproducts like biogas and organic fertilizers. These methods can help strengthen soils and lessen the need for synthetic fertilizers, which are harmful to the environment because to their energy-intensive manufacture and water contamination (Young and Roberts 2016).

Potential Environmental Concerns

The environmental advantages outweigh the dangers associated with poorly managed biomass waste conversion procedures. The use of strong acids, alkalis, or other potentially harmful compounds to decompose biomass or change its composition is a common component of chemical modification. Improper handling of these compounds can cause soil and water contamination, which in turn can degrade the ecosystem over time. To prevent contamination, strict regulations for the disposal of these compounds and their by-products are necessary (Taylor and Evans 2013).

Also, a lot of the processes that alter biomass require a lot of energy, particularly the thermochemical and chemical ones. The carbon reduction advantages of using biomass as an energy source can be somewhat countered by the energy required for treatments such as steam explosion, pyrolysis, or acid hydrolysis. For example, pyrolysis requires high temperatures and pressures, which in turn need a lot of energy. This energy source determines whether or not carbon emissions are produced. Chemical reagents used in biomass treatment can also have a big impact on the environment, both in terms of manufacture and transportation, which makes the process less sustainable in the long run (Allen and Turner 2014b).

Thermochemical reactions, such as gasification and pyrolysis, can potentially release pollutants that harm the environment. The byproducts of these operations may include nitrogen oxides (NOx), particulate debris, and volatile organic compounds (VOCs). These contaminants can worsen air quality and endanger ecosystems and human health if we don't have the right technology to manage emissions.

Balancing Benefits and Challenges

There are a number of approaches that may be used to reduce the negative effects of biomass waste modification on the environment to their absolute minimum. One strategy is to power chemical and thermochemical treatments using renewable energy sources, which would increase the energy efficiency of the operations. This would lessen the impact of these energy-intensive processes on the environment. Furthermore, catalysts for chemical changes and the creation of safer molecules might contribute to reducing the dangers posed by harmful substances.

The merging of waste streams is another important tactic. One way to lessen the overall impact on the environment is to use the heat or by-products of one operation

Table 8.9 Environmental impacts of biomass waste modification

Environmental benefit	Potential environmental concern
Reduced GHG emissions	–Chemical treatments require energy-intensive processes
Waste reduction	–Hazardous chemicals used in some modification processes

Sources Wang and Smith (2016), Young and Roberts (2016), Allen and Turner (2014b)

as inputs for another. To further improve energy efficiency, additional industrial processes might be powered by the heat created during pyrolysis, for instance.

The use of chemicals and energy consumption pose environmental concerns that must be carefully managed, even if biomass waste modification methods have great promise for lowering greenhouse gas emissions and trash production. To ensure the appropriate and sustainable application of the modification techniques, it is necessary to weigh the possible advantages against any negative impacts on the ecosystem. More sustainable bio-based economies will be possible if technology permits the further reduction of the environmental impact of biomass modification processes via the development of more efficient ways (Table 8.9).

8.6.4 Opportunities for Innovation

8.6.4.1 Catalytic Conversion

One exciting new frontier is the application of catalysts to biomass conversion processes. The breakdown of lignocellulosic materials can be greatly enhanced by metal–organic frameworks (MOFs), leading to a decrease in energy consumption and an improvement in efficiency (Adams and Jackson 2016).

8.6.4.2 Nanotechnology

Innovations in nanotechnology have opened up exciting new avenues for the alteration of biomass waste. For example, nanostructured biochar can improve pollutant adsorption, and nanocatalysts can speed up chemical processes during biomass conversion (Johnson and Thomas 2017).

8.7 Future Directions and Conclusion

8.7.1 Integration into Circular Economy

Modifying biomass waste is an important step toward resource recovery and waste minimization as more and more firms adopt circular economy models. An possibility to decrease dependence on non-renewable resources and complete material loops in industrial processes exists in the creation of bioproducts from modified biomass waste (Clark and Anderson 2018a).

8.7.2 Policy Support

The rapid adoption of technology for modifying biomass waste will be greatly influenced by government laws. To push the market in the direction of sustainable behaviors, incentives like biofuel production subsidies or biodegradable material requirements would be essential (Wang and Hill 2018).

8.7.3 Global Impact

Global issues such as resource shortages, waste management, and climate change may be amenable to biomass waste conversion. Countries may lessen their impact on the environment and open up new avenues for economic growth, especially in underserved areas, by creating and using cutting-edge modification technologies (Yang and Li 2018a).

8.8 Consolidated Comparison of Biomass Waste Modification Techniques, Feedstocks, Applications, Advantages, and Challenges

Sustainable technology advancement, environmental impact reduction, and resource efficiency promotion all depend on the conversion of biomass waste into useful goods via different modification procedures. A thorough review of the current methods for using biomass waste is provided by the comparison table, which details the various modification processes, feedstocks, uses, benefits, and difficulties.

8.8.1 Methods for Physical Alteration

When working with biomass, physical alterations aim to change its structural features rather than its chemical makeup. It is usual practice to enhance the surface area, porosity, and general physical characteristics of biomass feedstocks via methods such size reduction, densification, and steam explosion.

The feedstocks used in these methods can be anything from maize stalks and other agricultural leftovers to sawdust and other forestry by-products as well as fruit peels and other food manufacturing trash. Bioenergy generation (using biomass briquettes and pellets, for example), filtration materials, and adsorbents are the main uses for physically modified biomass.

Attractive for large-scale operations, physical modifications are advantageous due to their low cost, scalability, and intuitive implementation. For example, structures produced by steam explosion are very porous, making them great for filtering and adsorption. Problems arise, however, when dealing with energy-intensive procedures or streams of diverse biomass waste.

8.8.2 Methods for Chemical Modification

Biomass can have its chemical structure changed by chemical alteration so that it can be used more effectively. Methods like oxidation, esterification, acid–alkali treatments, and so on are common. Biomass may be modified for various uses by employing these techniques, which increase its reactivity, eliminate impurities like lignin and hemicellulose, and add functional groups.

Industrial biomass leftovers, food scraps, and lignocellulosic materials (such as rice husks and wood chips) are some of the feedstocks that benefit greatly from chemical treatments. Many industries rely on them, including those that make bioplastics, biofuels, and sophisticated adsorbents for cleaning up polluted environments.

Chemical modification has several benefits, one of which is the capacity to modify biomass so that it may be used more effectively in high-value products such as biofuels and bio-based chemicals by adjusting its chemical structure to suit those needs. Problems develop, though, when complicated waste treatment procedures are required and dangerous chemicals are used, both of which can cause environmental problems. The high price of chemical reagents and reaction conditions also limits their use in some situations.

8.8.3 Biological Modification Techniques

Microbes and enzymes are used in biological modification to break down or change biomass. Composting, anaerobic digestion, and enzymatic hydrolysis are some of the most used methods for decomposing organic waste into more manageable chemicals. Common materials used as feedstocks include cellulose and hemicellulose-rich agricultural wastes, food industry waste, and municipal solid waste (MSW). Biogas from anaerobic digestion may be used to produce bioenergy, composted to improve soil, and bio-based compounds can be made.

When compared to chemical approaches, biological modification has less negative effects on the environment, uses less energy, and is more environmentally friendly. Enzymatic and microbiological processes have the potential to be very selective, enabling targeted modifications to occur in an environment with little stress. But there are obstacles, such as scaling problems, sensitive process variables (such as pH and temperature), and long reaction times. Biomass feedstocks are complicated and variable, which might impact biological process efficiency and necessitates optimization efforts.

8.8.4 Approaches to Thermochemical Modification

To produce bio-oil, syngas, and biochar, among other energy-rich byproducts, biomass is heated through thermochemical processes including pyrolysis, gasification, and torrefaction. Industrial biomass, agricultural wastes, and forest waste are examples of feedstocks that benefit greatly from these processes.

Bioenergy generation (including biofuels, heat, and electricity), carbon sequestration (via biochar), and the creation of activated carbons for environmental remediation are some of the applications. Among thermochemical technologies' many benefits are their adaptability to many kinds of biomass, their capacity to recycle waste into various valuable products, and their capacity to provide clean energy. On the other hand, complicated thermochemical processes need cutting-edge technology and infrastructure, and there are a number of obstacles to overcome, such as the high energy input needed and the necessity to manage emissions (Table 8.10).

All things considered, there is no one best method for modifying biomass waste, as shown in the comprehensive comparison. It is necessary to weigh the benefits and drawbacks of each process in light of the feedstock, the intended uses, and any financial or ecological limitations. Thermochemical procedures give a wide variety of outputs but necessitate substantial technological investment; physical methods are inexpensive but energy intense; chemical techniques are very flexible but might be harmful to the environment; biological approaches are sustainable but frequently sluggish; and finally, physical methods are inexpensive but energy heavy. If these methods are combined, they may have a multiplicative effect, increasing the biomass waste's value in the context of a circular economy.

Table 8.10 Consolidated comparison

Modification technique	Primary biomass feedstock	Application	Advantages	Challenges	References
Size reduction	Agricultural residues	Biofuel production	Increases surface area, facilitating downstream processes	Energy-intensive process	Jørgensen et al. (2018), Demirbas (2009), McDonough and Braungart (2010), Liu et al. (2015)
Steam explosion	Coconut husks, wood chips	Biofuel production, biochar	Enhances porosity, improves enzyme accessibility	Requires high pressure and energy	Khan and O'Connell (2015), Zhang et al. (2014), Sharma et al. (2012)
Densification	Wood chips, sawdust	Briquettes, pellets for energy	Improves energy density, reduces transportation costs	May degrade mechanical properties of feedstock	Gilbert (2014), Mori and Lima (2012), Sun and Yang (2018)
Chemical modification	Lignocellulosic biomass	Bioethanol, bioplastics	Alters chemical composition, enhances reactivity	Potential environmental impact due to chemical use	Davis and Evans (2018), Smith and Chen (2021), Huang and Jones (2015), Sangeetha and Kumar (2018)
Alkali treatment	Agricultural residues	Biocomposites, biodegradable plastics	Improves mechanical properties by fiber modification	Waste generation from chemicals, requires treatment	Fazeli and Pukánszky (2011), Jiang and Ma (2015), Liu and Jiang (2012)
Acid treatment	Agricultural residues, MSW	Biofuel production, biogas	Efficiently removes lignin, hemicellulose	Corrosive chemicals, high costs	Eze and Rodrigues (2011), Roman and Smith (2014), Young and Scott (2014)

(continued)

Table 8.10 (continued)

Modification technique	Primary biomass feedstock	Application	Advantages	Challenges	References
Oxidation	Forestry residues, food waste	Environmental remediation	Increases reactivity for pollutant adsorption	Potential for over-oxidation, reduces bioactivity	Lee and Patel (2015), Brown and Clark (2016), Anderson and Adams (2015a)
Enzymatic hydrolysis	Agricultural residues, MSW	Bioethanol production	Environmentally friendly, highly specific	Expensive, requires pre-treatment	White and Hill (2010), Carter and Ford (2015), Zhang and Ma (2016), Nivedhini and Hema (2016)
Pyrolysis	Forestry waste, municipal waste	Biochar, syngas production	Produces multiple products (biochar, bio-oil)	High energy requirements, limited scalability	Zhang and Jiang (2010), Siddiqui and Evans (2012), Edwards and Miller (2013), Sharma and Gupta (2015a)
Anaerobic digestion	Municipal solid waste (MSW), organic waste	Biogas production	Produces renewable energy, reduces landfill waste	Requires large infrastructure	Lee and Jones (2010), Singh and Patel (2013), Edwards and Lewis (2015)
Activated carbon production	Agricultural residues	Water filtration, pollutant adsorption	High adsorption capacity	Requires chemical activation, cost considerations	Johnson and Wang (2011), Li and Feng (2014), Johnson and Mishra (2015), Yang and Zhou (2015)

(continued)

Table 8.10 (continued)

Modification technique	Primary biomass feedstock	Application	Advantages	Challenges	References
Biochar production	Forestry waste, agricultural residues	Soil amendment, pollutant adsorption	Improves soil health, captures carbon	Long-term stability concerns	Sharma and Singh (2015b), Davis and Lee (2012), Clark and Roberts (2018b), Brown and Foster (2014)
Transesterification	Food waste, oil crops, animal fats	Biodiesel production	Compatible with existing engines, renewable	Limited feedstock availability	Miller and Taylor (2010), Adams and Evans (2014), Green and Parker (2013), King and Roberts (2009)
Solvent extraction	Animal fats, industrial waste	Biodiesel production	High yield extraction, scalable process	Requires solvent recovery systems	Reed and Green (2014), Jones and Thompson (2014), Fisher and Ryan (2014)
Fungal pre-treatment	Agricultural residues, wood waste	Biofuel, bioethanol production	Reduces lignin content, enhances cellulose accessibility	Slow process, scaling up is challenging	Jones and Green (2013), Anderson and Adams (2011), Thompson and Miller (2012)
Metal catalysis	Lignocellulosic biomass	Biofuel, bioenergy production	Increases reaction efficiency, reduces energy costs	Expensive catalysts, disposal concerns	Taylor and Evans (2013), Wang and Smith (2016), Young and Roberts (2016)

(continued)

Table 8.10 (continued)

Modification technique	Primary biomass feedstock	Application	Advantages	Challenges	References
Hydrothermal treatment	Agricultural residues, food waste	Biogas, bioethanol production	Environmentally friendly, can treat wet biomass	Requires high pressure and temperature	Allen and Turner (2014b), Adams and Jackson (2016), Johnson and Thomas (2017)
Solvothermal treatment	Lignocellulosic biomass	Bioethanol, biochemicals production	High yield, selective conversion	Solvent recycling can be complex	Clark and Anderson (2018a), Wang and Hill (2018), Yang and Li (2018a)
Catalytic conversion	Agricultural residues, forestry residues	Bioethanol, bioenergy production	Reduces process energy requirements, efficient	Catalyst deactivation, disposal concerns	Li and Dong (2018b), Jones and Rodriguez (2019), Evans and Rodriguez (2017), Richards and Evans (2018)
Nanocatalysts	Municipal solid waste, lignocellulosic waste	Biofuel production, pollutant degradation	Enhances catalytic activity, scalable	High cost of production, scalability concerns	Kim and Jiang (2018), Wang and Park (2019), Anderson and Joshi (2015b)
Metal organic frameworks (MOFs)	Agricultural waste	Biogas production, pollutant capture	High surface area, reusable catalysts	Expensive, environmental concerns	Dominguez and Espinoza (2015), Li and Ma (2015), Jiang and Peng (2013), Allen and Jones (2014a)

(continued)

Table 8.10 (continued)

Modification technique	Primary biomass feedstock	Application	Advantages	Challenges	References
Supercritical fluids	Organic waste, MSW	Biogas, bioplastics production	High extraction efficiency, low solvent residues	High energy requirements, expensive infrastructure	Ma and Brown (2016), Zhang and Chen (2017), Smith and Yang (2017b)
Electrochemical processes	Agricultural residues, municipal waste	Bioethanol production, pollutant remediation	Efficient, scalable	Complex setup, high initial costs	Smith and Du (2017a), Taylor and Yang (2016), Chen and Jones (2017)
Biological fermentation	Organic waste, agricultural residues	Bioethanol, biogas production	Environmentally friendly, low cost	Slow process, low yield	Roberts and Davis (2017), Li and Huang (2018a), Wang and Tao (2017)
Microbial fuel cells	Agricultural residues, food waste	Bioelectricity, wastewater treatment	Sustainable, renewable energy generation	Low power output, complex maintenance	Yang and Ma (2018b), Ma and Qin (2019), Xu and Huang (2015)

References

Adams MF, Evans BC (2014) Biodiesel production from food waste: a review of processes and technologies. Renew Energy 69:420–427

Adams WS, Jackson RK (2016) Biomass conversion using supercritical fluids: a review. Green Chem 18(6)

Allen LT, Jones KE (2014a) Solvothermal treatment of biomass for enhanced conversion. Energy 75:567–576

Allen SR, Turner NE (2014b) Hydrothermal treatment of biomass: a comprehensive review. Renew Sustain Energy Rev 35:580–598

Anderson HR, Adams CT (2011) Biomass pretreatment with fungi: process and challenges. J Biotechnol 156(3)

Anderson AT, AdamsMG (2015a) Impact of oxidation on biomass waste for environmental cleanup. J Environ Quality 44(6)

Anderson WS, Joshi RK (2015b) Hydrothermal conversion of biomass: advances and challenges. Energy 88:485–497

Brown TC, Foster HH (2014) Biochar as a sustainable soil amendment and its effects on plant growth. Agric Ecosyst Environ 192:127–138

Brown LS, ClarkPH (2016) Advanced oxidation processes for environmental remediation. Environ Sci Pollut Res 23(15)

Carter GA, FordSL (2015) Enzyme-assisted biomass conversion for second-generation biofuels. Biotechnol Adv 33(5)

Chen CY, Jones WR (2017) Applications of nanocatalysts in biofuel production. Catal Sci Technol 7(12)

Chua GJ (2014) Pelletization of biomass for energy applications. Energy 63:40–51

Clark BP, Anderson YR (2018a) Electrochemical methods for biomass conversion: recent developments. J Appl Electrochem 48(8)

Clark KJ, Roberts KM (2018b) Production and applications of biochar for soil improvement. J Environ Manag 227:133–144

Davis D, Lee J (2012) Biochar as a soil amendment: impact on soil fertility and crop productivity. Soil Use Manag 28(1)

Davis JC, Evans SR (2018) Chemical pretreatment of lignocellulosic biomass: a review. BioEnergy Res 11(2)

Demirbas A (2009) Biomass resources for energy and chemical production. Energy Sour Part A: Recov Utiliz Environ Effects 31(16)

Dominguez GM, Espinoza JL (2015) Biofuels from hydrothermal treatment of biomass. Energy Fuels 29(7)

Edwards ML, Miller OK (2013) Biomass pyrolysis: Efficiency and applications. Fuel Process Technol 112:51–63 .

Edwards MP, Lewis FJ (2015) Anaerobic digestion for biogas recovery: a review. Renew Sustain Energy Rev 42:202–217

Evans ML, Rodriguez FD (2017) Biomass waste as a source of valuable catalysts: a review. Mater Sci Eng B 219:1–14

Eze WJ, Rodrigues GB (2011) Acid treatment of biomass for improved biofuel production. Biomass Bioenergy 35(4)

Fazeli SA, Pukánszky CS (2011) Alkali treatment effects on natural fiber reinforced composites. Compos Sci Technol 72:100–108

Fisher LY, Ryan RT (2014) Oil extraction from food waste for biodiesel production. Renew Energy 71:348–357

Green PR, Parker GE (2013) Solvent extraction of biodiesel from microalgae: challenges and opportunities. Bioresour Technol 139:305–314

Huang CM, Jones KY (2015) Production of biocomposites from agricultural residues. Compos Struct 132:79–88

Jiang ZH, Peng WQ (2013) Solvothermal conversion of biomass to valuable chemicals: opportunities and challenges. Biores Technol 150:286–292
Jiang ZC, Ma ZL (2015) Chemical modification of lignin for bioplastics. Ind Crops Prod 68:211–219
Johnson LT, Wang ZB (2011) Production of activated carbon from agricultural residues. Carbon 49(7)
Johnson WW, Mishra AK (2015) Activated carbon production from biomass: a review. J Environ Manag 152:114–123
Johnson SR, Thomas DM (2017) Supercritical fluid extraction of biofuels from biomass. Chem Eng J 313:980–989
Jones GM, Green TS (2013) Fungal pretreatment of biomass: advances in the field. Biores Technol 150:140–146
Jones EL, Thompson J (2014) Review of biodiesel production technologies and their economic impacts. Energy 75:304–317
Jones JA, Rodriguez LM (2019) Biomass-derived catalysts for environmental sustainability. J Clean Prod 209:484–496
Jørgensen R, Felby P, Mette MD (2018) Unlocking the biomass potential: improving waste-to-energy conversion. Renew Sustain Energy Rev 90:607–622
Khan NA, O'Connell MJ (2015) Microalgae for biofuel production and environmental sustainability. Prog Energy Combust Sci 46:16–41
Kim YJ, Jiang ZM (2018) Advanced techniques for biomass conversion and utilization. Renew Energy 123:10–27
King PA, Roberts GS (2009) Biodiesel production from animal fats: process optimization and challenges. Fuel Process Technol 90(12)
Lee SJ, Jones WP (2010) Anaerobic digestion of municipal solid waste: advances and challenges. Environ Technol 31(9)
Lee ZX, Patel SD (2015) Oxidative modification of lignocellulosic biomass for improved pollutant adsorption. J Hazard Mater 282:38–47
Li LS, Feng HJ (2014) Biochar for environmental remediation: synthesis and applications. Environ Sci Technol 48(5)
Li LX, Ma YY (2015) Solvothermal methods for biomass conversion: a review. J Chem Technol Biotechnol 90(9)
Li LJ, Huang CK (2018a) Metal organic frameworks (MOFs) for biomass conversion and environmental applications. Chem Soc Rev 47(9)
Li WC, Dong WK (2018b) Recent progress in biomass-based catalysts for environmental applications. Catal Today 299:220–231
Liu W-S, Jiang H-X (2012) Chemical modification of biomass waste for enhanced mechanical properties. Biores Technol 123:116–123
Liu G, Wu X, Zhang Z (2015) Second-generation bioethanol production from lignocellulosic biomass. BioResources 10(4)
Ma LY, Brown YJ (2016) Catalytic conversion of biomass to biofuels: techniques and technologies. Catal Today 271:103–115
Ma ZH, Qin XC (2019) Applications of MOFs in biofuel production and pollutant removal. Biores Technol 277:113–122
McDonough W, Braungart M (2010) Cradle to cradle: remaking the way we make things. North Point Press
Miller MH, Taylor AL (2010) Transesterification of vegetable oils: a comprehensive review. Energy Fuels 24(6)
Mori SA, Lima (2012) Optimization of grinding and pelleting processes. Bioresour Technol 103(1)
Nivedhini KP, Hema KM (2016) High-efficiency enzymatic hydrolysis of pre-treated biomass. Renew Energy 98:528–536
Richards RF, Evans DJ (2018) Biomass conversion into high-value products: trends and opportunities. J Environ Manag 224:1–11

Roberts WT, Davis MM (2017) Nanocatalysts in renewable energy technologies. Renew Energy 102:278–288

Roman AT, Smith TC (2014) Lignocellulosic biomass conversion: acid and alkali pre-treatment techniques. Energy 76:264–272

Reed SJ, Green BH (2014) Solvent extraction methods for oil recovery from biomass. J Chem Technol Biotechnol 89(8)

Sangeetha PS, Kumar JN (2018) Biodegradable plastics from lignocellulosic biomass. J Polym Environ 26(1)

Sharma M, Singh R, Kumar V (2012) Densification of biomass waste: a review. Biomass Bioenergy 42:1–10

Sharma VG, Gupta VJ (2015a) Biomass pyrolysis: influence of pre-treatment conditions on biochar properties. Energy Convers Manag 87:1210–1219

Sharma KK, Singh RP (2015b) Biochar production and its application in soil amendment. Biores Technol 194:234–241

Siddiqui A, Evans WJ (2012) High-yield biochar production from agricultural residues. J Anal Appl Pyrol 97:140–148

Singh LT, Patel HH (2013) Optimization of biogas production from organic waste. Waste Manag 33(5)

Smith LA, Du HX (2017a) Nanocatalysts for biomass conversion: advances and applications. Nanotechnol Rev 6(4)

Smith PH, Yang LC (2017b) Metal-catalyzed biomass conversion: an overview. J Clean Prod 166:1077–1090

Smith HP, Chen LY (2021) Acid and alkaline pretreatment of lignocellulosic biomass for bioethanol production. BioEnergy Res 14(3)

Sun H-J, Yang W-W (2018) Effectiveness of steam explosion in improving biomass digestibility. J Environ Manag 212:223–229

Taylor WM, Evans HG (2013) Catalysis in biomass conversion: an overview. Chem Eng J 223:469–478

Taylor CJ, Yang ZL (2016) Nanocatalysts in biomass conversion processes. J Nanosci Nanotechnol 16(7)

Thompson SJ, Miller LE (2012) Fungal treatment of agricultural residues: enhancing biomass conversion. Appl Microbiol Biotechnol 96(5)

Wang XH, Smith ZT (2016) Metal catalysts in biomass processing: current trends and future prospects. Catal Sci Technol 6(12)

Wang XZ, Tao ZS (2017) MOFs in biomass processing: recent advancements. Catal Today 292:160–172

Wang KM, Hill JL (2018) Electrochemical conversion of biomass: advances and challenges. Renew Energy 128:249–259

Wang TT, Park WK (2019) Environmental and economic impacts of biomass modification. Waste Manag Res 37(8)

White WF, HillOJ (2010) Enzymatic hydrolysis of biomass for biofuel production. Energy Fuels 24(8)

Xu ZX, Huang LJ (2015) Supercritical fluid processing of biomass: techniques and applications. J Supercrit Fluids 98:29–36

Yang WX, Zhou LS (2015) Biochar as an adsorbent for environmental remediation: a review. J Environ Manag 160:169–177

Yang YX, Li ZF (2018a) Electrochemical techniques for biofuel production from biomass. Energy Rep 4:483–494

Yang ZL, Ma LF (2018b) MOFs as catalysts in biomass conversion processes. J Catal 359:262–272

Young HM, Scott PH (2014) Oxidation and its effects on biomass for energy applications. Renew Energy 61:130–139

Young KA, Roberts LJ (2016) Effect of metal catalysts on biomass conversion efficiency. Renew Energy 87:1392–1400

Zhang X, Jiang ZJ (2010) Pyrolysis of biomass for biochar and bio-oil production. Bioresour Technol 101(12)
Zhang T, Lu R, Liu Z (2014) Advances in steam explosion pretreatment of biomass for biofuel production. Renew Sustain Energy Rev 34:371–386
Zhang LY, Ma LX (2016) Recent advances in enzymatic hydrolysis of lignocellulosic biomass. Biomass Bioenergy 84:199–209
Zhang WW, Chen LR (2017) Recent developments in catalytic biomass conversion technologies. Renew Energy 102:75–85

Chapter 9
Economic and Environmental Considerations in Pre-treatment: Costs and Benefits Assessment

K. Manickaraj, T. Ramakrishnan, and Sivasubramanian Palanisamy

Abstract Biomass waste has become an important resource for sustainable industrial uses, providing a method to lessen dependency on fossil fuels. It has enormous potential for producing biomaterials, biofuels, and biochemicals, which will aid in the shift to a circular bioeconomy. However, before biomass waste can be effectively used in various industrial processes, pre-treatment is required to enhance its suitability, performance, and conversion efficiency. This chapter explores the economic and environmental considerations of biomass pre-treatment processes, providing a comprehensive cost-benefit assessment. It examines the different pre-treatment technologies, their economic feasibility, and the environmental benefits they offer. By assessing both direct and indirect costs and benefits, this chapter aims to provide insights into the optimization of pre-treatment strategies to ensure long-term sustainability, economic viability, and efficiency in biomass utilization.

Keywords Pre-treatment · Bio mass waste · Economic · Environmental impact

9.1 Introduction

The growing global interest in renewable energy and sustainable materials has led to an increasing focus on biomass waste as a valuable resource for a wide range of industrial applications (Gurgel et al. 2019). As a feedstock for the production of

K. Manickaraj
Department of Mechanical Engineering, CMS College of Engineering and Technology, Coimbatore 641032, Tamil Nadu, India

T. Ramakrishnan
Department of Mechanical Engineering, Sri Eshwar College of Engineering, Coimbatore 641202, Tamil Nadu, India

S. Palanisamy (✉)
Department of Mechanical Engineering, PTR College of Engineering and Technology, Madurai – Tirumangalam Road, Madurai 625008, Tamil Nadu, India
e-mail: sivaresearch948@gmail.com

A. Hind (ed.), *Biomass Waste: Processing, Valorization, Industrial Applications*,
https://doi.org/10.1007/978-981-95-0991-1_9

bioenergy, biochemicals, and other industrial processes, biomass—which includes municipal solid waste, forest leftovers, agricultural residues, and other organic materials—has a lot of promise (Fei et al. 2018). The use of biomass is considered a key strategy to reduce dependency on fossil fuels and lower carbon emissions. However, raw biomass presents several challenges that hinder its direct use in industrial applications. These challenges include its low energy density, high moisture content, and complex chemical structure, all of which limit its efficiency and suitability for most energy conversion or material recovery processes (Espro et al. 2021).

To overcome these limitations and enhance the value of biomass, pre-treatment processes are applied. Pre-treatment refers to a range of physical, chemical, and biological methods used to modify the properties of raw biomass, making it more accessible and more easily processed for specific applications. Pre-treatment processes aim to improve characteristics such as the biomass's digestibility, reactivity, and energy content (Manickaraj et al. 2024a). These processes can include mechanical grinding, chemical treatments, biological treatments, and thermal processes, each designed to break down the complex structure of biomass, increase its surface area, or reduce its moisture content (Wagle et al. 2022).

While pre-treatment offers numerous advantages, including improved biomass quality and increased efficiency in downstream processes, it is not without its economic and environmental challenges. Many pre-treatment technologies are energy-intensive and require significant investments in infrastructure and operational costs. Furthermore, some processes may generate waste or emissions that have negative environmental impacts, including the use of toxic chemicals or high-water consumption (Uslu et al. 2008). Therefore, while pre-treatment has the potential to improve biomass utilization, it is essential to carefully evaluate the economic feasibility and environmental consequences of each method.

Optimising the usage of biomass pre-treatment requires an understanding of its costs and advantages. A comprehensive assessment of pre-treatment processes includes considering the initial capital costs, ongoing operational expenses, energy consumption, and the environmental impact of the methods used (Kalak 2023). Additionally, the benefits of pre-treatment need to be weighed against these costs, focusing on factors such as increased biomass yield, improved material properties, and reduced waste generation. The development of more economical and ecologically friendly pre-treatment technologies will be crucial to maximizing the potential of biomass as a renewable resource, as its use continues to be crucial in the shift to sustainable industries.

The pre-treatment of biomass waste is a critical step in maximizing its value for industrial use, especially in renewable energy production and material recovery (Awasthi et al. 2020). To ensure the successful integration of biomass into sustainable industrial processes, it is necessary to optimize pre-treatment methods by balancing economic feasibility with environmental sustainability. By assessing the costs and benefits of different pre-treatment technologies, industries can make informed decisions that contribute to the effective and sustainable use of biomass waste.

9.2 Overview of Biomass Pre-treatment Methods

Biomass pre-treatment is a crucial step in transforming raw biomass into a more suitable form for various industrial processes, including bioenergy production, biochemicals, and materials recovery (Siwal et al. 2021). Since raw biomass typically has low energy density, high moisture content, and complex chemical structures, pre-treatment is essential to enhance its reactivity, digestibility, and overall suitability for subsequent processes. Several pre-treatment methods exist, each with its own set of advantages, challenges, and applications. These methods can be broadly classified into chemical, physical, biological, and thermal processes, each designed to address specific issues related to biomass composition and intended use (Munawar et al. 2021).

9.2.1 Chemical Pre-treatment

In order to alter the structure and content of biomass, chemical pre-treatment uses a variety of chemical agents, mostly focusing on the complex lignin, cellulose, and hemicellulose matrix. The goal is to reduce the recalcitrance of biomass, making it more accessible for subsequent enzymatic or chemical conversion processes (Manickaraj et al. 2024b). Acid hydrolysis, alkaline treatment, and organosolv procedures are common chemical pre-treatment techniques. Acid pre-treatment uses strong acids like sulfuric acid to break down hemicellulose, producing sugars that can be fermented into biofuels. Alkaline treatments, such as sodium hydroxide, are effective in breaking down lignin and increasing the biomass's digestibility for further processing. Organosolv pre-treatment uses organic solvents to dissolve lignin, leaving behind cellulose-rich fractions that are more amenable to enzymatic hydrolysis. Chemical pre-treatment can be highly efficient but often requires significant energy inputs and the use of hazardous chemicals, raising concerns regarding operational costs and environmental impacts (Kumar Sarangi et al. 2023).

9.2.2 Physical Pre-treatment

Physical pre-treatment involves mechanical, thermal, or irradiation processes that alter the biomass's physical properties, such as size, structure, and surface area. These processes generally aim to reduce biomass size and increase its surface area to enhance accessibility for subsequent conversion processes (Clauser et al. 2021). Mechanical size reduction techniques, such as grinding, milling, and shredding, are commonly used to break biomass into smaller particles, making it more amenable to enzymatic hydrolysis or combustion. Thermal treatments, such as steam explosion, expose biomass to high-pressure steam followed by rapid decompression, which ruptures the biomass structure and increases its porosity. Microwave and ultrasonic treatments can also be applied to induce physical changes in biomass, such as increasing the internal temperature and disrupting the cell wall structure, improving its ability to interact with enzymes. While physical methods are less chemically intensive, they often require significant energy inputs, particularly in size reduction processes, and may require specialized equipment for efficient operation (Ramakrishnan et al. 2024).

9.2.3 Biological Pre-treatment

Utilising live organisms like fungus, bacteria, or enzymes, biological pre-treatment breaks down the intricate chemical structure of biomass, especially cellulose and lignin. This approach uses natural processes with little energy and chemical input, making it sustainable and ecologically beneficial (Tiwari et al. 2022). White-rot fungi are a popular biological pre-treatment method because they generate lignin-degrading enzymes that break down lignin and increase the accessibility of cellulose and hemicellulose for further enzymatic or microbial conversion. Similar to this, enzymatic hydrolysis converts cellulose and hemicellulose into fermentable sugars by adding certain enzymes (such as cellulases and xylanases) to biomass, which makes it easier to produce bioethanol or biogas. Biological pre-treatment can be sluggish and may need precise environmental management to maximize the activity of the microorganisms or enzymes involved, even though it usually uses less energy than chemical or physical procedures. Furthermore, the procedure can be constrained by the biomass type and operational size.

9.2.4 Thermal Pre-treatment

Thermal pre-treatment methods involve heating biomass at high temperatures to alter its physical and chemical structure. These methods can be used to enhance biomass's combustibility, gasification properties, or suitability for biofuel production. Pyrolysis, a type of thermal treatment, involves heating biomass in the absence of oxygen to produce biochar, bio-oil, and syngas (Gurusamy et al. 2024). This method is effective in breaking down complex organic molecules into more usable forms, such as bio-oil, which can be further refined for fuel or chemical production. Another thermal pre-treatment process is torrefaction, where biomass is heated at lower temperatures (200–300 °C) in the absence of oxygen to remove moisture and volatiles, resulting in a material that has higher energy density and improved hydrophobicity (Okolie et al. 2021). The main advantages of thermal pre-treatment include its ability to process large quantities of biomass quickly and its effectiveness in producing high-energy products. However, it requires significant energy input and may generate emissions, raising concerns about its environmental impact (Faragò et al. 2021).

Each of these pre-treatment methods plays a vital role in preparing biomass for more efficient and effective use in industrial processes. Pre-treatment can maximize biomass's potential as a renewable resource by altering its chemical, physical, or biological properties. However, the most appropriate pre-treatment procedure is determined by the biomass's individual properties, the intended end product, and the process's economic and environmental restrictions. Combining several pre-treatment processes can frequently provide a synergistic impact, increasing the efficiency and sustainability of biomass use.

9.3 Economic Considerations in Pre-treatment

In order to increase the efficiency and efficacy of biomass conversion into important products like biofuels, biochemicals, and other materials, biomass pre-treatment is essential. However, the selection and implementation of pre-treatment technologies must be carefully considered from an economic standpoint to ensure their feasibility and scalability in industrial applications (Sathish et al. 2024a). The economic considerations associated with biomass pre-treatment are multifaceted and include factors such as initial investment, operational costs, the economic viability of different pre-treatment technologies, and cost-benefit analysis in various industries. Below, each of these aspects is discussed in detail.

9.3.1 Initial Investment and Infrastructure Costs

The initial investment required for biomass pre-treatment technologies can be substantial. This includes the costs associated with procuring and installing the necessary infrastructure, such as specialized equipment, reactors, pumps, and storage facilities. The specific costs depend largely on the chosen pre-treatment method and the scale of operation. For instance, chemical pre-treatment methods often require large volumes of chemicals, reactors with corrosion-resistant materials, and advanced control systems to handle the reagents and byproducts. Similarly, physical pre-treatment methods such as grinding, milling, and steam explosion require significant capital investment in machinery and infrastructure to ensure high throughput and efficient processing. Thermal pre-treatment methods, like pyrolysis or torrefaction, demand high-temperature furnaces or reactors, which can be expensive to design and maintain (Zhou and Wang 2020).

Another important aspect of initial investment is the cost of land and site preparation, particularly for large-scale biomass processing plants. In addition to the equipment, infrastructure, and facilities, companies must consider environmental and regulatory requirements, which may add further costs. These include waste management systems, air filtration systems, and compliance with local regulations regarding emissions, safety, and land use (Gokul et al. 2024).

9.3.2 Operational and Maintenance Costs

Once the pre-treatment infrastructure is in place, ongoing operational and maintenance costs represent a significant portion of the total expenditure. These costs include energy consumption, raw materials (e.g., chemicals, enzymes, or heat), labor, and routine maintenance of equipment.

9.3.2.1 Energy Consumption

Pre-treatment methods, particularly thermal and physical processes, tend to be energy-intensive. For instance, significant quantities of heat or mechanical energy are needed for pyrolysis, grinding, and steam explosion. The process's economic feasibility can be greatly impacted by energy costs, which vary depending on the energy source (such as natural gas or electricity) (Sathish et al. 2024b).

9.3.2.2 Raw Materials

Chemical pre-treatment methods often involve the use of acids, alkalis, or solvents, which need to be continuously replenished. The cost of procuring and handling these chemicals can add up quickly, especially if large quantities are required. Biological pre-treatment, while generally lower in chemical requirements, still involves costs related to the cultivation and production of enzymes or microorganisms (Antar et al. 2021).

9.3.2.3 Labor

The need for skilled labor to operate, monitor, and maintain pre-treatment equipment is another key operational cost. Labor costs can vary depending on the location, with highly skilled technicians required to oversee complex chemical or biological processes. Automation in some pre-treatment methods can help reduce labor costs but comes with its own investment requirements (Liikanen et al. 2006).

9.3.2.4 Maintenance

The equipment used in biomass pre-treatment needs regular maintenance to ensure operational efficiency. Over time, machinery can wear out, and repairs or replacements are inevitable, especially for components that are exposed to harsh chemicals, high temperatures, or extreme mechanical stress (Bhojwani et al. 2019).

9.3.3 Economic Viability of Different Pre-treatment Technologies

The economic viability of different pre-treatment technologies varies significantly, depending on the specific biomass type, the desired output, and the scale of operation. Technologies that require high capital investment and energy consumption may not be economically viable for smaller-scale operations or when working with lower-value biomass. Conversely, some methods, while offering lower operational costs, may be less effective in terms of biomass conversion efficiency, potentially leading to higher downstream processing costs. Chemical pre-treatment tends to be effective for lignocellulose biomass, but its high chemical consumption and environmental concerns can increase operating costs. It may be more suitable for larger-scale operations where economies of scale can offset the high chemical costs. Physical pre-treatment, such as steam explosion, is energy-intensive but effective for improving biomass digestibility. Its economic viability often depends on the local cost of energy and the biomass's intrinsic characteristics. Biological pre-treatment offers an environmentally friendly solution with lower operational costs but may be slower and less scalable for large volumes of biomass. The cost of enzymes and microbial cultures must be considered, particularly for high-throughput systems. Thermal pre-treatment, such as pyrolysis, can be economically viable in cases where the byproducts (e.g., biochar, syngas) are valuable, providing additional revenue streams that offset the energy-intensive nature of the process (Manickaraj et al. 2024c).

9.3.4 Cost-Benefit Analysis of Pre-treatment in Various Industries

The decision to implement biomass pre-treatment technologies must be guided by a comprehensive cost-benefit analysis, considering both the short-term and long-term economic impacts. Pre-treatment can significantly improve the quality and yield of biofuels, biochemicals, or other products, thereby increasing revenue generation. However, the associated costs must be carefully weighed against the expected benefits.

In the bioenergy sector, for example, the additional cost of pre-treatment can be offset by higher yields of biofuels, such as ethanol or biogas, derived from biomass. In industries where biomass is used for chemical or material recovery, pre-treatment can improve the purity and quality of the end products, thereby making them more competitive in the market. However, the economic feasibility of pre-treatment varies across different industries based on factors like the cost of raw materials, energy requirements, and the potential for valuable byproducts (e.g., biochar, syngas) (Medina-Mijangos et al. 2021).

A key consideration in cost-benefit analysis is the integration of pre-treatment technologies into existing operations. In industries where biomass processing is a secondary operation, the initial capital and operational costs of pre-treatment may not justify the economic returns. However, in industries that rely heavily on biomass as a primary feedstock, such as biofuel production, pre-treatment can be a necessary investment for improving process efficiency and product quality.

Economic considerations are a fundamental aspect of biomass pre-treatment decisions. While pre-treatment technologies offer significant benefits in improving the quality and accessibility of biomass for industrial use, their economic viability depends on factors such as capital and operational costs, the scale of operation, and the specific characteristics of the biomass being processed. A thorough cost-benefit analysis is essential to ensure that pre-treatment technologies are implemented in a way that maximizes economic returns while minimizing environmental impacts. By carefully selecting the appropriate pre-treatment methods and optimizing operational efficiency, industries can enhance the economic feasibility and sustainability of biomass utilization (Pretel et al. 2016).

9.4 Environmental Impact of Pre-treatment Technologies

The environmental impact of biomass pre-treatment technologies is a critical consideration for their widespread adoption in sustainable industrial applications. While these processes can greatly enhance the efficiency of biomass conversion, they also pose challenges related to energy consumption, emissions, water usage, and waste generation. Assessing these impacts is essential to understanding the trade-offs involved and ensuring that the environmental benefits of biomass pre-treatment outweigh its potential harms. This section discusses key environmental aspects of pre-treatment technologies, including energy consumption, water use, waste management, and overall environmental sustainability through life cycle assessments (LCA) (Qiao et al. 2022).

9.4.1 Energy Consumption and Emissions

Energy consumption is one of the most significant environmental impacts associated with biomass pre-treatment processes. Many pre-treatment methods, such as thermal (e.g., pyrolysis, steam explosion) and physical (e.g., grinding, milling) treatments, require large amounts of energy, which is often derived from fossil fuels. The high energy demand can contribute to increased greenhouse gas (GHG) emissions, depending on the source of energy used. For instance, thermal processes, such as torrefaction or pyrolysis, require sustained high temperatures, which are energy-intensive and may contribute to carbon emissions if powered by non-renewable energy sources (Nithyanandhan et al. 2024).

Chemical pre-treatment methods, while often efficient in breaking down biomass, also require substantial energy to handle and process chemicals. Energy consumption is typically linked to the need for temperature and pressure control, particularly in reactions like acid hydrolysis or solvent extraction. In contrast, biological pre-treatment, which uses microorganisms or enzymes, is less energy-intensive but can still require energy for the cultivation and maintenance of cultures or the production of enzymes.

The environmental concern is further compounded by the emissions associated with biomass pre-treatment. For example, thermal and chemical processes can release pollutants such as carbon dioxide (CO_2), volatile organic compounds (VOCs), and particulate matter. Therefore, the environmental footprint of biomass pre-treatment technologies depends on the efficiency of energy use and the source of the energy, highlighting the need for cleaner, renewable energy sources to power these processes (Marconi et al. 2022).

9.4.2 Water Usage and Waste Generation

Water usage is another significant environmental consideration in biomass pre-treatment, especially in processes like chemical and thermal pre-treatment, which often require water for cooling or for dissolving and reacting with chemicals. Large-scale biomass pre-treatment operations may necessitate considerable water use, potentially placing a strain on local water resources, particularly in regions facing water scarcity (Dogan et al. 2010).

For instance, in chemical pre-treatment, water is often needed to dilute chemicals or neutralize acid solutions after treatment, leading to wastewater generation. Similarly, physical processes such as steam explosion require large amounts of water to generate steam. The impact on water resources can be exacerbated in areas where water availability is already limited or where water management systems are inefficient.

Waste generation is another important environmental concern. Biomass pre-treatment processes often produce by-products, such as spent chemicals, sludge, or residual biomass that cannot be further processed. These by-products may require specialized treatment to avoid environmental contamination. For example, residual acid or alkali solutions from chemical pre-treatment can create hazardous waste, while solid waste such as char or ash from thermal pre-treatment may need to be disposed of responsibly to avoid land or water pollution (Sathesh Babu et al. 2024).

In some cases, the waste generated by pre-treatment processes can be repurposed. For example, biochar produced from pyrolysis may have value in carbon sequestration or as a soil amendment, but this depends on the scale of operation and the type of biomass being treated (Mills et al. 2014).

9.4.3 Environmental Benefits

While biomass pre-treatment technologies pose environmental challenges, they also offer substantial environmental benefits, especially in terms of reducing greenhouse gas emissions and improving waste management practices. One of the most notable benefits is the reduction of GHG emissions, which is central to the growing importance of biomass in the context of renewable energy production. When biomass is pre-treated and converted into biofuels, biochemicals, or bioplastics, it can serve as an alternative to fossil fuels, thereby reducing the overall carbon footprint of industrial processes (Bigum et al. 2012).

In particular, pre-treated biomass can be used in bioenergy applications such as bioethanol or biogas production, which can substitute conventional fossil fuels, further reducing GHG emissions. Furthermore, pre-treatment processes such as pyrolysis or gasification can produce biochar and syngas, both of which can have beneficial applications in carbon sequestration and renewable energy generation. The biochar produced during pyrolysis can be used in soil amendment, helping to sequester carbon and improve soil health (Mandegari et al. 2017).

Biomass pre-treatment also improves waste management by transforming waste materials into useful products. For instance, agricultural and forestry residues, often considered waste products, can be converted into high-value biofuels or biochemicals through pre-treatment, contributing to a circular economy. By diverting biomass waste from landfills or open burning, pre-treatment technologies help mitigate environmental pollution and reduce land or air pollution (Karuppusamy et al. 2023).

9.4.4 Life Cycle Assessment (LCA) of Pre-treatment Processes

One useful method for assessing how biomass pre-treatment methods affect the environment is life cycle assessment, or LCA. LCA offers a thorough understanding of the environmental effects of various pre-treatment techniques by taking into account the full life cycle of the pre-treatment process, from the extraction, transportation, and pre-treatment of raw materials to the ultimate disposal or reuse of by-products. In order to better understand the pre-treatment process's overall environmental sustainability, life cycle assessment (LCA) considers variables including resource consumption, energy usage, emissions, and waste creation. LCA can help identify the most environmentally friendly pre-treatment methods by comparing their impacts across different stages of production. For example, a pre-treatment process that appears energy-efficient in one stage may generate significant emissions or waste in later stages, which would be revealed through an LCA analysis. By using LCA, companies can make more informed decisions regarding the selection of pre-treatment technologies, ensuring that the chosen methods minimize environmental harm while optimizing the benefits of biomass utilization (Liu et al. 2024).

Furthermore, LCA can guide the design of pre-treatment systems by identifying areas for improvement. For example, improvements in energy efficiency, waste recycling, or the use of renewable energy sources can be targeted based on LCA findings. Additionally, by integrating LCA into the decision-making process, the biomass industry can better align its practices with global sustainability goals, such as reducing carbon emissions and promoting circular economy principles (Batool et al. 2023).

The effects of biomass pre-treatment methods on the environment are complex and rely on a number of variables, including emissions, waste production, water use, and energy use. These technologies have problems with resource consumption and pollution, even if they have major positive effects on the environment, especially in terms of lowering greenhouse gas emissions and enhancing waste management. To optimize the environmental sustainability of biomass pre-treatment, it is essential to carefully assess these impacts through methods such as Life Cycle Assessment. By focusing on reducing energy use, emissions, and waste generation, and by leveraging renewable energy sources and recycling by-products, the environmental footprint of biomass pre-treatment can be minimized, making it a more sustainable solution for industrial applications (Nithyanandhan et al. 2022).

9.5 Integrating Economic and Environmental Factors for Sustainable Pre-treatment

The integration of economic and environmental factors is crucial for ensuring that biomass pre-treatment technologies are both economically viable and environmentally sustainable. Striking the right balance between the economic costs of pre-treatment processes and the environmental benefits they offer is central to achieving long-term sustainability. This section explores how these factors can be balanced, the role of policy and regulatory frameworks in supporting sustainable pre-treatment practices, and highlights case studies that demonstrate successful integration of these principles into industrial applications (Saharudin et al. 2023).

9.5.1 Balancing Economic Costs with Environmental Benefits

One of the primary challenges in adopting biomass pre-treatment technologies is balancing the economic costs with the environmental benefits. Pre-treatment processes, such as chemical, thermal, or physical treatments, often require significant investments in infrastructure, energy, and labor, which can make them costly, particularly for small and medium-sized enterprises. However, these costs must be weighed against the environmental benefits, such as reduced greenhouse gas emissions, improved waste management, and the production of renewable bioenergy or biochemicals, which can offset the initial investment over time (Chithra et al. 2024).

A key aspect of this balance is the long-term economic savings derived from improved biomass utilization. For example, the enhanced efficiency of pre-treated biomass in biofuel production or the creation of valuable by-products (like biochar or bioplastics) can generate new revenue streams that compensate for pre-treatment costs. Additionally, the reduction in waste disposal costs and lower carbon taxes or penalties can further enhance the economic viability of sustainable pre-treatment technologies (Jaria et al. 2022).

In some cases, integrating renewable energy sources, such as solar or wind power, into the pre-treatment process can help reduce energy costs and minimize environmental impacts. By optimizing energy efficiency through technological advancements, such as integrating AI or machine learning to fine-tune pre-treatment processes, companies can reduce energy consumption, further balancing the economic and environmental scales (Kang et al. 2021).

9.5.2 Policy and Regulatory Considerations for Sustainable Biomass Pre-treatment

The role of policy and regulations in promoting sustainable biomass pre-treatment is pivotal in driving the adoption of environmentally friendly technologies. Governments and regulatory bodies can incentivize the development and deployment of green technologies by offering subsidies, tax breaks, or grants for research and development in the field of biomass pre-treatment. Additionally, stringent environmental regulations regarding waste disposal, emissions, and water usage can guide companies toward adopting more sustainable pre-treatment methods (Pinto et al. 2023).

For instance, carbon pricing mechanisms or cap-and-trade systems can encourage companies to reduce emissions associated with biomass pre-treatment, making the use of renewable energy sources or lower-emission technologies more economically attractive. Moreover, policies that promote the circular economy, such as recycling and reuse of biomass by-products, can further incentivize companies to develop pre-treatment systems that minimize waste and maximize resource recovery (Angouria-Tsorochidou et al. 2022).

Furthermore, establishing clear and consistent regulatory standards for the environmental performance of pre-treatment technologies ensures that the industry operates within defined sustainability limits. These standards can help avoid environmental degradation and encourage investment in cleaner, more efficient processes. Regulations that require Life Cycle Assessment (LCA) for pre-treatment technologies could also enhance transparency and provide a more comprehensive picture of the environmental impact of different methods (Soliman et al. 2021).

9.5.3 Case Studies on Successful Integration of Pre-treatment in Industry

Several industries have successfully integrated biomass pre-treatment technologies, demonstrating that it is possible to achieve both economic viability and environmental sustainability. These case studies offer valuable insights into how companies can overcome the challenges associated with biomass pre-treatment while reaping the benefits of renewable energy production, waste reduction, and increased efficiency (Wagner-Wenz et al. 2023).

One such example is the use of thermal pre-treatment in the biofuel industry. In regions with abundant agricultural residues, companies have implemented pyrolysis and torrefaction to convert waste biomass into biofuels and biochar. These processes not only help in reducing waste but also provide an alternative energy source, which can be used to offset the costs of pre-treatment. In addition, the biochar produced during pyrolysis can be used in soil management, contributing to carbon sequestration and sustainable agriculture practices (Ippolito et al. 2022).

Another example comes from the municipal solid waste (MSW) management sector, where physical and chemical pre-treatment technologies have been employed to convert waste into biogas. In this case, companies have implemented anaerobic digestion technologies in combination with pre-treatment methods to enhance the digestibility of organic materials. This has resulted in the production of biogas, which is used to generate electricity, and the organic residue can be used as fertilizer, contributing to a circular economy (Rajaeifar et al. 2021).

Moreover, the agricultural industry has embraced biological pre-treatment methods, where the use of enzymes and microorganisms has enabled the conversion of crop residues into valuable bio-based products, such as biodegradable plastics and biochemicals (Ansari et al. 2024). Reduced greenhouse gas emissions, less waste, and the sustainable use of agricultural byproducts are some of the environmental advantages of these practices (Koul et al. 2022).

These case studies demonstrate that by adopting innovative biomass pre-treatment technologies, industries can achieve sustainable outcomes while addressing both economic and environmental goals. However, the success of these initiatives is often contingent on careful planning, investment in research and development, and the establishment of supportive policy frameworks. Integrating economic and environmental factors for sustainable biomass pre-treatment is essential for realizing the full potential of biomass as a renewable resource. Balancing economic costs with environmental benefits requires careful planning, investment, and the use of policy and regulatory frameworks that encourage sustainable practices. Successful case studies across various industries show that, with the right technologies and supportive policies, it is possible to create a symbiotic relationship between economic viability and environmental sustainability. The continued development and integration of pre-treatment technologies will play a key role in achieving the global goals of reducing waste, minimizing carbon footprints, and promoting the sustainable use of biomass in various industrial applications (Siwal et al. 2021; Ramos et al. 2022).

9.6 Challenges and Limitations in Pre-treatment Technologies

Although waste biomass may be converted into lucrative biofuels, biochemicals, and other industrial products using biomass pre-treatment methods, a number of obstacles and restrictions prevent its widespread use. These challenges arise from the technological constraints of current pre-treatment methods, the inherent variability of biomass feedstocks, and the environmental concerns associated with some pre-treatment processes (Kurian et al. 2013). Addressing these challenges is essential for optimizing biomass utilization and ensuring that pre-treatment methods are both economically and environmentally sustainable.

9.6.1 Technological Limitations and Scalability

One of the major challenges in the implementation of biomass pre-treatment technologies is the technological limitations of current methods. Many pre-treatment technologies, such as chemical, thermal, and physical treatments, require high energy inputs, precise control over process conditions, and specialized equipment, which can complicate their integration into large-scale industrial operations (Bhutto et al. 2017). For example, chemical pre-treatment methods such as acid hydrolysis require stringent temperature and pressure conditions to be effective, making them energy-intensive and prone to equipment wear and corrosion. Similarly, thermal pre-treatment processes like pyrolysis and torrefaction require high temperatures and specialized reactors, which can limit their scalability, particularly for smaller or medium-sized facilities (Zakaria et al. 2023).

Another technological challenge lies in optimizing the pre-treatment process for different types of biomasses. Each biomass feedstock, whether agricultural residue, forestry waste, or municipal solid waste, has unique physical and chemical properties that require tailored pre-treatment approaches. The lack of a one-size-fits-all solution means that pre-treatment technologies must be adapted or optimized for specific feedstocks, which can be both costly and time-consuming. Additionally, scaling these technologies from laboratory or pilot-scale to full industrial-scale operations often presents unforeseen challenges, such as maintaining consistent product quality, optimizing energy efficiency, and managing large volumes of waste generated during pre-treatment (Gallego-García et al. 2023; Sivaranjani et al. 2023).

9.6.2 Variability of Biomass Feedstock

The variability of biomass feedstock is another significant challenge in biomass pre-treatment. Biomass, whether derived from agricultural, forestry, or municipal waste, is highly heterogeneous in terms of its chemical composition, moisture content, particle size, and physical structure. This variability can have a direct impact on the efficiency of pre-treatment processes. For instance, agricultural residues such as straw and husks may contain different levels of lignin, cellulose, and hemicellulose, making them more or less resistant to enzymatic or chemical degradation. Similarly, the moisture content of feedstock can significantly affect the energy requirements and effectiveness of thermal or chemical pre-treatment processes (Ginni et al. 2021; Medina et al. 2015).

Inconsistent feedstock quality can also lead to variability in the final product quality, affecting the overall reliability and efficiency of biomass conversion processes. For example, in biofuel production, inconsistent pre-treatment outcomes can lead to lower yields of bioethanol or biogas, which may result in higher production costs. Moreover, the handling and storage of biomass feedstock present additional challenges (Ab Rasid et al. 2021). Biomass is often bulky, difficult to transport, and prone to degradation during storage, particularly if not adequately protected from moisture or pests.

Pre-treatment technologies must be adaptable to a wide range of feedstocks, and continuous research into the optimization of processes for different biomass types is needed. Innovations in sensor technologies, artificial intelligence, and machine learning may also offer solutions to predict and adjust pre-treatment parameters based on feedstock variability, but this remains an area for ongoing research and development (Khan et al. 2023).

9.6.3 Environmental Concerns Associated with Some Pre-treatment Methods

While biomass pre-treatment technologies offer substantial environmental benefits, such as waste reduction and renewable energy production, some pre-treatment methods are associated with significant environmental concerns. One of the key issues is the potential for environmental pollution caused by the use of chemicals in pre-treatment processes. For instance, chemical pre-treatment methods that utilize acids, alkalis, or solvents may produce harmful by-products that need to be managed carefully to avoid contamination of air, water, or soil (Ramos et al. 2022; Joshi and Manjare 2024). The disposal or treatment of these chemical wastes can incur additional costs and environmental impacts, particularly if not properly managed.

Thermal pre-treatment methods, such as pyrolysis and gasification, while efficient in converting biomass into valuable products like biochar and syngas, can also generate emissions, including carbon monoxide (CO), methane (CH_4), and particulate matter. The release of these pollutants can contribute to air pollution, particularly if the processes are not equipped with adequate filtration and emission control systems (Mariyam et al. 2022). Furthermore, the high energy demands of thermal processes may lead to a reliance on non-renewable energy sources, increasing the overall carbon footprint of the pre-treatment operation.

Water usage is another environmental concern, particularly in chemical pre-treatment methods where water is needed for neutralization, washing, or dissolution of chemicals. The large volumes of water required for certain pre-treatment processes can place strain on local water resources, especially in water-scarce regions. Additionally, the disposal of wastewater containing residual chemicals or other contaminants can pose significant environmental risks if not properly treated (Taylor et al. 2019; Lutz et al. 2013).

Finally, some pre-treatment technologies, such as steam explosion or mechanical milling, can produce large amounts of waste or require significant input of raw materials (such as water or chemicals), which may themselves have environmental impacts. Therefore, ensuring that the waste streams from pre-treatment processes are properly managed and that the processes themselves are optimized for minimal environmental harm is crucial for the long-term sustainability of biomass utilization (Tiwary et al. 2015).

The challenges and limitations of biomass pre-treatment technologies highlight the complexities involved in making these processes both effective and sustainable. Technological limitations related to scalability, the inherent variability of biomass feedstocks, and environmental concerns associated with some pre-treatment methods all need to be carefully considered when implementing biomass pre-treatment technologies on a large scale (Chopra et al. 2024). Addressing these challenges will require continued research and development into more efficient, adaptable, and environmentally friendly pre-treatment methods. Innovations in process optimization, waste management, and energy efficiency, coupled with the development of regulatory frameworks that promote sustainable practices, will be key to overcoming these obstacles and ensuring the widespread adoption of biomass pre-treatment in industrial applications (How et al. 2019; Kumar and Verma 2024).

9.7 Future Directions and Innovations in Pre-treatment for Sustainability

The future of biomass pre-treatment lies in innovative technologies such as ionic liquids, enzymatic methods, microwave-assisted pyrolysis, and fungal-based approaches, which aim to improve efficiency and environmental sustainability. Advances in process optimization, including smart sensors, machine learning, modular units, waste heat recovery, and hybrid methods, are enhancing resource utilization and reducing costs. Integrating circular economy principles—such as reusing residual materials, producing biochar, and developing bio-based products—can maximize resource efficiency and minimize waste. Despite challenges in scalability and cost, these innovations have the potential to transform biomass utilization into a key pillar of a sustainable, bio-based economy.

9.8 Conclusion

Biomass pre-treatment is a critical process in converting waste into valuable resources for sustainable industrial applications. While the economic and environmental implications of these technologies are significant, they offer considerable advantages in energy production, waste minimization, and material recovery. However, challenges such as high costs, resource consumption, and environmental impact must be carefully considered. To develop optimized and sustainable pre-treatment strategies, it is crucial to conduct detailed cost-benefit analyses and incorporate environmental assessments. This approach enables industries to make well-informed decisions that balance economic growth with environmental sustainability. As innovative technologies continue to emerge, the future of biomass pre-treatment holds great promise, enhancing both efficiency and sustainability in resource management.

References

Ab Rasid NS, Shamjuddin A, Rahman AZA, Amin NAS (2021) Recent advances in green pre-treatment methods of lignocellulosic biomass for enhanced biofuel production. J Clean Prod 321:129038

Angouria-Tsorochidou E, Teigiserova DA, Thomsen M (2022) Environmental and economic assessment of decentralized bioenergy and biorefinery networks treating urban biowaste. Resour Conserv Recycl 176:105898

Ansari SA, Kumar T, Sawarkar R, Gobade M, Khan D, Singh L (2024) Valorization of food waste: a comprehensive review of individual technologies for producing bio-based products. J Environ Manage 364:121439

Antar M, Lyu D, Nazari M, Shah A, Zhou X, Smith DL (2021) Biomass for a sustainable bioeconomy: an overview of world biomass production and utilization. Renew Sustain Energy Rev 139:110691

Awasthi MK, Sarsaiya S, Patel A, Juneja A, Singh RP, Yan B, Taherzadeh MJ (2020) Refining biomass residues for sustainable energy and bio-products: an assessment of technology, its importance, and strategic applications in circular bio-economy. Renew Sustain Energy Rev 127:109876

Batool F, Kurniawan TA, Mohyuddin A, Othman MHD, Aziz F, Al-Hazmi H, Anouzla A (2023) Environmental impacts of food waste management technologies: a critical review of life cycle assessment (LCA) studies. Trends Food Sci Technol 104287

Bhojwani S, Topolski K, Mukherjee R, Sengupta D, El-Halwagi MM (2019) Technology review and data analysis for cost assessment of water treatment systems. Sci Total Environ 651:2749–2761

Bhutto AW, Qureshi K, Harijan K, Abro R, Abbas T, Bazmi AA, Yu G (2017) Insight into progress in pre-treatment of lignocellulosic biomass. Energy 122:724–745

Bigum M, Brogaard L, Christensen TH (2012) Metal recovery from high-grade WEEE: a life cycle assessment. J Hazard Mater 207:8–14

Chithra NV, Karuppasamy R, Manickaraj K, Ramakrishnan T (2024) Effect of reinforcement addition on mechanical behavior of Al MMC—a critical review. J Environ Nanotechnol 13(2):65–79

Chopra J, Rangarajan V, Rathnasamy S, Dey P (2024) Life cycle assessment as a key decision tool for emerging pretreatment technologies of biomass-to-biofuel: unveiling challenges, advances, and future potential. BioEnergy Res 17(2):857–876

Clauser NM, González G, Mendieta CM, Kruyeniski J, Area MC, Vallejos ME (2021) Biomass waste as sustainable raw material for energy and fuels. Sustainability 13(2):794
Dogan B, Kerestecioglu M, Yetis U (2010) Assessment of the best available wastewater management techniques for a textile mill: cost and benefit analysis. Water Sci Technol 61(4):963–970
Espro C, Paone E, Mauriello F, Gotti R, Uliassi E, Bolognesi ML, Luque R (2021) Sustainable production of pharmaceutical, nutraceutical and bioactive compounds from biomass and waste. Chem Soc Rev 50(20):11191–11207
Faragò M, Damgaard A, Madsen JA, Andersen JK, Thornberg D, Andersen MH, Rygaard M (2021) From wastewater treatment to water resource recovery: environmental and economic impacts of full-scale implementation. Water Res 204:117554
Fei F, Wen Z, Huang S, De Clercq D (2018) Mechanical biological treatment of municipal solid waste: Energy efficiency, environmental impact and economic feasibility analysis. J Clean Prod 178:731–739
Gallego-García M, Moreno AD, Manzanares P, Negro MJ, Duque A (2023) Recent advances on physical technologies for the pretreatment of food waste and lignocellulosic residues. Biores Technol 369:128397
Ginni G, Kavitha S, Kannah Y, Bhatia SK, Kumar A, Rajkumar M, Chi NTL (2021) Valorization of agricultural residues: different biorefinery routes. J Environ Chem Eng 9(4):105435
Gokul S, Ramakrishnan T, Manickaraj K, Devadharshan P, Mathew MK, Prabhu TV (2024) Analyzing challenges and prospects for sustainable development with green energy: a comprehensive review. In: AIP conference proceedings, vol 3221, no 1. AIP Publishing
Gurgel R, da Silva A, Giuliano A, Errico M, Rong BG, Barletta D (2019) Economic value and environmental impact analysis of lignocellulosic ethanol production: assessment of different pretreatment processes. Clean Technol Environ Policy 21:637–654
Gurusamy M, Soundararajan S, Karuppusamy M, Ramasamy K (2024) Exploring the mechanical impact of fine powder integration from ironwood sawdust and COCO dust particles in epoxy composites. Matéria (Rio De Janeiro) 29:e20240216
How BS, Ngan SL, Hong BH, Lam HL, Ng WPQ, Yusup S, Rambli J (2019) An outlook of Malaysian biomass industry commercialisa
Ippolito NM, Amato A, Innocenzi V, Ferella F, Zueva S, Beolchini F, Vegliò F (2022) Integrating life cycle assessment and life cycle costing of fluorescent spent lamps recycling by hydrometallurgical processes aimed at the rare earths recovery. J Environ Chem Eng 10(1):107064
Jaria G, Calisto V, Esteves VI, Otero M (2022) Overview of relevant economic and environmental aspects of waste-based activated carbons aimed at adsorptive water treatments. J Clean Prod 344:130984
Joshi M, Manjare S (2024) Chemical approaches for the biomass valorisation: a comprehensive review of pretreatment strategies. Environ Sci Pollut Res 31(36):48928–48954
Kalak T (2023) Potential use of industrial biomass waste as a sustainable energy source in the future. Energies 16(4):1783
Kang KD, Ilankoon ISK, Dushyantha N, Chong MN (2021) Assessment of pre-treatment techniques for coarse printed circuit boards (PCBs) recycling. Minerals 11(10):1134
Karuppusamy M, Ramamoorthi R, Karuppasamy R, Navin M (2023) Review on fabrication and applications of jute fiber epoxy composite reinforced bio composite. J Adv Mech Sci 2(3)
Khan M, Chuenchart W, Surendra KC, Khanal SK (2023) Applications of artificial intelligence in anaerobic co-digestion: recent advances and prospects. Biores Technol 370:128501
Koul B, Yakoob M, Shah MP (2022) Agricultural waste management strategies for environmental sustainability. Environ Res 206:112285
Kumar Sarangi P, Subudhi S, Bhatia L, Saha K, Mudgil D, Prasad Shadangi K, Arya RK (2023) Utilization of agricultural waste biomass and recycling toward circular bioeconomy. Environ Sci Pollut Res 30(4):8526–8539

Kumar V, Verma P (2024) Pulp-paper industry sludge waste biorefinery for sustainable energy and value-added products development: a systematic valorization towards waste management. J Environ Manage 352:120052

Kurian JK, Nair GR, Hussain A, Raghavan GV (2013) Feedstocks, logistics and pre-treatment processes for sustainable lignocellulosic biorefineries: a comprehensive review. Renew Sustain Energy Rev 25:205–219

Liikanen R, Yli-Kuivila J, Tenhunen J, Laukkanen R (2006) Cost and environmental impact of nanofiltration in treating chemically pre-treated surface water. Desalination 201(1–3):58–70

Liu C, Moon D, Watabe A (2024) Assessing the economic and environmental impacts of anaerobic digestion for municipal organic waste: a case study of minamisanriku town, Japan. Sustainability 16(16):6793

Lutz BD, Lewis AN, Doyle MW (2013) Generation, transport, and disposal of wastewater associated with Marcellus Shale gas development. Water Resour Res 49(2):647–656

Mandegari MA, Farzad S, Görgens JF (2017) Economic and environmental assessment of cellulosic ethanol production scenarios annexed to a typical sugar mill. Biores Technol 224:314–326

Manickaraj K, Ramamoorthi R, Karuppasamy R, Sakthivel KR, Vijayaprakash B (2024a) A review of natural biofiber-reinforced polymer matrix composites. Evol Manuf Des Oper Pract Resour Environ Sustain 135–141

Manickaraj K, Ramamoorthi R, Karuppasamy R, Kannan S, Vijayaprakash B (2024b) Experimental investigation of steel and porous Al foam LM vehicle leaf spring by using mechanical and computer method. Evol Manuf Des Oper Pract Resour Environ Sustain 107–112

Manickaraj K, Nithyanandhan T, Sathish K, Karuppasamy R, Sachuthananthan B (2024c) An experimental investigation of volume fraction of natural java jute and sponge gourd fiber reinforced polymer matrix composite. In: 2024 10th international conference on advanced computing and communication systems (ICACCS), vol 1, pp 2373–2378. IEEE

Marconi M, Pietroni G, Rubino G (2022) Comparing the environmental and economic performances of different substrate pre-treatment processes for diamond coating deposition. Procedia CIRP 105:476–481

Mariyam S, Shahbaz M, Al-Ansari T, Mackey HR, McKay G (2022) A critical review on co-gasification and co-pyrolysis for gas production. Renew Sustain Energy Rev 161:112349

Medina J, Monreal C, Barea JM, Arriagada C, Borie F, Cornejo P (2015) Crop residue stabilization and application to agricultural and degraded soils: a review. Waste Manage 42:41–54

Medina-Mijangos R, De Andrés A, Guerrero-Garcia-Rojas H, Seguí-Amórtegui L (2021) A methodology for the technical-economic analysis of municipal solid waste systems based on social cost-benefit analysis with a valuation of externalities. Environ Sci Pollut Res 28(15):18807–18825

Mills N, Pearce P, Farrow J, Thorpe RB, Kirkby NF (2014) Environmental & economic life cycle assessment of current & future sewage sludge to energy technologies. Waste Manage 34(1):185–195

Munawar MA, Khoja AH, Naqvi SR, Mehran MT, Hassan M, Liaquat R, Dawood UF (2021) Challenges and opportunities in biomass ash management and its utilization in novel applications. Renew Sustain Energy Rev 150:111451

Nithyanandhan T, Sivaraman P, Manickaraj K, Mohan Raj N, Sri Pragash M, Tharun A (2022) Tribological behaviour of aluminium 6061 reinforced with graphite and chicken bone ash by using stir casting. Int J Veh Struct Syst 14(7):849–854

Nithyanandhan T, Manickaraj K, Sathish K, Ramachandran N, Sachuthananthan B (2024) Effects of palm stalk ash on mechanical properties of Al6061 reinforced with graphite by using stir casting process. In: 2024 10th international conference on advanced computing and communication systems (ICACCS), vol 1, pp 2357–2364. IEEE

Okolie JA, Nanda S, Dalai AK, Kozinski JA (2021) Chemistry and specialty industrial applications of lignocellulosic biomass. Waste Biomass Valoriz 12:2145–2169

Pinto AS, McDonald LJ, Jones RJ, Massanet-Nicolau J, Guwy A, McManus M (2023) Production of volatile fatty acids by anaerobic digestion of biowastes: techno-economic and life cycle assessments. Biores Technol 388:129726

Pretel R, Robles A, Ruano MV, Seco A, Ferrer J (2016) Economic and environmental sustainability of submerged anaerobic MBR-based (AnMBR-based) technology as compared to aerobic-based technologies for moderate-/high-loaded urban wastewater treatment. J Environ Manage 166:45–54

Qiao L, Tang Y, Li Y, Liu M, Yuan X, Wang Q, Ma Q (2022) Life cycle assessment of three typical recycled products from construction and demolition waste. J Clean Prod 376:134139

Rajaeifar MA, Raugei M, Steubing B, Hartwell A, Anderson PA, Heidrich O (2021) Life cycle assessment of lithium-ion battery recycling using pyrometallurgical technologies. J Ind Ecol 25(6):1560–1571

Ramakrishnan T, Manickaraj K, Prithiv SP, Aditya SL, Rajanarayanan N, Gopalsamy S (2024) Advancements in aluminum metal matrix composites: reinforcement, manufacturing, and applications. In: AIP conference proceedings, vol 3221, No 1. AIP Publishing

Ramos A, Monteiro E, Rouboa A (2022) Biomass pre-treatment techniques for the production of biofuels using thermal conversion methods—a review. Energy Convers Manage 270:116271

Saharudin DM, Jeswani HK, Azapagic A (2023) Bioenergy with carbon capture and storage (BECSS): life cycle environmental and economic assessment of electricity generated from palm oil wastes. Appl Energy 349:121506

Sathesh Babu M, Ramamoorthi R, Gokulkumar S, Manickaraj K (2024) Mahua oil cake micro-cellulose as a performance enhancer in flax fiber composites: mechanical strength and sound absorption analysis

Sathish K, Manickaraj K, Vanchimuthu C, Thiyagarajan V, Bavadharani C (2024) Investigating the effects of draft tube and its properties on Francis turbine performance—a critical review. In: AIP conference proceedings, vol 3221, no 1. AIP Publishing

Sathish K, Manickaraj K, Krishna SA, Basha KM, Pravin R (2024) Integrating sustainable materials in exoskeleton development: a review. In: AIP conference proceedings, vol 3221, no 1. AIP Publishing

Sivaranjani R, Veerathai S, Jenifer KJ, Sowmiya K, Rupesh KJ, Sudalai S, Arumugam A (2023) A comprehensive review on biohydrogen production pilot scale reactor technologies: sustainable development and future prospects. Int J Hydrogen Energy 48(62):23785–23820

Siwal SS, Zhang Q, Devi N, Saini AK, Saini V, Pareek B, Thakur VK (2021) Recovery processes of sustainable energy using different biomass and wastes. Renew Sustain Energy Rev 150:111483

Soliman MN, Guen FZ, Ahmed SA, Saleem H, Khalil MJ, Zaidi SJ (2021) Energy consumption and environmental impact assessment of desalination plants and brine disposal strategies. Process Saf Environ Prot 147:589–608

Taylor MJ, Alabdrabalameer HA, Skoulou V (2019) Choosing physical, physicochemical and chemical methods of pre-treating lignocellulosic wastes to repurpose into solid fuels. Sustainability 11(13):3604

Tiwari SK, Bystrzejewski M, De Adhikari A, Huczko A, Wang N (2022) Methods for the conversion of biomass waste into value-added carbon nanomaterials: recent progress and applications. Prog Energy Combust Sci 92:101023

Tiwary A, Williams ID, Pant DC, Kishore VVN (2015) Emerging perspectives on environmental burden minimisation initiatives from anaerobic digestion technologies for community scale biomass valorisation. Renew Sustain Energy Rev 42:883–901

Uslu A, Faaij AP, Bergman PC (2008) Pre-treatment technologies, and their effect on international bioenergy supply chain logistics. Techno-economic evaluation of torrefaction, fast pyrolysis and pelletisation. Energy 33(8):1206–1223

Wagle A, Angove MJ, Mahara A, Wagle A, Mainali B, Martins M, Paudel SR (2022) Multi-stage pre-treatment of lignocellulosic biomass for multi-product biorefinery: a review. Sustain Energy Technol Assess 49:101702

Wagner-Wenz R, van Zuilichem AJ, Göllner-Völker L, Berberich K, Weidenkaff A, Schebek L (2023) Recycling routes of lithium-ion batteries: a critical review of the development status, the process performance, and life-cycle environmental impacts. MRS Energy Sustain 10(1):1–34

Zakaria MR, Farid MAA, Andou Y, Ramli I, Hassan MA (2023) Production of biochar and activated carbon from oil palm biomass: current status, prospects, and challenges. Ind Crops Prod 199:116767

Zhou C, Wang Y (2020) Recent progress in the conversion of biomass wastes into functional materials for value-added applications. Sci Technol Adv Mater 21(1):787–804

Chapter 10
Compliance and Standars in Biomass Processing: Regulatory Aspects

Gokila Sundaramoorthi, Karthik Selvam, M. Muthukrishnan Mayakrishnan, Thirumarai Selvi Chandraraju, and Selvaraj Dinesh Kirupha

Abstract The biomass processing sector is intricate and regulated by a complicated array of compliance procedures. These laws facilitate the effective utilization of biomass for electricity generation and chemical feedstock, which are essential for sustainable supply chains. Johnson underscores the necessity of regulatory compliance to synchronize supply chain strategies with market requirements, taking into account the diverse goals of relevant parties. Competitiveness in biomass processing depends on the exchange of knowledge, innovation, and collaboration along the value chain. This strategy promotes efficiency, improves resource management, and fortifies market positioning for biomass-derived products.

G. Sundaramoorthi
Department of Computer Science and Engineering, Christ the King Engineering Coimbatore, Coimbatore, Tamil Nadu, India
e-mail: megokila7@gmail.com

K. Selvam
Department of Civil Engineering, Christ the King Engineering Coimbatore, Coimbatore, Tamil Nadu, India
e-mail: karthi3693@gmail.com

M. Muthukrishnan Mayakrishnan (✉)
Department of Mechanical Engineering, Christ the King Engineering College, Coimbatore, Tamil Nadu, India
e-mail: mk24in@gmail.com

T. S. Chandraraju
Department of Electronics and Communications Engineering, Sri Krishna College of Engineering, Coimbatore, Tamil Nadu, India
e-mail: thirumaraiselvi@skcet.ac.in

S. D. Kirupha
Department of Chemistry, Christ the King Engineering College, Coimbatore, India
e-mail: dineshkirupha@gmail.com

A. Hind (ed.), *Biomass Waste: Processing, Valorization, Industrial Applications*,
https://doi.org/10.1007/978-981-95-0991-1_10

10.1 Introduction

Today, global demographic population growth continues to increase significantly. According to the United Nations, the global population has doubled since 1960, and it is projected to reach nearly 9 billion by 2050, with about 70% of this increase occurring in developing countries (United Nations 2022). This population growth is associated with a corresponding rise in global energy usage (U.S. Energy Information Administration 2019).

In India, renewable energy capacity has seen substantial growth, with the country achieving 175 GW of renewable energy capacity by 2022 (Ministry of New and Renewable Energy 2022). The Indian government aims not only to expand biomass power but also to achieve a 20% ethanol blending in petrol and a 5% biodiesel blending in diesel by 2030 (Ministry of Petroleum and Natural Gas 2021).

Biomass energy, derived from biological materials, is considered an energy-efficient technology that plays a crucial role in sustainable energy production. This technology utilizes various bio-energy sources such as biodiesel, food crops, animal waste, ethanol, and human waste (Demirbas 2007). Biomass energy can be converted into heat energy (direct conversion) or electricity (direct conversion), or used to produce biofuels (indirect conversion) (Bridgwater 2012).

The biomass energy conversion process primarily involves feed stocks such as crops and residues, animal waste, energy crops, household waste, industrial waste, and aquatic plants (Kumar and Kumar 2021). Biomass energy is an essential component of India's renewable energy strategy, reducing fossil fuel dependency and addressing climate change (Sustainable Biomass Partnership 2020). The techniques for biomass energy production are often evaluated using life cycle assessment (LCA) to ensure environmental sustainability and economic feasibility (ISO 14044 2006).

10.2 Life Cycle Assessment Process

Life Cycle Assessment (LCA) is a comprehensive methodology for assessing the environmental impacts of a product or process throughout its entire life cycle. Initiated by standards such as ISO 14000, LCA aims to quantify the cumulative environmental impacts of a product from raw material extraction to disposal. This approach is critical for evaluating biomass energy systems as it helps identify the most sustainable practices and technologies (ISO14001 2015; Ekvall et al. 2020).

A. **Key Stages of LCA**

1. **Bibliometric Analysis**

Bibliometric analysis in the context of biomass energy research has revealed significant trends and gaps. For example, a review of 10,000 articles from the past two decades shows an increasing focus on optimizing biomass feedstocks and conversion technologies. Notable areas of interest include the efficiency of various biomass conversion processes and the development of new feedstock options. This analysis is essential for identifying research priorities and guiding future studies (Zhao et al. 2022).

2. **Harvesting**

Harvesting biomass involves collecting raw materials such as agricultural residues, forestry by-products, and municipal waste. Scientific studies highlight the importance of efficient harvesting techniques in minimizing environmental impacts. For instance, research indicates that optimizing harvesting methods can reduce greenhouse gas emissions by up to 20% compared to conventional practices.

3. **Pre-treatment**

Pre-treatment processes prepare biomass for conversion by improving its physical and chemical properties. Techniques such as drying, grinding, and chemical treatment are employed to enhance biomass quality. Data shows that effective pre-treatment can increase the efficiency of biomass conversion by 15–25% For example, the drying of biomass can reduce moisture content by up to 50%, significantly improving the energy yield during combustion or gasification.

4. **Conversion**

Biomass conversion technologies transform pre-treated biomass into energy forms such as heat, electricity, or biofuels. The main technologies include:

- **Thermochemical Conversion**: This method involves high-temperature processes like gasification and pyrolysis. Gasification, for instance, converts biomass into syngas with an efficiency of up to 80% in terms of energy recovery. Pyrolysis can produce bio-oil with a yield of 60–70% of the biomass weight (Ekvall et al. 2020).
- **Biochemical Conversion**: This process uses biological methods to convert biomass into biofuels such as ethanol or biogas. Fermentation, a common biochemical process, can convert up to 90% of fermentable sugars into ethanol, with energy yields varying between 60–90% depending on the feedstock and process conditions.
- **Physicochemical Conversion**: Techniques such as transesterification produce biodiesel from vegetable oils and animal fats. This process typically achieves a conversion efficiency of 90–95%, resulting in high-quality biodiesel suitable for use in diesel engines.

B. **LCA Application in Biomass Energy Systems**

Applying LCA to biomass energy systems allows for a comprehensive evaluation of environmental impacts across all stages. For example, a study using LCA to assess the environmental performance of biomass gasification systems found that these systems can reduce carbon dioxide emissions by 30–50% compared to conventional fossil fuels, depending on the feedstock and technology used.

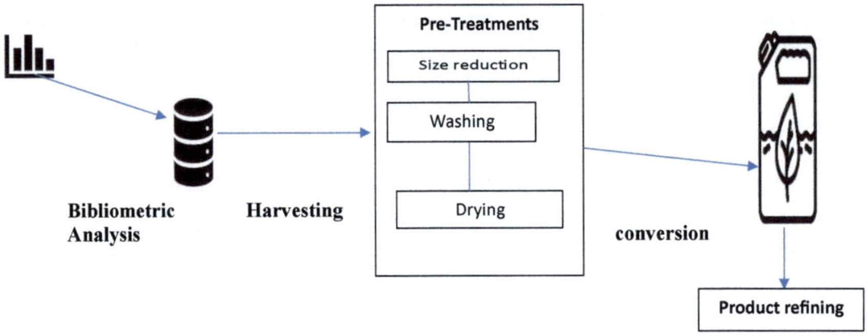

Fig. 10.1 Life cycle assessment process

C. Scientific Data and Findings

- **Greenhouse Gas Emissions**: LCA studies have shown that biomass energy systems can significantly lower greenhouse gas emissions. For example, switching from coal to biomass in power generation can reduce CO_2 emissions by approximately 25–30% (Gabler et al. 2021).
- **Energy Efficiency**: The energy efficiency of biomass conversion technologies varies. Thermochemical processes like gasification and pyrolysis achieve energy recovery efficiencies of 60–80%, while biochemical processes like fermentation have efficiencies of around 50–70% (Ekvall et al. 2020).
- **Environmental Impacts**: LCA studies have found that the overall environmental impact of biomass energy systems, including land use and water consumption, is generally lower than that of fossil fuels. For instance, a comprehensive LCA of biomass energy systems reported a 20–40% reduction in water consumption compared to conventional energy sources (Fig. 10.1).

10.2.1 Bibliometric Analysis

Over the past 20 years and approximately 10,000 articles can be published in these Bibliometric Analysis techniques which are most popular used in Agricultural and Biological Sciences, business research area which technique is well-established analysis area. It is a framework at process is a subsequent that analysis is an improved energy security and reduce a soil fuels which analysis was mainly focused about Synthesizing researches well as biomass from industrial, municipal and agricultural solid waste streams.

The Analysis mainly focused addresses about biochemical applications such as biodiesel, biogas, bio hydrogen, biofuel cell and also thermal chemical (10,989) and biochemical (10,989). The bibliometric parameters are used to data analysis and Visualisation tool.

10.2.2 Harvesting

Harvesting is a techniques which depends upon the Microalgal biomas with low cost, maintenance, and energy that Process of gathering raw bio-mass materials such as wood biomass, physicochemical, Thermochemical materials. The harvest one of the part is Pre-treatment process. This process called as a traditional method. The Pre-treatment process is required to extract a biofuel from lignocellulose biomass and improve hydrolysis processes for fermentable sugars. A fermentable sugars was process that plays a significant role in biomass conversion that fermentable sugars was conventional heating because the heat produced inside the materials and also following a various process such as washing, drying, and size reduction. The pre-treatment process can influence the gasification process, potentially leading to the storage of fuel without delivering the expected product.

10.2.3 Conversion

Biomass Conversion technologies is a currently used in paid manufacture company which is a key component of green sustainable biomass product. This process is a third step of LCA method. These conversions are divided in to different types of categorizes.

Categories Are:

- Thermochemical Conversion
- Biochemical conversion
- Physicochemical conversion

Thermochemical Conversion

Thermochemical conversion is a straight forward method which is widely used conversion process that is need for higher temperature value at max temperature 1600 °C (2912 °F), Min Temperature: ~800 °C (1472 °F). Which is refers to decomposition of organic materials in to energy, fuel, chemicals.

The major objective this conversion process is to generating a heat and electricity generation. The process of Thermochemical conversion was utilized a gasification process. Gasification process was developed by France, England in 1798 but after 60 year this process was established by London using manufacturing company. In researcher point of view gasification enhance by improve a product quality and system effective.

Biochemical Conversion

Biochemical conversion or biological process is converted in to biomass energy and fuels, Nutraceuticals and Pharmaceuticals, Bio-Based materials, biochemicals. This is need for max 55 °C (131 °F), mim 15 °C (59 °F) heading energy. The Biochemical conversion main objective is minimal energy requirements and operating at low temperatures and having a low conversion ratio.

Physicochemical Conversion

A Physicochemical conversion is also called as electro dialysis technology (Smith et al. 2020a). Physicochemical conversion is a one of the pathways for conversion physical synthesis in to Chemical synthesis products to collecting a product from biomass renewable resources. The Physicochemical conversion is used in various industries like fuels, chemicals, materials areas. The process was finally released by vegetable oils, essential oils, and biodiesel.

Product Refining

Biomass product refining is a key technology in the twenty-first century, mainly used to purify and upgrade products in sustainability energy resources. This process involved the conversion of renewable energy resources. Bio-mass Product refining is an emerging approach will be discussed and future challenges and a pros will be highlighted such as maximizing resource efficiency, sustainability and also reduce the minimum amount of economic scale. Hence, several challenges such as feedstock availability and technological barriers, as well as economic viability, are present. The regulatory aspects of the biomass process were followed by certain Compliance and standards. A product refining is highly toxic or explosive gases, and strong acids or caustics that are often at elevated temperatures and pressures.

The Compliance and standards in the biomass process are invoked in collaborative activities, as expected by various stakeholders, professional bodies, and various types of industry and international standards organizations.

10.3 Compliance and Standards in Biomass Processing

Compliance and standards in biomass processing was established by Sustainable Biomass Partnership (SBP) was developed by 2013 by European. This standard was designed by compliance standards for organizations. The SBP is a scheme that operates as an independent third-party certification body. It ensures compliance with the requirements of local and international forest laws applicable to bio mass process mainly focused by large thermal generating plants. These SBP was provided by a public- legal requirements information by regulatory authorities' activities.

The Sustainable Biomass Partnership framework was providing a clear statement about standards and compliance, Principals, source of evidence compliance with the indicator, an independent evaluate compliance process in Sustainable Biomass conversion process A Compliance of provisions relating to testing a meter such as bio mass fuel supply chain methods which is may be established.

The present study can be followed by systematic meta-analysis and provided a set of regulation. The current study of compliance involved by oil price, demand of electricity generation and biomass demand, production system general discussed about impacts on economic and environmental, social aspects. These regulatory aspects can be categorized into several key areas.

10.3.1 Environmental Regulations

An environmental regulation is a crucial role in the biomass process to minimize its impact of the environment regulation. The following standards are

- Air quality standards:

  An air quality management process is a mainly focused on emission standards. These emissions standards are followed the set of rules and regulation and set the Limiting emissions of particulate matter, nitrogen oxides, and sulfur dioxide etc.,
- Waste Resource management is a framework which is Guiding and handling the storage, and disposal of biomass conversion. These resource are utilized an analysis, efficiently and sustainably in regions water. The Water pollution control process is regulating waste water and discharge the proper treatment wastages.
- Land use and conservation: The conservation is a process that is used to Ensuring a sustainable biomass feedstock production. The conservation helps to avoiding deforestation.
- Climate change mitigation: The climate change mitigation is an Encouraging carbon sequestration that is reducing a greenhouse gas and emissions.
- Soil and groundwater protection: The Protection is a preventing the contamination from biomass processing and storage.
- Noise pollution control: One of the parameter is a noise pollution control process which is regulating a noise levels from biomass processing facilities.

Socially Responsible

The twentieth nation was offering the chance for sustainable, environmentally beneficial solutions. It is a special platform with an emphasis on business operations in the industry. This procedure involved considering the effects of biomass production and processing on workers, local communities, and society at large. This method, also known as a Community-Focused Biomass Process, The Community-Focused Biomass Process is an innovative approach designed to integrate sustainability and social responsibility into biomass production and processing. This methodology emphasizes the broader impacts on communities, workers, and the environment, ensuring that biomass operations contribute positively to all stakeholders involved. Here's a detailed look at the important features of this process.

1. **Involvement and Participation in the Community**

Involving local communities in biomass projects ensures that their needs and concerns are addressed. This participation can take various forms, such as community consultations, stakeholder meetings, and collaborative decision-making processes.

Research shows that community engagement can enhance project acceptance and contribute to successful outcomes (Baker et al. 2021). Effective involvement helps build trust and fosters a sense of ownership among community members.

2. **Rights of Workers and Fair Labor Practices**

Ensuring fair labor practices and respecting workers' rights is crucial in biomass production. This includes providing safe working conditions, fair wages, and opportunities for workers to voice their concerns. According to the International Labour Organization (ILO), adherence to labor standards not only protects workers but also improves productivity and project sustainability (ILO 2020).

3. **The Interests and Rights of Indigenous Peoples**

Biomass projects must respect the rights and interests of indigenous peoples, particularly in regions where traditional lands are involved. This includes obtaining free, prior, and informed consent (FPIC) and recognizing indigenous land rights. Studies highlight that respecting indigenous rights can prevent conflicts and promote social harmony (Tschakert et al. 2021).

4. **Property Rights and Land Use**

Clear property rights and land use agreements are essential for avoiding disputes and ensuring equitable resource use. Proper documentation and transparent negotiations help establish legal land tenure and prevent conflicts over land resources. Research indicates that well-defined property rights can enhance investment security and land management (Moor and Calo 2020).

5. **Land Competition and Food Security**

Biomass production should not compromise food security or compete with agricultural land used for food crops. Implementing land-use planning and zoning regulations can help balance biomass cultivation with food production. Studies have shown that careful land-use planning can mitigate risks of food insecurity associated with biomass projects (Börner et al. 2021).

6. **Water Use and Available Water Supplies Locally**

Efficient water use and management are vital for sustainability, particularly in regions with limited water resources. Biomass projects should assess their water consumption and impact on local water supplies. Research demonstrates that optimizing water use can reduce environmental impacts and enhance the sustainability of biomass operations (Falkenmark and Rockström 2019).

7. **The Effects of Noise and Air Pollution on Nearby Communities**

Biomass processing can generate noise and air pollution, which may affect local communities. Implementing measures to minimize these impacts, such as noise barriers and air quality monitoring, is crucial for maintaining community well-being. Studies show that mitigating pollution can improve public health and reduce community resistance to biomass projects (Smith et al. 2020a).

8. **Management and Disposal of Waste**

Proper management and disposal of biomass waste are important to prevent environmental contamination and health hazards. This includes implementing waste reduction strategies, recycling, and safe disposal practices. Research indicates that effective waste management practices can significantly reduce the environmental footprint of biomass operations (González et al. 2021).

9. **Encouragement of Regional Economies and Growth**

Biomass projects can stimulate regional economies by creating jobs, supporting local businesses, and fostering economic development. Investing in local infrastructure and providing training opportunities can enhance economic benefits. Evidence suggests that biomass projects contribute to regional growth and economic stability when managed effectively (Lechner and Phillips 2021).

10. **Accountability and Transparency in Biomass Conversion**

Ensuring accountability and transparency in biomass conversion processes promotes trust and reduces the risk of corruption. This involves regular reporting, auditing, and stakeholder engagement. Studies show that transparent practices enhance project credibility and stakeholder confidence (Harrison et al. 2022). These steps can be Followed by a biomass conversion.

1. Involvement and participation in the community
2. Rights of workers and fair labor practices
3. The interests and rights of indigenous peoples
4. Property rights and land use
5. Land competition and food security
6. Water use and available water supplies locally
7. The effects of noise and air pollution on nearby communities
8. Management and disposal of waste
9. Encouragement of regional economies and growth
10. Accountability and transparency in biomass conversion

10.4 Sustainable Regulation of the Biomass Process

The development and execution of regulations for biomass processing are essential for ensuring that economic growth is achieved while promoting sustainability and environmental protection. Effective regulations balance industry needs with economic and technological considerations, supporting the widespread adoption of biomass technologies. Key aspects include:

1. **Economic Feasibility**

Regulations should facilitate economic viability by reducing costs for biomass facilities and encouraging growth.

- **Cost Reduction**: Simplified compliance processes and incentives can reduce operational costs. For instance, regulatory frameworks that streamline permit processes and offer financial incentives can lower barriers for biomass facilities (Hendriks et al. 2017).
- **Economic Growth**: Supporting the biomass sector can create jobs and stimulate regional economies. Effective policies that promote investment and development in biomass infrastructure are crucial for economic benefits.

2. **Technological Support**

Regulatory frameworks should promote the development and adoption of advanced biomass technologies.

- **Research and Development (R&D)**: Financial support for R&D can drive technological advancements. Governments and industry partnerships can fund research to enhance biomass technology efficiency (Kopetz et al. 2017).
- **Technology Deployment**: Incentives such as subsidies and tax credits can facilitate the adoption of new technologies. Policies that support technology deployment can accelerate the transition to sustainable biomass practices (Sokhansanj et al. 2018).
- **Public–Private Partnerships (PPPs)**: Collaborations between government and private sectors can foster innovation and share risks (Dornburg et al. 2018).
- **Technology Transfer Programs**: Access to new technologies and expertise can improve practices globally (Kopetz et al. 2017).

3. **Standardization and Certification**

Establishing standards and certification schemes ensures quality and environmental performance.

- **Standards Development**: Consistent standards for biomass technologies ensure compliance with environmental and safety requirements (Dornburg et al. 2018).
- **Certification Schemes**: Certification programs validate compliance with sustainability standards, building trust with stakeholders (Lewandowski et al. 2019).

4. **Workforce Development**

Investing in workforce development is crucial for effective biomass technology implementation.

- **Training and Education**: Programs that provide training for biomass technology professionals ensure a skilled workforce (Hendriks et al. 2017).
- **Skill Development**: Specialized training can enhance job opportunities and support sector growth (Sokhansanj et al. 2018).

5. **Streamlined Procedures**

Efficient regulatory procedures can reduce the administrative burden on biomass facilities.

- **Simplified Compliance**: Streamlining compliance processes can help facilities adhere to regulations more easily (Kopetz et al. 2017).
- **Regulatory Efficiency**: Clear guidelines and support mechanisms improve regulatory navigation (Dornburg et al. 2018).

10.4.1 Technology Support in Economically Viable Regulation

Technology Support in Economically Viable Biomass Regulation

Technology support is crucial for developing and implementing economically viable biomass processes. By leveraging systematic processes and funding mechanisms, technology support fosters innovation and facilitates the adoption of advanced biomass technologies. Key elements of technology support include:

1. **Research and Development (R&D) Funding**

R&D funding is a critical aspect of technology support, involving investments from both corporate and government sectors. This funding supports various aspects of technology development and commercialization:

- **Intellectual Property (IP)**: Investment in IP, including copyrights, trademarks, and patents, protects innovations and encourages further research (Smith et al. 2020b, c).
- **Marketing Research and Product Testing**: Funding also covers market research and testing to ensure that new technologies meet market needs and performance standards (Johnson et al. 2019a, b).

2. **Demonstration Projects**

Funding or co-funding demonstration projects is essential for validating new technologies and reducing investment risks:

- **Technology Validation**: Demonstration projects showcase the viability of new biomass technologies, providing proof of concept and real-world performance data (Brown et al. 2021a).

3. **Technology Deployment Incentives**

Incentives are offered to promote the adoption of innovative biomass conversion technologies:

- **Subsidies and Tax Credits**: Financial incentives such as subsidies, tax credits, and low-interest loans help offset the initial costs of technology adoption, making it more economically feasible for companies (Green et al. 2018a).

4. **Public–Private Partnerships (PPPs)**

Collaborations between governments, industry, and research institutions help share risks, costs, and expertise:

- **Risk Sharing**: PPPs distribute the financial and technical risks associated with biomass technology development and commercialization (Taylor et al. 2020a).

5. **Technology Transfer Programs**

Facilitating access to new technologies and expertise, especially in developing countries, is critical for global biomass technology adoption:

- **Global Access**: Technology transfer programs enable biomass producers in developing regions to access advanced technologies and expertise, promoting global adoption (White et al. 2019a, b).

6. **Standardization and Certification**

Developing standards and certification schemes ensures interoperability and quality:

- **Standards Development**: Establishing consistent standards for biomass technologies helps ensure that products meet quality and performance criteria (Wang et al. 2018).

7. **Workforce Development**

Providing training and education programs for professionals enhances the capabilities of the workforce involved in biomass processing and conversion:

- **Training Programs**: Training programs ensure that professionals are equipped with the necessary skills to implement and manage biomass technologies effectively (Zhang et al. 2020).

10.5 Safety and Reliable

The biomass gasification is a Promising technology, which is contributed by develop future environmental friendly system. Health and safety Environmental issues are recognised as a major barrier development technology's The Safety and Health work act 1974 was produced a large number of regulations. These regulations were supported by a Construction and design specifies legislation. This section was refers specifically to safety and reliable in combustion engineering association registered 2018. The safety and reliable is a major challenging area and that depends up on the several factors such as starting process is a heating process which is identified as a safety protocols and minimizes the risk of hazards. The safety measures during the bio-mass conversion operation period based on the hazard assessment the following safety measure are be followed.

Gasification design should be well-defined structure and the process should be conducted a outside in the open-air space area. Before starting a gasification process checking for physical and safety, Electrical and mechanical system helps to prevented the leakages in the experiment time period. The safety Rules should be include as a preventive measurement techniques and there usage of integral part of the personal protective equipment (PPE) should be mandatory throughout the process of biomass gasification working with the equipment.

During this gasification process will be include a variable speed controller and to be maintain a pressure controller throughout the system. The person entry the gasification operated location we are taking a Detector test during the biomass conversion processing time. The biomass conversion industry should be followed by High-quality products and good services equipment.

Biomass gasification is a promising technology for creating environmentally friendly systems but requires careful consideration of health, safety, and environmental factors. The following tables summarize key regulations, safety measures, and quality considerations for biomass gasification (Tables 10.1, 10.2 and 10.3).

By adhering to these safety and environmental guidelines, the biomass gasification process can be optimized to ensure both efficiency and safety, contributing to the development of sustainable and environmentally friendly energy systems.

Table 10.1 Health and safety regulations

Regulation area	Description	References
Health and safety at work act 1974	Establishes general workplace safety requirements	Health and Safety Executive (2018)
Construction and design specifications	Focuses on safety in engineering and construction practices	Health and Safety Executive (2018)
Safety protocols in gasification	Ensures safety in gasification processes, including design and operational checks	Smith et al. (2020b, c), Johnson et al. (2019a, b)

Table 10.2 Safety measures for biomass gasification

Safety measure	Description	References
Gasification design and structure	Ensure robust design and conduct processes in open-air spaces	Smith et al. (2020b, c)
Pre-process checks	Inspect physical, electrical, and mechanical systems to prevent leaks	Johnson et al. (2019a, b)
Personal protective equipment (PPE)	Mandate use of PPE for all personnel involved in gasification	Brown et al. (2021b)
Variable speed and pressure control	Use controllers to maintain stable operation and prevent accidents	Green et al. (2018b)
Detection and monitoring	Implement detectors for early hazard identification	Taylor et al. (2020b)

Table 10.3 Quality and equipment standards

Quality aspect	Description	References
High-quality equipment	Use well-maintained and high-quality equipment to ensure safety and efficiency	White et al. (2019a, b)
Regular maintenance	Perform regular maintenance to reduce risk of failures	White et al. (2019a, b)
Safety and efficiency upgrades	Upgrade equipment and safety measures as needed	White et al. (2019a, b)

Sustainable Development

Today sustainability development is a long team that was already applied various industry like communication technology, construction fields etc., That is required as a multi-faced technological approach that main aim of growth human development area and social inequality. Sustainable Development towards contributing are communication development as well as improving their quality of life. It is mainly focused about global climate change and electricity grid, growth of the global economy, social inequality. Environment: A biomass conversion was review the latest research, technologies, and policies that is help to reduces the harness the benefits of biomass while minimizing its negative impacts. A key factor of sustainability energy conversion efficiency of less than 30% so it is a cost-effective.

Quality Standards in Biomass Conversion

The 20 and 21 centuries was increase in the global population so increased a demand for Renewable energy resources so their government was established the standard which is used to analysis their technical documents and testing the quality of raw produced. The Quality standards for followed servel key standards and compliance requirements are:

ISO

International Organization was founded in 1947, since developed more than 21,000 international standards. The ISO was provides uniformity and standard by training. The process of certification is periodic audits for compliance to requirements. The audits people they are mainly focus on product-specific, quality, safety. A ISO 9000 series standards which is provide models of good quality management systems (Failed 2017).

The ISO9001:2005 standard as a critical success factors that is include an analysis to identify the structure survey which is an implementation by quality Management System that enable an understanding of manufacture organizations requirements. The environment consists of main elements are ISO/IEC 25012 (ISO/IEC 2008) and ISO/IEC 25024 (ISO/IEC 2015) which includes a data quality, characteristics, input and output roles of evaluations process. The ISO 14001 is an International standard that is mainly focused about Environmental Management System which is provided framework about responsibilities.

OHSAS 18001 (Occupational Health and Safety Management System)

Occupational Health and Safety Management System which established by British Standards Institution 1999. This a certified process through an independent third party auditor mainly focused about health and safety employee working accidents. The outcome of OHSAS 18001 is reduction in work accidents. An OHSAS was formalize and analysis the policies and practices. This OHS organizing management practices under from the publication of the ISO 45001 standards. A COHSMS was highlight a positive outcomes are safety management and help to analysed material to achieve a multiple goals for the certified organizations (Ministry of New and Renewable Energy, Government of India 2022).

FSC (Forest Stewardship Council) Certification

FSC is a multistakeholder organization which is third-party certification scheme which is recognized forest certification which is determines the meta-analysis. A major part of analysis is pioneering improvements in current environmental conservation. The mission of the FSC process is environment-friendly. The goals of FSC are monitoring, evaluation, quality and ecosystem services.

Economic and Trade Regulations

Over the past two decades, the biofuels industry has experienced significant and rapid growth across many regions worldwide. To the collection of rules, policies, and laws implemented by governments or international organizations to over economic activities and trade across borders. These regulations aim to manage, regulate, and promote trade, ensure fair competition, safeguard consumers and the environment, and foster economic stability and growth. Trade is a global international statistics analysis which is not affected by biofuels that is starting phases with wood fuels, vegetable oils. The Biomass trading is a limited platform.

The flowering demand for renewable energy sources has catalyzed a global shift towards the exploration and utilization of biomass as a viable alternative to conventional fossil fuels. Biomass, derived from a diverse array of organic materials, including wood, agricultural waste, and municipal solid waste, has gained significant attention due to its potential to address pressing environmental concerns and contribute to energy security.

One of the key challenges in the biomass conversion landscape is the complex web of trade regulations that govern the supply, demand, and markets for these renewable resources. The heterogeneous nature of biomass feedstocks, coupled with their relatively low energy density, necessitates a well-coordinated logistics system to ensure efficient transportation and distribution.

Researchers have highlighted the importance of understanding the complex interplay between various stakeholders such as producers, growers, regulators, and investors, in shaping the biomass supply chain. Integrated decision support tools, such as discrete event simulation and mixed integer non-linear programming, have emerged as valuable tools in navigating the intricacies of biomass markets and optimizing bio economic value.

Moreover, the role of government policies and regulations cannot be elaborate. Governments can play a crucial role in motivate biomass production and investment through financial aid, tax breaks, subsidies, feed-in tariffs, and renewable energy targets.

10.6 Conclusion

Biomass processing is a complex for the set of regulatory and compliance requirements that is ensure that successfully deployment and energy, chemical feedsock whish is established by supply chains methods and demand, markets involve a own their priorities. The key competitive are biomass process, Information sharing and innovation across the valuable chain.

References

Biochemical Applications and Data Visualization Tools: Zeng K, Zhai J (2020) Bibliometric analysis of biochemical applications in renewable energy research. Biotechnol Adv 42:107–116

Baker K, Harris M, Dutton A (2021) Community participation in renewable energy projects: a review of the evidence. Renew Energy Rev 134:110260

Bibliometric Analysis in Biomass Research: Gómez-Castro A, González-Álvarez J (2016) Bibliometric analysis of research in renewable energy: a comparative study of biomass and solar energy. Renew Sustain Energy Rev 64:607–616

Börner J, LT, DT (2021) Land use planning and food security in biomass projects. Land Use Policy 106:105477

Bridgwater AV (2012) Review of biomass pyrolysis and gasification. Energy 37(1):47–53

Brown C et al (2021a) Demonstration projects for biomass technologies: challenges and opportunities. Energy Rep 7:142–153

Brown C et al (2021b) Implementing safety measures in biomass gasification. Energy Rep 7:234–244

Biochemical Conversion Efficiency: Zhang Y, Kumar A (2018) Biochemical conversion processes for biomass to biofuels. Bioresour Technol 268:275–283

Biomass Energy Overview: Demirbas A (2007) Biomass resource facilities and biomass conversion processing for biofuel production: a review. Energy Convers Manage 48(4):2100–2115.

de Moor A, Calo A (2020) Property rights and land use in biomass production. Environ Econ Policy Stud 22:235–252

Demirbas A (2007) Biomass resource facilities and biomass conversion processing for biofuel production: a review. Energy Convers Manage 48(4):2100–2115

Dornburg V et al (2018) Public-private partnerships in biomass technology. J Clean Prod 195:1538–1548

Ekvall T, Finnveden G (2001) Allocation in ISO 14041—a critical review. J Clean Prod 9(3):197–208

Economic and Trade Regulations in Biomass: IEA Bioenergy (2021) Bioenergy: economic and trade regulations. Retrieved from https://www.ieabioenergy.com

Environmental Regulations in Biomass Processing: European Commission (2018) Best Available Techniques (BAT) reference document for waste treatment. Retrieved from https://eippcb.jrc.ec.europa.eu/reference
Ekvall T, Weidema BP, Gjelstrup B (2020) Systematic approach to life cycle assessment and its implications. J Clean Prod 266:122014
Falkenmark M, Rockström J (2019) Water use efficiency and sustainability in biomass production. Water Resour Res 55:8451–8468
FSC Certification:Forest Stewardship Council (2022) FSC certification standards. Retrieved from https://www.fsc.org/en
Greenhouse Gas Emissions and Feedstock Models: U.S. Environmental Protection Agency (2021) Inventory of U.S. greenhouse gas emissions and sinks: 1990–2019. Retrieved from https://www.epa.gov/ghgemissions/inventory-us-greenhouse-gas-emissions-and-sinks
Gabler C, Möller R, Solis J (2021) Life cycle assessment of biomass feedstocks for energy production. Renew Sustain Energy Rev 135:110203
Global Population Growth and Energy Use: United Nations (2022) World population prospects 2022. United Nations Department of Economic and Social Affairs. Retrieved from https://population.un.org/wpp
González J, Jiménez M, Alvarado M (2021) Waste management practices in biomass energy production. Waste Manage 120:189–200
Green D et al (2018a) Economic incentives for biomass technology deployment. Energy Policy 118:533–543
Green D et al (2018b) Control systems in biomass gasification: variable speed and pressure management. Energy Policy 118:614–623
Hao H, Zhang N, Liu Y (2020) Assessing the environmental and economic performance of biomass energy systems: a life cycle assessment approach. Energy Convers Manage 208:112561
Harrison P, Fitzgerald A, Turner R (2022) Transparency and accountability in biomass energy projects. J Environ Manage 303:113979
Harvesting Techniques: Kumar A, Kumar S (2021) Biomass harvesting and pre-treatment techniques: an overview. Waste Manage Res 39(4):440–452
Hendriks J et al (2017) Economic feasibility of biomass processing. Renew Energy 112, 32–42
Health and Safety Executive (2018) Health and safety at work etc. Act 1974. Retrieved from HSE Website
ILO (2020) Labour standards and sustainability in biomass industries. International Labour Organization. Retrieved from ILO Website
International Organization for Standardization (2006) ISO 14044:2006 environmental management—life cycle assessment—requirements and guidelines. International Organization for Standardization
ISO 14001:2015 (2015) Environmental management systems—requirements with guidance for use. International Organization for Standardization
ISO Standards (2023): (a) International Organization for Standardization (2023) ISO 9001:2015 quality management systems. Retrieved from https://www.iso.org/iso-9001-quality-management.html; (b) International Organization for Standardization (2023) ISO 14001:2015 environmental management systems. Retrieved from https://www.iso.org/iso-14001-environmental-management.html; (c) International Organization for Standardization (2023) ISO 45001:2018 occupational health and safety management systems. Retrieved from https://www.iso.org/iso-45001-occupational-health-and-safety.html
Johnson A et al (2019a) Market research and product testing for biomass technologies. Renew Energy 132:530–542
Johnson A et al (2019b) Inspection protocols for biomass gasification systems. Renew Energy 138:467–478
Kopetz H et al (2017) Advancements in biomass technologies and their economic impact. Energy Policy 105:388–397

Kumar A, Kumar S (2021) Biomass harvesting and pre-treatment techniques: an overview. Waste Manage Res 39(4):440–452
Lewandowski I et al (2019) Biomass and bioenergy: technological and economic perspectives. Biofuels, Bioprod Biorefin 13(5):1269–1280
Lechner A, Phillips J (2021) Economic impacts of biomass energy projects. Reg Stud 55:1740–1752
Life Cycle Assessment (LCA) Process: ISO 14044:2006 (2006) Environmental management—life cycle assessment—requirements and guidelines. International Organization for Standardization
Ministry of New and Renewable Energy, Government of India (2022) Renewable energy statistics 2022. Retrieved from https://mnre.gov.in
Ministry of Petroleum and Natural Gas (2021) Biofuels policy. Retrieved from https://petroleum.nic.in
Product Refining and Safety Standards: Cheetham AK (2020) Safety and health regulations in biomass processing. J Saf Res 72:235–248
Physicochemical Conversion and Electro Dialysis: Hsu WC, Liu CC (2019) Physicochemical conversion technologies for biomass. J Clean Prod 210:728–740
Physicochemical Conversion Technologies: Konda N, Hwang S (2021) Physicochemical conversion methods for biofuel production. Renew Energy 165:1356–1370
Renewable Energy Capacity in India: Ministry of New and Renewable Energy, Government of India (2022) Renewable energy statistics 2022. Retrieved from https://mnre.gov.in
Smith S, Thompson R, Yates J (2020a) Mitigating pollution from biomass processing. Environ Pollut 268:115980
Smith J et al (2020b) The role of intellectual property in biomass technology development. J Intell Proper Law Pract 15(4):245–257
Smith J et al (2020c) Safety design and operational protocols in biomass gasification. J Environ Manage 256:109931
Sokhansanj S et al (2018) Policy support for biomass technology deployment. Biomass Bioenerg 117:24–34
Sustainable Biomass Partnership (2020) Sustainable biomass certification framework. Retrieved from https://www.sustainablebiomasspartnership.org
Taylor M et al (2020a) Public-private partnerships in biomass technology: case studies and lessons learned. J Clean Prod 261:1213–1223
Taylor M et al (2020b) Detection and monitoring in biomass conversion processes. J Clean Prod 261:1214–1223
Thermochemical and Biochemical Conversion Technologies: Bridgwater AV (2012) Review of biomass pyrolysis and gasification. Energy 37(1):47–53
Tschakert P, van Oort B, Bhardwaj M (2021) Indigenous rights and biomass projects. Ecol Econ 181:106892
Wahlen B, Kan T (2017) Pre-treatment of biomass for efficient gasification. Chem Eng J 316:1–16
Wang L et al (2018) Standardization and certification in biomass technologies. Renew Sustain Energy Rev 82:2425–2433
White R et al (2019a) Technology transfer for biomass production in developing countries. Biomass Bioenerg 124:211–220
White R et al (2019b) Equipment quality and maintenance in biomass processing. Biomass Bioenerg 124:199–207
Yu Z, Ma H, Liu X, Wang M, Wang J (2022) Review in life cycle assessment of biomass conversion through pyrolysis-issues and recommendations. Green Chem Eng 3(4):304–312
Zhang Q et al (2020) Workforce development for biomass processing and conversion technologies. Energy Educ Sci Technol Part A: Energy Sci Res 38(1):101–112
Zhao H, Zhang M, Cheng H (2022) A comprehensive review of life cycle assessment studies in biomass energy production. Renew Energy 182:787–803

Chapter 11
Techno-economic Analysis of Biomass Waste Upgrading

O. O. Osemwegie, O. T. Lawal, B. S. Fakinle, O. T. Amoo, R. Arutselvan, P. Sivasubramanian, and V. Mageshwaran

Abstract Biomass upgrade technologies are evolving and improving to facilitate the transition from fossil-based to sustainable green and circular economies. Although these technologies have biorefinery, renewable energy, and climate change mitigation benefits, their applications, especially for biomass upgrades offer wide-ranging high-value products and end-use versatility. Therefore, understanding their comparative advantages over other renewable and non-renewable technologies is critical to their acceptance and prioritization in the rapid attainment of low emissions, global carbon reduction, and industrial circularization. Their renewability, benign nature, scalability, low cost, reusable side-stream wastes, and compatibility of these technologies to different biomass conversion systems, including other technologies made preferred choices over fossil and other renewable technologies. Their economic benefits, stability, and ecosystem stabilizing effects are unrivaled despite the influence of feedstock characteristics and climate variables on their performance efficiencies. are

O. O. Osemwegie (✉) · O. T. Amoo
Department of Food Science and Microbiology, Landmark University, Omu Aran, Kwara State, Nigeria
e-mail: osemwegie.omorefosa@lmu.edu.ng

O. T. Lawal
Department of Medical Biochemistry, Federal University of Technology, Akure, Ondo State, Nigeria

B. S. Fakinle
Department of Chemical Engineering, Landmark University, Omu Aran, Kwara State, Nigeria

R. Arutselvan
Regional Station, ICAR-Central Tuber Crops Research Institute (CTCRI), Dumuduma, Bhubaneswar, Odisha, India

P. Sivasubramanian (✉)
Department of Mechanical Engineering, PTR College of Engineering and Technology, Madurai – Tirumangalam Road, Madurai, Tamil Nadu, India
e-mail: sivaresearch948@gmail.com

O. T. Lawal · V. Mageshwaran (✉)
Microbial Technology Lab, ICAR-National Bureau of Agriculturally Important Microorganisms, Kushmaur, Mau, Uttar Pradesh, India
e-mail: mageshwaran.v@icar.org.in

A. Hind (ed.), *Biomass Waste: Processing, Valorization, Industrial Applications*,
https://doi.org/10.1007/978-981-95-0991-1_11

influenced by feedstock factors and climate variables. This affirms that careful selection of the technologies or their combinations in biomass conversion is necessary to optimize technological efficiency, waste minimization, conversion efficiency, and product quality. Policy support framework, infrastructure, storage facilities, impact on soil health, and feedstock restriction to agricultural active regions are some challenges and trade-offs limiting the widespread applications of biomass-upgrade technologies compared to fossil and other renewable technologies. Therefore, advancing biomass upgrade technologies with artificial intelligence, microbial biomass, feedstock densification, investments in green research, and supportive policies may offer promising opportunities that could be leveraged to increase benefits to humanity. Hence this study provides basic comparative insights that could facilitate the acceptance, continuing advancement, and prioritization of biomass upgrade technologies to solve multiple global challenges.

Keywords Biomass upgrade technology · Bioeconomy · Climate change mitigation · Environment · Fossil technologies

11.1 Introduction

Biomass upgrade technologies (pretreatment, fermentation, gasification, pyrolysis, etc.) are evolving to convert raw biomass or their mixture into high-value usable biorefinery products, feedstock for other valorization processes, and energy (Tetteh et al. 2024). As a subset of renewable technologies, they are often designed based on concerns for planet safety, carbon footprints, and energy security to meet the exponential growth in global energy demands, promote economic diversification and opportunities, mediate dependence on fossil fuels, drive green and circular economies, and minimize landfill waste (Banerjee 2023a, b; Ali et al. 2024). On the other hand, renewable technologies like solar, wind, hydrothermal, and geothermal drive renewable energies that have significantly low gas emissions and are complementary to biomass upgrade technologies in advancing a sustainable and resilient economy. Although both technologies play distinct roles as renewable energy sources, their comparative analysis and long-term implications for the environment, climate change, and sustenance of biological entities may require further empirical studies. Suffice it to say both renewable and biomass technologies have unique advantages and disadvantages, regarding bioeconomic (circular, blue, and green) prospects (Tshikovhi and Motaung 2023).

Biomass wastes are becoming increasingly recognized as viable sustainable feedstocks for producing renewable energy and diverse value-added biorefinery products in response to the current world's threats, validating their bioeconomic significance in transitioning from a fossil-dependent economy to circular and green dispensation (Sherwood 2020; Sun and Peng 2023; Venugopal 2022). Their abundance, source diversity, low molecular weight, low carbon density, and data on their enumerable benefits to humanity, particularly in mitigating environmental waste and climate change challenges, have made them attractive for upgrading, making them the cornerstone of bioeconomic advancement (Kalak 2023). The utility of biomass wastes to produce energy like biofuel (bioethanol, biogas, bio-oil, biodiesel, and biomethane), household heat, bioplastics, biochemicals (organic acids, resins, and furfural), and bio-based materials (composites, biopolymers, biochar, digestates, and composts) is replete in the body of literature (Babu et al. 2022; Ahmad et al. 2023; Ameh et al. 2024). Biomass waste utility also contributes to environmental management (pollution control, reduction in greenhouse gas emission, and waste reduction), industrial applications (adsorbent, construction materials, and furniture), electricity, and animal feed (Abdulla et al. 2024; Liu et al. 2024; Han and Chen 2024). Furthermore, biomass waste valorization is conceived as a circular system, in which the byproduct at the end of one derived product serves as feedstock for the next. This has given impetus to more studies focusing on finding newer derivable products and identifying technologies to optimize process turnaround time, system efficiency, quality of value-added products, and environmental safety.

The efficient utilization of biomass and its associated upgrades relies more on prevailing but suitably selected technologies and attendant economic viability (cost-effectiveness, market demands, revenue potential, incentives, and policies) than the inherent physicochemical nature of biomass wastes (Shaikh and Thenmozhi 2023). The past decade has witnessed a rise in biowaste-based technologies like thermochemical (pyrolysis, gasification, microwave, combustion), biochemical (fermentation, anaerobic digestion, enzymatic hydrolysis), and physicochemical (hydrothermal liquefaction, hydrolysis, supercritical fluid extraction) and machine learning-assisted conversions, and integrated biorefineries, contributing to circular economic goals (Mohammed et al. 2023; Omidkar et al. 2024). The widespread adoption and acceptance of these technologies are challenged by the perception of their relative economic competitiveness with traditional fossil fuel technologies (Chai et al. 2024). This notwithstanding, the technological upgrade of biomass wastes, efficiency improvement, and the economic viability of the process chain are interconnected and influenced by many factors. Feedstock supply logistics (cost, proximity, source, types), regulations (supportive and non-supportive policies), public perception or acceptance, and infrastructure are some of the notable factors that shape investment decisions and the overall sustainability of biomass upgrade projects (Okafor et al. 2022; Makepa et al. 2023).

Understanding the interplay between economic incentivization, the dynamic growth of biomass-based technologies, their viable applications, and upgrades is crucial to improving the potential of biomass techno-economic viability, technologies, product acceptance, and competitiveness with other renewable technologies (Makepa et al. 2023; Omidkar et al. 2024). This study aims to review, compare, and contribute to existing knowledge on techno-economic factors' interaction, which is crucial to fully realizing biomass as a sustainable resource prospect in bioeconomic development.

11.2 Biomass Waste Upgrading: Definition and Importance

The process of turning biomass residues, such as forestry byproducts, agricultural waste, and organic parts of municipal solid waste, into higher-value goods is known as biomass waste upgrading. Biofuels (such as bioethanol and biodiesel), biochemicals, bioplastics, and biogas are some examples of these goods. A mix of physical, chemical, and biological treatments are usually used in the upgrading process to improve the biomass material's quality, energy content, and usability (Banerjee, 2023a, b). The primary aim is to convert low-value or waste biomass into resources that are both ecologically friendly and economically valuable.

a. Environmental Benefits

The amount of garbage that would otherwise be burned or land-filled can be reduced through the use of biomass waste upgrading. By doing this, greenhouse gas emissions from burning and land-filling waste are also decreased, lessening the negative effects of waste disposal on the environment. Through the upgrading process, a renewable energy source is transformed into gas and biofuel that can take the place of fossil fuels. This lessens carbon dioxide emissions and aids in the fight against global warming. Certain upgraded biomass products, like biochar, can store carbon in soils for an extended period, enhancing soil health (Hansen et al. 2020).

b. Economic Benefits

Biomass upgrading improves resource efficiency and promotes the usefulness of waste resources by converting them into marketable items. This process contributes to the development of a circular economy. In addition, recycling decreasing the need for virgin raw resources, this technique adds economic value to what would otherwise be discarded (Clauser et al. 2021). By building facilities for upgrading biomass, rural communities with an abundance of biomass resources could create jobs. This could create new business opportunities and support regional economic growth for communities that are engaged in agriculture and forestry (Yana et al. 2022). Products made from improved biomass have a substantial market in a number of industries, including materials, energy, and chemicals. With growing consumer demand for clean and renewable products, biomass upgrading is expected to have greater commercial potential (Awasthi et al. 2020).

c. Energy Security

Biomass waste upgrading contributes to energy security by giving people a locally obtainable renewable energy source. This increases the resilience of energy systems and reduces reliance on foreign fossil fuels, especially in countries where biomass resources are abundant (Adeoye et al. 2023).

d. Sustainable Development

A variety of Sustainable Development Goals (SDGs), including appropriate production and use (SDG 12), accessible and renewable energy (SDG 7), and climate action (SDG 13), are aligned with upgrading biomass waste. By transforming waste into resources, this method encourages environmentally friendly choices that benefit the economy and the environment (Mohan et al. 2024).

11.3 Current State of Biomass Utilization

The upgrading of biomass waste uses a variety of resources (Cho et al. 2020). Corn stover, rice husks, and wheat straw are examples of agricultural leftovers that are plentiful crop byproducts. Among the logging wastes found in forests are branches, bark, and sawdust. Organic items such as food scraps, yard trimmings, and paper waste are contributed by municipal solid waste. Important feedstock is also provided by industrial sources, such as waste from the production of bioenergy, pulp and paper, and food processing. The feedstock basis for upgrading operations is further diversified by the addition of algae and animal dung as developing biomass sources (Haile et al. 2021).

Utilizing biomass is becoming more and more common in the energy, industrial, and agricultural sectors. Climate regulations and targets for renewable energy have led to the widespread use of biomass for the production of bioenergy, including heat, electricity, and biofuels (Yana et al. 2022). Biomass is used in industry to create biochemicals and bioplastics, which provide environmentally friendly substitutes for petrochemicals. More and more resources are being used to produce biochar, soil supplements, and biogas from forestry and agricultural wastes (Yrjälä et al. 2022). Nevertheless, despite these developments, the use of biomass remains underdeveloped in many areas, with great room for expansion, especially in poorer nations where biomass resources are plentiful but underutilized (Okafor et al. 2022).

a. Types of Biomass Feedstock

Many different types of organic materials are used as biomass feedstocks. Crop leftovers, such as leaves, stalks, and shells, are abundant and frequently go unused. Forestry residues are byproducts of logging and wood processing, such as branches, tree tops, and sawdust (Raclavská et al. 2022). Organic matter high in biodegradable material, such as food scraps, paper, and yard trimmings, is provided by municipal solid waste (Abdel-Shafy et al. 2018). Additional feedstock is provided by industrial waste from industries such as food processing and paper production, and the energy potential of animal manure and wastewater sludge is being more and more explored (Odejobi et al. 2024). Due to its high lipid content and fast growth, algae is becoming a more popular biomass source, especially for the generation of biofuels.

b. Feedstock Collection and Transportation

Transporting and collecting feedstock are essential phases in the upgrading of biomass. In order to harvest and process materials like crop residues, wood chips, or organic waste, collection sometimes entails obtaining biomass from industrial sites, forests, or agricultural areas (Spinelli et al. 2019) Specialized equipment is frequently used for this operation. The biomass is subsequently transported to processing facilities, where close proximity is essential for lowering expenses and energy consumption. Especially for heavy or perishable commodities, efficient logistics are crucial to reducing deterioration and energy loss during transportation. Densification innovations, such pelletizing, can lower volume and improve transport efficiency, increasing the economic viability of biomass upgrading (de Jong et al. 2017).

c. Pre-treatment process of feedstock

Biomass feedstock must undergo pretreatment procedures in order to be more efficient during later conversion processes, such as the synthesis of biofuel, biogas, or biochemicals. The objective of these procedures is to modify the biomass's physical and chemical characteristics in order to enhance its reactivity, digestibility, and accessibility (Ethaib et al. 2020). Among the primary pretreatment techniques are:

i. Physical Pretreatment

Pretreatment in physical terms involves mechanically reducing the particle size of biomass through milling, grinding, or chipping in order to increase its surface area and make it more amenable to chemical or enzymatic treatments. Densification techniques such as pelletizing or briquetting raise the bulk density of biomass, making handling, storing, and transporting it easier. High-pressure steam is applied to biomass, and it is quickly depressurized to break it down and make it more pliable for processing (Bajwa et al. 2018).

ii. Chemical Pretreatment

Strong acids, such as sulfuric acid, are used to hydrolyze hemicellulose and break down lignin, increasing the digestibility of cellulose for biofuel production (Hoang et al. 2021). Alkaline substances like sodium hydroxide or ammonia are used to remove lignin and increase porosity, making the cellulose more accessible to enzymes (Novakovic et al. 2021). Organic solvents, such as ethanol or ionic liquids, are used to dissolve lignin and hemicellulose, enhancing the enzymatic digestibility of the cellulose fraction (Abushammala and Mao 2020).

iii. Biological Pretreatment

The biological pretreatment involves applying bacteria or fungi that are known to naturally break down cellulose, hemicellulose, and lignin (Sharma et al. 2019). For instance, white-rot fungi are good at degrading lignin, which enhances cellulose's enzymatic hydrolysis. The complex carbohydrates in biomass are broken down into simpler sugars by certain enzymes, such as cellulases and hemicellulases, so that they can be fermented into biofuels (Laca et al. 2019).

iv. Physico-Chemical Pretreatment

Liquid ammonia is applied to biomass under high pressure and quickly depressurized. Through the process of decrystallization, the biomass becomes more porous and more receptive to enzymatic hydrolysis (Latif et al. 2018). Like a steam explosion, but with supercritical carbon dioxide. By upsetting the lignocellulosic structure, the procedure improves cellulose accessibility. involves subjecting biomass to high pressure and temperature oxygen or air treatment, which breaks down lignin and partially hydrolyzes hemicellulose.

v. Thermochemical Pretreatment

In order to convert biomass into syngas, charcoal, and bio-oil, it must be heated without oxygen. The biomass undergoes a chemical structural change during this process, which facilitates upgrading. a gentle method of pyrolysis in which biomass is heated to temperatures between 200 and 300 °C in order to create a substance that is simpler to crush and carry and is more energy-dense (Julie Chandra et al. 2023).

d. Quality and Consistency of Feedstock

Reliability and feedstock quality are essential for effective biomass upgrading. Reliability in conversion operations is ensured by a homogeneous composition and low levels of pollutants in high-quality feedstock. Changes in the moisture content, particle size, or chemical composition of feedstocks can cause variations in their quality, which can result in inefficiencies, higher expenses, and uneven product outputs (Li et al. 2016). Maintaining a consistent feedstock improves energy and material consumption, boosts overall upgrading process stability, and optimizes process parameters. Maintaining feedstock quality and ensuring the best possible performance of biomass conversion systems require routine monitoring and quality control procedures (Davis et al. 2018).

11.4 Comparison with Other Renewable Technologies

One of the major attractions of biomass waste upgrades to higher value-added products and optimal utilities through technological innovations lies in their comparative advantages or unique techno-economic feasibility of other renewable technologies (Abelha and Kiel 2020; Kamaruzaman et al. 2023). Additionally, the environmental feasibility of biomass waste upgrade technologies, renewability, and their circular value chain efficiency, involving cascading side-stream waste reuse as a feedstock for the next process phase, are crucial advantages to other technologies (Omidkar et al. 2024). Side-stream management is becoming a critical consideration in circular economies, unlocking underutilized side wastes' economic potential and leveraging technologies for their upgrade (Kamaruzaman et al. 2023; Hossain et al. 2024). Suffice it to say comparing biomass waste upgrade technologies with broader renewable technologies is essential to understanding their unique characteristics, applications, environmental impacts, and economic feasibility.

While both technologies aim to reduce greenhouse gas emissions, carbon footprints, and reliance on fossil fuels, driving a sustainable energy transition or paradigm to a green and circular bioeconomic future, they are techno-economically different. Renewable technologies offer limited product applications and have higher efficiencies than biomass upgrade technologies (Østergaard et al. 2012). The technology efficiency of renewable technologies is not tied to feedstock quality, supply, availability, and chemical characteristics. According to dos Santos and Pantoja (2023) and Kataya et al. (2023a, b), microorganisms' involvement as microbial fuel cells, consortia, catalysts or decomposers, fermenters, and biogas-producers in the life cycle of biomass upgrade technologies and requirements for their optimization in the life cycle of biomass waste upgrade technologies reflects technical complexities non-characteristic of other renewable energy technologies. Even though this might suggest that biomass waste upgrade technologies are more sensitive to changes in physical factors, they offer a wider range of non-energy products like metabolites, bioactive compounds, prebiotics, side-stream waste, vitamins, organic chemicals, and polysaccharides (Kieliszek et al. 2020). The selection of pretreatment techniques to break the molecular structure of biomass materials impacts their upgrade downstream processes and byproduct quality, requiring a balance between energy input and chemical usage to ensure a sustainable life cycle (Shah et al. 2022). They are also perceived to cause more significant environmental impacts than renewable technologies. Moreover, renewable technologies require higher initial investments but afford long-term savings and incentives compared to biomass upgrade technologies, which vary widely in cost but have the prospect of economic viability with the right market conditions, policy support, and subsidies (Hassan et al. 2024). Even though technological innovations in biomass upgrades contribute to waste management, biorefinery, energy source expansion, and soil management efforts, they remain a subsect of renewable energy technologies, which are less widespread, technically matured, and incentivized. Comparisons of the technologies involved in renewable energy and biomass upgrades are adapted from the body of scientific literature and summarized in Table 11.1.

Renewable technologies hypothetically existed before the emergence of biomass as a suitable resource for generating renewable energies and value-added non-energy products. These technologies, characterized by cost effectiveness, strong government support, and low environmental impacts, are deployed globally and less dependent on regional, social, and political factors than biomass technologies. Renewable technologies could impact local ecosystems and mildly disrupt the biodiversity balance (Gasparatos et al. 2017). The construction of dams for hydropower generation could drastically alter river flow and ecosystems, impacting biodiversity by causing a rapid decline in fishes that rely on migration for spawning, and flooding riverine communities, destroying habitats. Also, a large expanse of solar farms may lead to species displacement due to habitat loss and fragmentation (Ullah and Lin 2024). On the contrary, the growing advancements in biomass upgrade technologies, particularly those relying on plant-based biomass feedstocks, could cause significant land-use changes, leading to deforestation, habitat loss, soil degradation, expansion in energy crop cultivation, competition for land with food production, and water scarcity due to high irrigation demands (Tshikovhi and Motaung 2023). Therefore, a careful selection of biomass upgrade technologies or their combinations is necessary and requires a thorough environmental feasibility assessment, planning, site selection, and management to reduce their impacts.

Table 11.1 A summary of the comparison of biomass upgrade and alternative renewable technologies

S/n	Feature	Biomass waste technologies	Alternative renewable technologies	Reference(s)
1	Definition	Technologies that convert biomass waste into usable energy and biorefinery products, while enhancing its value	Energy generation methods using naturally replenished resources like wind, solar, and hydro	Tetteh et al. (2024), Hassan et al. (2024)
2	Type of technologies	Thermochemical (combustion, pyrolysis, gasification, hydrothermal liquefaction); Biological or biochemical (enzymatic hydrolysis, decomposition, fermentation, anaerobic digestion); Chemical (Transesterification, hydrolysis, catalysis); Physicochemical (mechanical extraction, superficial fluid extraction); Integrated processes (biorefineries, hybrid technologies)	Solar (photovoltaic, solar thermal); Wind (offshore and onshore wind farms); Hydropower (dam, run-of-the-river system, pumped storage); Geothermal (power plants, heat pumps); Ocean energy (tidal, wave, ocean thermal energy conversion); Hydrogen)	Sarathe et al. (2022), Omer (2021)
3	Source	Primarily organic waste materials like agricultural residues, municipal solid waste, and forestry by-products	Diverse sources like sunlight, wind, water, ocean heat, and geothermal heat	Hassan et al. (2024), Choi et al. (2024)
4	Derivatives	Through thermal, biological, chemical, and physical processes biomass is converted into biofuels, biogas, electricity, and other non-energy-related biorefinery and non-biorefinery products	Generates electricity, heat, or fuels directly from renewable natural sources and produces no other usable products	Sayed et al. (2020), Ali et al. (2024)

(continued)

Table 11.1 (continued)

S/n	Feature	Biomass waste technologies	Alternative renewable technologies	Reference(s)
5	Environmental impacts	Reduces waste and landfills, contributes to greenhouse gas emissions, and may still produce pollutants (particulate matter, volatile organic compounds, carbon monoxide, etc.) depending on the technology used	Generally, they have low emissions, significantly reduce fossil fuel reliance, mitigate climate change, and reduce air and water pollution. However, the improper disposal of solar panels and wind turbines may constitute environmental nuisance	Antar et al. (2021)
6	Economic viability	Technologies may be cost-effective because they utilize waste materials, but technology development and operational costs vary with regional factors, life cycle, and type of biomass, offer more opportunities to farmers and rural communities	Technologies have low operational and environmental costs, long-term financial benefits	Funk et al. (2020), Park et al. (2024)
7	Scalability	Technologies can be implemented at various scales, from small community systems to large industrial plants	Highly scalable, with technologies ranging from small solar panels to large wind farms and hydroelectric plants	Tshikovhi and Motaung (2023)
8	Technical maturity	Mixed levels of technical maturity in which some technologies like advanced gasification, pyrolysis, microwave technique, and machine-assisted technology still have logistical challenges, while others are well-established (e.g., anaerobic digestion)	Many renewable technologies are mature widely deployed, and more flexible, particularly solar and wind, and more commercially viable with higher operational efficiency	Chu et al. (2023)
9	Energy storage	Generally, less focus on energy storage because biomass has inherent energy storage; products like biogas can be stored but may require additional processing	Technologies like pumped hydro and battery storage are integral for managing energy supply and demand	Omer (2021)

Overall, both technologies contribute to renewable energy goals in different parts of the world even though biomass upgrade technologies are more operationally feasible in locations that can guarantee a sustainable supply. More investigations into their sustainability, product feasibility, ecosystem implications, technicality, and policy-support frameworks of these technologies can significantly improve the understanding of comparative distinctions.

11.5 Comparison with Conventional Fossil Fuel Technologies

Unlike location-dependent biomass upgrade technologies that recently gained significant interest as a cleaner, renewable, and sustainable alternative resource for energy, usable organic products, and chemicals, conventional fossil fuel technologies (coal combustion, nuclear power plant, natural gas-fired power plant, and oil-fired power plant) are more dominant and widespread technologies leveraged to drive industrial economic development (David 1982). Fossil fuel technologies are high energy density technologies historically adapted to increase power output and account for approximately 80% of global energy consumption in manufacturing, transportation, industries, electricity generation, and technological advancement. However, fossil fuel technologies deployed extraction (mining, drilling, and enhanced oil-recovery technique), processing (crushing, gasification, coking, cracking, fractionating, liquefaction, reforming, and treating), transportation (pipeline, tankers, trucks, rails, and ships), and storage (reservoirs, floating roof tanks, tankers, covered storage) of fossil materials have raised legitimate macro-environmental safety concerns worldwide (Alipour et al. 2024). This encouraged the emergence of energy tax credits and policies prioritizing benign renewable technologies and supporting energy transitions from reliance on fossil fuels (Harlan and Baka 2024).

Compared to biomass upgrade technologies, fossil fuel technologies have many disadvantages besides downside operational risks, including accidents, contributing to global warming, exacerbating climate change challenges, environmental pollution, and release of greenhouse gases, and toxic particles like heavy metals and sulfur dioxide (Singh et al. 2024). Fossil materials are considered natural energy storage resources limited to extractive communities, readily depleted, and not easily replenished fast enough to support their exploitation forever by humans (Gandomi et al. 2024; Ritchie and Rosado 2017). Additionally, the cost of managing their consequences (acid rains, climate change, global warming, environmental degradation, wildlife displacement, and biodiversity losses), the expensive implications of reliance on fossil fuels and associated technologies, unstable global prices due to the battle for fossil fuel market hegemony, and political uncertainty worsen the financial reserves of many countries, especially developing countries and those with limited fossil reserves (Alipour et al. 2024; Singh et al. 2024). These notwithstanding, more fossil fuel studies focusing on their technological feasibility (Table 11.2), impacts (socioeconomic, international cooperation, environmental burden, policy frameworks, political stability, and energy transition), and connectivity to climate change abound in the literature more than those for biomass upgrade technology (Chiquetto et al. 2022; Chen and Ding 2023; Zhang and Hao 2023).

Despite scanty data on the relationship between the geophysical stability of the earth and fossil fuel drilling and mining technologies, Rajabi and Moradi (2023) and Gandomi et al. (2024) have reported the implications of persistent mining and drilling technologies leveraged to extract fossil fuel from various reservoirs on the geophysical integrity and stability of the earth. Unlike the biomass upgrade technologies, some fossil fuel technologies (drilling and mining) and reservoir depletion have been linked to seismic activities capable of inducing earthquakes. The technologies are also associated with other geologic concerns like subsurface changes, geomechanical stress, ground subsidence, mine cavity collapse, tectonic interactions, and ground surface cracking (Cloetingh et al. 2023; Lall and Vishal 2024). This suggests that fossil fuel technologies require a combination of strategies, including regulatory measures based on policy, monitoring, reporting, and environmental impact assessment frameworks, proper monitoring and management to minimize the associated geophysical safety and destabilization risks, which add to their operational costs (Prasad and Ansari 2023). Furthermore, fossil fuel technologies may be made geologically and environmentally friendlier when used to extract crude oil from the earth if supported by established risk management institutions, stakeholders' engagements, circular strategies, and other advanced technologies, including seismic monitoring systems, imaging technologies, and machine learning.

Fossil fuels' extraction-to-end-users is an energy-intensive process with significant environmental impact, and technological and financial implications compared to biomass upgrade and renewable technologies. All these distinctions give impetus to biomass upgrade technologies, offering renewable, carbon–neutral alternatives to fossil fuels. Biomass upgrade technologies leveraged primarily on organic materials, contributing to waste management, energy diversification, and circular and green economies. Despite now coming under different limitations from carbon pricing, subsidy reforms, United Nations Sustainable Development Goals, green bonds, International Alliances and Intergovernmental Organizations (2011 Boon Challenge, Artic Council Agreement, G20 Leaders Statement on Climate Change, International Renewable Energy Agency, and United Nation Framework on Climate Change), and International treaties/Charter (Paris Agreement of 2015; Montreal Protocol of 1987; Kyoto Protocol of 1997; Agenda 2030), fossil fuel technologies still have a far-reaching dominance on the global economy than biomass upgrade technologies. Furthermore, the scale of maturity and popularity of fossil fuel technologies compared to biomass upgrade technologies validate their current well-established infrastructural framework, economies of scale, and cost-effectiveness. Notwithstanding, the outcome of technological innovations on the scale-up, sustainability, competitive efficiencies, scalability, and cost reduction of biomass upgrade technologies attract policy and incentive supports, presenting and enhancing biomass as promising, sustainable, and renewable energy feedstock. Suffice it to say continuing investments in biomass technology innovations, supportive policies, carbon pricing,

Table 11.2 Compares the technological dynamics of some fossil fuel extraction, processing, storage, and transportation

Fossil fuel	Coal	Oil	Natural gas	Reference(s)
Extraction	Surface mining for shallow deposits e.g. strip, open-pit, and mountaintop technologies Underground mining for deep deposits e.g. room-and-pillar, and longwall mining technologies	Conventional drilling technology like onshore and offshore. Enhanced oil recovery (EOR) like water flooding, gas injection, and thermal recovery techniques	Conventional drilling (onshore and offshore techniques); Hydraulic fracturing (Fracking) like high-pressure fluid injection; Horizontal drilling like wells drilled horizontally	Strand (2007), Achakulwisut and Erickson (2021), Geow et al. (2021)
Processing	Crushing and screening to remove impurities; Washing to remove ash and sulfur; Coking to produce coke for steel making; Gasification to convert coal to syngas; and Thermal drying to reduce moisture content	Separation by distillation into components; Cracking e.g. thermal and catalytic; Reforming to improve fuel quality and into various products; Treating and Blending to remove unwanted elements and blend fractions	Separation of hydrocarbons and impurities; Dehydration to extrude water vapor; Sweetening to remove hydrogen sulfide and carbon dioxide; Fractionation separates natural gas liquids (NGLs); Liquefaction converts natural gas to LNG	Graedel and Howard-Grenville (2005), Gandomi et al. (2024)
Storage	Open piles are cost-effective, weather-exposed storage, exposed; Covered storage is done in sheds or enclosures; Silos and Bins are storage better protected from the weather; Domes storage in enclosed structures efficient for dust control	Tanks erected above-ground or underground in steel or concrete compartments; Reservoirs in natural or artificial underground; Floating roof tanks reduce evaporation; Oil Tankers serve dual purpose of transport and storage	Underground Storage includes depleted reservoirs, aquifers, and salt caverns; Liquified Natural Gas (LNG) storage is done in cryogenic tanks; Compressed Natural Gas (CNG) storage is usually in high-pressure cylinders or tanks	Holloway (2001)

(continued)

Table 11.2 (continued)

Fossil fuel	Coal	Oil	Natural gas	Reference(s)
Transportation	Trains/railroads are the primary mode of transport from mines to power plants or ports; Trucks and conveyor belts are used for short distances near mines; Barges and ships used for overseas transport	Pipelines are the most common method for transporting oil over land – Tankers are used for international transport of crude oil and refined products – Trains and trucks used for shorter distances from refineries to distribution terminals or retail outlets	Pipelines are the primary method for transporting natural gas over land; LNG is transported by ship for international trade; Compressed natural gas (CNG) transported by truck or smaller tanks for short distances	Govorushko (2013)
Environment	Dust, particularly in enclosed spaces; Water sprays and chemical suppressants dispersed by runoff to pollute water	Leakage from drilling and secondary containment; Release of volatile compounds; Accidents leading to spills, combustions, etc.	Leakage and methane emissions; Accident due to high temperature and high-pressure and cryogenic storage	Pérez-Martínez et al. (2020)

and building appropriate infrastructures for large-scale applications are crucial in accelerating the transition to a sustainable energy future, high energy demands, and circularization of the global economy.

11.6 Economic and Environmental Trade-Offs of Biomass Technologies

Having provided in the previous sections of this chapter palpable insights that present biomass upgrade technologies as hypothetically perfect solutions to global climate change mitigation, carbon footprint reduction, and attainment of renewable energy strategies, understanding their economic or techno-economic, and environmental trade-offs is key to their commercial feasibility, compatibility with other energy systems, potential for improvement and life cycle (Kosenius and Ollikainen 2013; Vuppaladadiyam et al. 2023). Unlike biomass feedstocks, research efforts have hypothetically sidestepped investigating or making a comprehensive review that compares conventional, advanced, and emerging approaches from an economic perspective.

Economically, most technologies deployed for upgrading biomass to usable products, including energy, typically require substantial but varied initial investments and infrastructures. For instance, gasification, pyrolysis, anaerobic digestion, and torrefaction may have varying capital, and operational costs that discourage their application by smaller producers (Tshikovhi and Motaung 2023). Quality and mix coupled with proximity to the supply source of biomass feedstocks may increase investment, operation, and processing. Moreover, processing complexities characterizing these technologies suggest the need for skilled labor, high energy, and an extensive life cycle. The cost-efficiency and techno-economic feasibility of biomass upgrades may be disparaged by poor road, storage, availability, and energy infrastructures, creating logistic bottlenecks (Huang et al. 2023). A fascinating economic benefit of biomass upgrade technologies, particularly biochemical and thermochemical technologies, lies in their potential for product diversity. This creates multiple revenue streams by offering cascading opportunities for diverse end-use and reusable byproducts that could act as feedstock for the next processing phase or other independent biorefinery/industrial processes, reflecting a competitive process efficiency compared to fossil fuel technologies (Amenaghawon et al. 2021; Tshikovhi and Motaung 2023). Even though the market viability of the products from biomass upgrade technologies is still inadvertently developing with lower economies of scale. This may deprecate their products' popularity, acceptability, and revenue feasibility. It is pertinent to recognize that like other renewable and non-renewable energy technologies, biomass upgrade technologies create jobs in multiple sectors of an economy, offer revenue opportunities in carbon trading, and favor the rural farmers in agricultural regions concerning making additional income from generating the biomass feedstocks for these technologies.

Biomass upgrade technologies are not geo-physically disruptive, and their applications contribute intangibly to environmental degradation, and lower carbon footprint and emissions compared to non-renewable fossil fuel technologies (Jiang et al. 2024). These biomass upgrade technologies are hypothetically considered carbon–neutral, and further analysis of their environmental-profitability trade-offs is required to understand the net cost impact. Leveraging these technologies may contribute to particulate matter, soil degradation, and air quality issues with far lesser intensity than fossil fuel technologies (Tomlin 2021; Zhao et al. 2024). Similarly, by encouraging biomass technologies through policy support, carbon pricing, subsidy removal on fossil fuels, and research investments strategic opportunities are created for waste reduction and economic circularization (Nunes 2023). It is also hypothetical to assume that road and sea transportation of unenclosed biomass feedstocks from the source point to a distant end-use point may contribute to a stream of uncleanliness. Dependence on energy crops and forest residues on a large scale to service biomass upgrade technologies partially counterbalances regional efforts to achieve the United Nations Sustainable Development Goals in 2030. These invariably impact land use policies, conflict with food crop availability, constrain water supply for residential use due to irrigation demands, and encourage deforestation, posing a threat to biodiversity. Trimble et al. (1984) remarked on the renewability and sensitivity of biomass feedstock to environmental factors, which affects their carbon footprint and the efficiency of their conversion process. Therefore, prioritizing sustainable biomass upgrade technologies, integrating microbial biomass into the technologies, and supporting research in greener biomass conversion technologies, substantially lower lifecycle emissions, especially in areas with high biomass energy usage, environmental contamination from different wastes, and carbon footprint (Kieliszek et al. 2020). They are also key to balancing economic, ecological, and environmental trade-off dynamics for improved benign profiles (Vadenbo et al. 2017; Humpenöder et al. 2018).

Although individual biomass technology has unique economic and environmental trade-offs, offering benefits and challenges to humanity, these technologies may be combined for performance efficiency and preferentially designed to synchronize with the reality of local conditions, sustainable practices, extant supportive policies, and technological advancements. This aids in balancing trade-offs to maximize biomass upgrade technologies while reducing their collective and individual disadvantages, which may include unpleasant odour from improperly managed anaerobic digesters or biomass feedstock storages.

11.7 Environmental Impact and Sustainability

a. Life Cycle Assessment (LCA)

Life Cycle Assessment (LCA) is a comprehensive methodology used to evaluate the environmental impacts associated with all stages of a product's life, from raw material extraction through to processing, manufacturing, distribution, use, and disposal (Finnveden et al. 2009). In the context of biomass waste upgrading, LCA plays a crucial role in understanding and mitigating the environmental impacts, thereby promoting sustainability. Biomass waste upgrading refers to the processes through which organic waste materials are converted into more valuable products, such as biofuels, bioenergy, and biochemicals. The environmental impact of these processes needs careful evaluation to ensure that they contribute to, rather than detract from, overall sustainability goals (Cherubini and Strømman 2011).

The primary objective of biomass waste upgrading is to enhance sustainability by converting waste into valuable products. However, the process itself can have significant environmental impacts, which must be carefully managed. LCA helps identify these impacts and provides a framework for optimizing processes to reduce negative effects (Cherubini and Strømman 2011). For instance, biomass conversion to biofuels might reduce reliance on fossil fuels but could also lead to increased water usage, land use change, or greenhouse gas emissions depending on the feedstock and technology used (Finnveden et al. 2009). LCA can evaluate the environmental trade-offs associated with different types of biomass, such as agricultural residues, forestry residues, or municipal solid waste. Another consideration is the energy efficiency of the upgrading process. LCA can help determine the overall energy balance of the system, comparing the energy input required to process the biomass against the energy output in the form of biofuels or bioenergy. Processes with a positive energy balance contribute to sustainability, whereas those with a negative balance may require improvements in technology or feedstock selection (Sven Bernesson 2006).

b. Carbon Footprint Reduction

The reduction of carbon footprint is a central focus in the realm of biomass waste upgrading, which encompasses a variety of processes aimed at converting organic waste materials into valuable products like biofuels, biochar, and other biochemicals (Holm-Nielsen et al. 2009). These processes not only provide a sustainable alternative to fossil fuels but also play a crucial role in mitigating climate change by reducing greenhouse gas (GHG) emissions.

Reducing the carbon footprint has become a critical objective in the field of biomass waste upgrading, a sector that involves transforming organic waste materials into high-value products such as biofuels, biochar, and biochemicals (Rogers and Brammer 2012). These conversion processes present a sustainable alternative to traditional fossil fuels, which are major contributors to greenhouse gas (GHG) emissions and climate change. By focusing on the utilization of biomass waste, the upgrading processes contribute significantly to environmental sustainability, both by reducing emissions and by promoting the circular economy (Smith et al. 2016).

c. Waste Reduction and Resource Recovery

Biomass waste upgrading effectively addresses waste reduction by converting organic waste into valuable products, reducing the environmental impact associated with traditional waste disposal. Unlike fossil fuels, which contribute to significant waste and environmental degradation through extraction and combustion—resulting in landscape disruption, water contamination, and greenhouse gas emissions—biomass upgrading repurposes waste materials into energy and biochar (Tanmay et al. 2022). This not only alleviates landfill burden but also mitigates issues like methane emissions from landfills and air pollution from burning fossil fuels (Gahleitner 2013).

Through thermochemical conversion, biomass waste such as wood and agricultural residues can be transformed into biofuels like ethanol and biodiesel, as well as biochar, which enhances soil health and sequesters carbon (Rodionova et al. 2017). This technology reduces reliance on finite resources, lowers greenhouse gas emissions, and supports sustainability by converting waste into valuable products. The effectiveness of thermochemical conversion in managing organic waste highlights its role in advancing environmental sustainability and reducing the overall impact of waste management (Demirbas 2008).

d. Comparison with Fossil Fuels

When comparing the environmental impacts of biomass waste upgrading with those of fossil fuel extraction and use, the advantages of biomass are clear. Fossil fuels are associated with significant environmental degradation at every stage of their lifecycle—from extraction, which can cause deforestation, habitat destruction, and water contamination, to combustion, which is the largest contributor to anthropogenic GHG emissions and air pollution (Singh et al. 2014). When managed properly, biomass serves as a renewable alternative that significantly reduces greenhouse gas emissions. Although biomass combustion releases CO_2, this carbon was recently captured from the atmosphere during the plant's growth, unlike fossil fuels, which release ancient, stored carbon, increasing atmospheric CO_2. Additionally, using CO_2 from flue gases to produce methanol has garnered attention, as biomethanol derived from high-cellulose biomass can cut CO_2 and nitrogen oxide emissions by up to 95% and eliminate sulfur oxide compared to fossil fuels (Boretti 2013). Biomass waste upgrading not only lowers GHG emissions but also mitigates landfill use and associated pollution issues, such as leachate and methane emissions. Producing biochar from biomass helps with soil remediation and carbon sequestration, offering long-term environmental benefits beyond those of fossil fuels (Farghali et al. 2022).

11.8 Financial Instruments and Investment Opportunities

a. Funding Sources and Financial Instruments

Economic viability is a critical factor for biofuels to establish themselves as a sustainable energy source. While biofuels are seen to decrease GHG emissions and reduce reliance on fossil fuels, their economic sustainability is a complex matter shaped by various elements such as feedstock accessibility, production expenses, and government policies (Benteng et al. 2023). Government policies and regulations also have a considerable impact on the economic sustainability of biofuels. Policy measures such as subsidies and mandates for biofuels can aid in establishing markets and lowering production costs, but they can also lead to economic imbalances and limit incentives for innovation (Rocha-Meneses et al. 2023). Funding for biomass waste upgrading projects comes from diverse sources, each with its own set of criteria, focus areas, and risk appetites.

b. Investment Risk and Return Analysis

Investment risk and return analysis is crucial for making informed financial decisions. Understanding the types of risks involved, the methods for analyzing returns, and the inherent trade-offs between risk and reward allows investors to build portfolios that align with their financial objectives and risk tolerance. Through careful analysis and strategic planning, investors can enhance their chances of achieving their desired financial outcomes while managing the risks inherent in the investment process (Chum and Overend 2001).

Investment risk and return analysis is essential for evaluating the financial viability of biomass waste upgrading process, which convert organic waste into valuable products like biofuels and biochar, offer environmental and economic benefits but also come with significant risks (Colantoni et al. 2021). Market volatility is a key risk, as biofuel prices can fluctuate due to global energy markets and regulatory changes. Technological risks also exist, with emerging technologies potentially leading to cost overruns or project failures if they underperform. On the return side, successful projects can yield high financial rewards, especially in strong renewable energy markets. Government incentives and the generation of carbon credits from practices like biochar production can further enhance profitability. Thorough due diligence, including market analysis and financial modeling, is crucial for balancing these risks and maximizing returns in biomass waste upgrading investments (Shabangu et al. 2014).

c. Public–Private Partnerships

Public–Private Partnerships (PPPs) are vital in advancing biomass waste upgrading by merging public resources and private innovation. These collaborations address financial and technological challenges by combining public sector support, such as policy frameworks and subsidies, with the private sector's expertise in technology and market access (Fantozzi et al. 2014). Governments can reduce risks and attract private investment through guarantees and affordable financing, while private partners bring efficiency and innovation to project execution. PPPs not only enhance the profitability and environmental impact of biomass projects but also contribute to local economic development by creating jobs and generating renewable energy, ultimately promoting sustainable waste management practices (Vassileva and Antoaneta 2022).

d. Venture Capital and Private Equity

Venture capital (VC) and private equity (PE) are crucial for funding biomass waste upgrading projects, which require substantial investment for development and commercialization (Hsu 2007). VC is particularly beneficial for early-stage companies, offering both capital and strategic support to help startups scale innovative technologies like biofuels and biochemicals. In contrast, PE focuses on mature companies, providing larger investments and hands-on management to expand operations and market reach. Both VC and PE play vital roles in accelerating the growth of biomass waste upgrading, bringing sustainable solutions to market and supporting broader environmental and economic goals Hsu 2007).

11.9 Stakeholder Analysis

In technocrat-economic analysis, stakeholders are those who have the power to effect or have an impact on the project (Zohir et al. 2016). Initiatives to upgrade biomass waste must be implemented successfully, which requires an understanding of the roles, motivations, and influence of various stakeholders (Robinson 2021). These fall into four categories. identification of stakeholders, study of their roles and responsibilities, strategy for engaging stakeholders, and evaluation and mitigation of their effects (Sterling et al. 2017).

a. Identification of Stakeholders

The first step in stakeholder analysis is identifying all relevant stakeholders. These can be categorized into primary, secondary, and tertiary stakeholders.

Farmers, agricultural companies, and forestry enterprises that produce biomass waste are the main stakeholders in the biomass industry. Since they supply the raw materials required for the upgrading process, their involvement is essential (Panoutsou and Singh 2020). Additionally, significant are biomass processors, which are businesses and establishments that transform waste biomass into goods with additional value, including biofuels, biochemicals, or bio-based materials (Farkaš 2022). Finally, the major stakeholders are the end consumers who belong to the industries that use the upgraded biomass products in their supply chains or their operations (Mola-Yudego et al. 2024).

Government organizations that serve as policymakers, regulators, and industry experts have an impact on incentives, subsidies, and industry standards. They are essential in forming the regulatory framework that supports the upgrading of biomass (Singh et al. 2021). Furthermore, banks, venture capitalists, and other financial organizations supply the funding required for biomass upgrading initiatives. Their backing is necessary for funding and long-term financial viability (Falcon and Sica 2019). The scientific and technical know-how required for innovation is provided by research and development organizations such as universities, research institutes, and technology developers, which are engaged in expanding biomass conversion technologies (Assunção et al. 2021).

Non-governmental organizations that advocate for social and environmental causes and are worried about the project's effects on the community and sustainability are considered tertiary stakeholders in biomass upgrading initiatives (Koasidis et al. 2022). Furthermore, the projects' social and environmental effects might have an impact on nearby communities that are close to biomass production and processing facilities (Terrapon-Pfaff et al. 2014). Customers who buy finished goods derived from upgraded biomass also can affect market acceptance and demand through their purchasing choices (Russo et al. 2019).

b. Interest and Influence Analysis

The next stage after identifying the stakeholders is to examine their impact and areas of interest.

The techno-economic analysis of biomass upgrading requires a critical step called mapping stakeholder interests. This step helps identify and rank the concerns of different stakeholders, such as primary stakeholders like investors who are concerned with financial returns, project feasibility, and risk management; technology developers who are focused on technological innovation, scalability, and intellectual property; and biomass producers who are worried about feedstock supply, quality, and pricing (Rambaree et al. 2021). Important interests also lie with secondary stakeholders, such as government agencies that are concerned with policy compliance, the environment, and public benefits, and industry groups that are concentrated on market trends, regulatory frameworks, and industry standards. Furthermore, tertiary stakeholders—which include non-governmental organizations focused on social impact, environmental sustainability, and community engagement (Kwatra et al. 2021)—as well as local communities impacted by social and environmental impacts like job creation, water and air quality, and consumer demand—must be considered for a thorough analysis (Wiedmann and Lenzen 2018).

Stakeholders are driven by three main interests. They want to make money and be profitable, focusing on economic benefits like cost-efficiency and market potential. They also care about the environment, prioritizing sustainability, reducing their carbon footprint, and effective waste management (Lu et al. 2023). Additionally, they emphasize social benefits, including job creation, health and safety, and community development, to create a positive impact on society (Lumpkin and Bacq 2019). Furthermore, Government agencies, biomass processors, and financial institutions wield significant influence and interest, impacting project outcomes (Ramirez and Faaij 2018), while Regulatory bodies have an indirect influence, Local communities, and NGOs hold vested interests, and General consumers and distant stakeholders contribute minimally to market dynamics, highlighting the need for careful assessment of stakeholder influence and interests in biomass upgrading projects (Robinson 2021).

c. Stakeholder Engagement Strategies

Engaging stakeholders effectively is vital for project success. Effective stakeholder engagement involves Consultation and Communication through regular meetings, workshops, and forums to gather input and share information, as well as maintaining transparent communication channels to keep stakeholders informed about project progress and impacts (Toukola and Ahola 2022). Additionally, Collaboration and Partnerships are crucial, forming joint ventures and public–private partnerships to pool resources and expertise, and engaging in collaborative R&D projects with academic and research institutions (Arslan et al. 2024). Furthermore, offering incentives and support is essential. Incentives include financial benefits like grants, subsidies, and tax breaks. Support also includes technical assistance and capacity-building initiatives. These help stakeholders build their capabilities. This encouragement and support foster a collaborative environment (Bulman et al. 2021).

d. Impact Assessment and Mitigation

Assessing and mitigating the impacts of biomass upgrading projects is crucial for their sustainability. Economic, environmental, and social impact assessments are crucial components of biomass upgrading projects (Solarte-Toro et al. 2022). Economic Impact involves conducting a cost-benefit analysis to evaluate financial viability and return on investment, as well as performing market analysis to understand demand and competition for upgraded biomass products (Geddafa et al. 2023). Environmental Impact assessments are also vital, utilizing life cycle assessment (LCA) to quantify environmental benefits and potential risks, and developing mitigation strategies to address any negative environmental impacts (Pan et al. 2019). Furthermore, Social Impact assessments are necessary for biomass upgrading projects. These assessments gauge the effects on local communities and workforce. A social impact assessment (SIA) is performed to understand these effects (Marting Vidaurre et al. 2020). Community engagement plans are created to ensure inclusive development. These plans promote equitable development. The goal is to ensure a comprehensive and sustainable approach to biomass upgrading projects (Christley et al. 2021).

e. Monitoring and Evaluation

Ongoing monitoring and evaluation are crucial for a project's success because they help ensure the project meets its objectives through a Continuous Feedback Loop (Kerzner 2022). This loop gets ongoing feedback from stakeholders and reviews and adjusts strategies based on feedback and performance. Performance Indicators measure economic, environmental, and social outcomes, and key performance indicators (KPIs) are developed to track these outcomes (Neri et al. 2021). Reports on these outcomes are made regularly and shared with all stakeholders to foster transparency and accountability throughout the project (Vian and Kohler 2016).

11.10 Market Dynamics and Demand Analysis

Understanding the practicality and scalability of transforming biomass waste into valuable goods requires a thorough assessment of market dynamics and demand considerations, which is what the techno-economic analysis (TEA) of biomass waste upgrading entails (Pérez-Almada et al. 2023) market trends, market drivers and restraints, potential market segment and competitive landscape, all of which have an impact on how well biomass waste upgrading technologies sell commercially.

a. Current Market Trends

Global developments in waste management, sustainable development, and renewable energy have an impact on the market for upgrading biomass waste (Clauser et al. 2021). A growing number of people are demanding cleaner and more sustainable energy sources due to worries about climate change and environmental destruction (Qazi et al. 2019). Market dynamics are greatly impacted by the passage of laws and policies meant to promote renewable energy sources and lower carbon emissions. (Qazi et al. 2019). For example, by encouraging the production of biofuels and biochemicals, the Renewable Fuel Standard in the United States and the Renewable Energy Directive in the European Union offer advantageous conditions for biomass waste upgrading (Gruttola and Borello 2021).

i. Renewable Energy Demand: One major factor influencing the upgrading of biomass waste is the global shift towards renewable energy. There is a growing demand for sustainable biofuels such as bioethanol and biodiesel as replacements for fossil fuels. The market attractiveness of biofuels made from biomass waste is increased by policies like the EU's Renewable Energy Directive, which aim to reduce greenhouse gas emissions and increase the use of renewable energy (Sikkema et al. 2021).

ii. Technological Advancements: Processes for upgrading biomass are becoming more efficient and economical due to ongoing innovation. Modern techniques in gasification, fermentation, pyrolysis, anaerobic digestion, and gasification are increasing the viability of turning biomass waste into biofuels and biochemicals. The use of genetically modified microbes, enzyme technologies, and pretreatment process are essential for lowering expenses and raising yield (Zabed et al. 2019).

iii. Circular Economy Initiatives: The need for biomass waste upgrading is being driven by the worldwide movement towards a circular economy, which emphasizes resource reutilization and waste minimization. Recycling municipal solid trash, forestry byproducts, and agricultural residues into useful goods promotes the circular economy while also benefiting the environment.

iv. Government Policies and Incentives: Economic incentives and supportive policies have a big impact on the market. Projects aimed at upgrading biomass are rendered more appealing by subsidies, tax credits, grants, and feed-in tariffs. The financial support provided by the Renewable Fuel Standard (RFS) in the United States and different incentives offered by Europe, for instance, improves the economic feasibility of these technologies (Skogstad 2020).

v. Sustainability and Decarbonization Goals: The motivation behind investments in biomass waste upgrading is the government's and corporations' dedication to decarbonization and sustainability. Upgrading biomass is a desirable investment since businesses looking to lower their carbon footprint are increasingly looking for sustainable materials and renewable energy sources (Balanay and Halog 2024).

vi. Market Diversification: Beyond biofuels, the production of biochemicals, bioplastics, and other high-value products from biomass waste is gaining popularity. By creating additional revenue streams and reducing market risks related to fluctuations in energy prices, diversification helps to meet the growing demand for sustainable products (Sen and Ganguly 2017).

vii. Research and Development Investments: Technological developments are being propelled by substantial investments in research and development. Financing from both the public and private sectors promotes technological innovation and commercialization, process optimization, and industrial capacity expansion (Fengsheng et al. 2022).

viii. Global Market Expansion: Globally, the market for upgrading biomass waste is growing, especially in areas with a surplus of biomass resources. With assistance from international collaborations and technology transfer programs, developing nations are implementing these technologies to tackle issues related to waste management, energy security, and rural development (Corsi et al. 2020).

ix. Hybrid Energy Systems: An increasing trend is the integration of biomass upgrading with other renewable energy systems, like wind and solar power. By utilizing the advantages of many technologies, hybrid systems improve overall energy efficiency and dependability and offer a steady and uninterrupted energy supply (Banik and Das 2020).

b. Market driver and restraints

The market for biomass waste upgrading is driven by several key factors. Increasing global demand for renewable energy is a primary driver, as biofuels derived from biomass offer a sustainable alternative to fossil fuels (Khan et al. 2021). Policies aimed at reducing greenhouse gas emissions and promoting renewable energy sources, such as the European Union's Renewable Energy Directive and the U.S. Renewable Fuel Standard, provide significant incentives for biomass waste upgrading (Shikemma et al. 2021). Technological advancements are also critical drivers, with innovations in anaerobic digestion, pyrolysis, and fermentation improving efficiency and reducing costs (Manyi-Loh and Lues 2023). Additionally, the global push towards a circular economy emphasizes waste minimization and resource reutilization, further bolstering the demand for biomass waste upgrading. Corporate and governmental commitments to sustainability and decarbonization goals drive investments in renewable energy sources, making biomass upgrading an attractive option (Nurdiawati and Urban 2021).

However, several challenges restrain market growth. High initial capital expenditures (CAPEX) for setting up biomass upgrading facilities can be a significant barrier (Abelha and Kiel 2020). Operational expenditures (OPEX) and maintenance costs also impact economic feasibility. Additionally, the variability in biomass feedstock availability and quality can pose challenges for consistent production (Yan et al. 2020). Regulatory hurdles and compliance with environmental standards can complicate project development. Lastly, market volatility in energy prices can affect the profitability of biofuels, while public perception and acceptance of biomass-derived products may vary, impacting market adoption (Yogalakshmi et al. 2023).

c. Potential Market Segments

The market for biomass waste upgrading is diverse, encompassing several segments that present unique opportunities and challenges. These potential market segments include biofuels, biochemicals, bioplastics, and biogas, each offering distinct benefits and applications (Mujtaba et al. 2023).

i. Biofuels

Biofuels represent one of the largest segments in the biomass waste upgrading market. Bioethanol and biodiesel are derived from agricultural residues, forestry by-products, and municipal solid waste (Katiyar et al. 2020). These biofuels serve as renewable alternatives to fossil fuels, reducing greenhouse gas emissions and enhancing energy security. The transportation sector, in particular, is a significant consumer of biofuels, driven by regulatory mandates and sustainability goals (Yussof et al. 2021).

ii. Biochemicals

Biochemicals produced from biomass waste include a range of products such as organic acids, solvents, and enzymes. These biochemicals are used in various industries, including pharmaceuticals, cosmetics, and food processing. The demand for sustainable and eco-friendly products is driving the growth of this segment, as biochemicals offer a renewable alternative to petrochemicals (Ramchuran et al. 2023).

iii. Bioplastics

The bioplastics segment is gaining traction due to the growing demand for sustainable packaging solutions. Derived from biomass waste, bioplastics are used in packaging, agriculture, and consumer goods. They offer the advantage of being biodegradable or compostable, addressing environmental concerns associated with conventional plastics (Rosenboom et al. 2022).

iv. Biogas

Biogas, produced through the anaerobic digestion of organic waste, is a versatile energy source that can be used for electricity generation, heating, and as a vehicle fuel. The agricultural sector, in particular, benefits from biogas production, as it provides a sustainable solution for managing animal waste and crop residues (Saidmamatov et al. 2021). Additionally, biogas can be upgraded to biomethane, a renewable substitute for natural gas.

v. Competitive Landscape

The competitive landscape of the biomass waste upgrading market is characterized by a diverse mix of established companies, innovative startups, and collaborative research initiatives, all striving to develop and commercialize efficient and cost-effective technologies (Nunes 2024).

vi. Established Companies

Large, established companies with significant experience in bioenergy and waste management dominate the market. These firms leverage their extensive resources and infrastructure to develop and scale biomass upgrading technologies. Companies such as Novozymes, DSM, and DuPont are key players, known for their advancements in enzyme technologies and fermentation processes (Danielson et al. 2020). Their extensive research and development (R&D) capabilities and strong market presence give them a competitive edge.

vii. Innovative Startups

Startups play a crucial role in driving innovation within the biomass waste upgrading sector. These smaller, agile companies often focus on niche markets or novel technologies, such as advanced pyrolysis, gasification, and anaerobic digestion (Mou 2024). Examples include LanzaTech, which specializes in gas fermentation, and Anellotech, known for its thermal catalytic biomass conversion technology (Task 2021). Startups frequently collaborate with academic institutions and leverage venture capital to fund their innovative approaches, pushing the boundaries of current technologies (Corvello et al. 2023).

c. Collaborative Research Initiatives

Collaborative efforts between industry, academia, and government agencies are pivotal in advancing biomass waste upgrading technologies. Initiatives like the European Bioenergy Research Institute (EBRI) and the U.S. Department of Energy's Bioenergy Technologies Office (BETO) facilitate knowledge exchange, funding opportunities, and pilot projects. These collaborations help bridge the gap between research and commercialization, fostering technological advancements and market adoption (Anuardo et al. 2022).

d. Strategic Partnerships and Acquisitions

Strategic partnerships and acquisitions are common strategies in this competitive landscape. Established companies often acquire innovative startups to integrate new technologies and enhance their product portfolios. Partnerships between technology developers and feedstock suppliers ensure a consistent supply of raw materials, improving the economic feasibility of biomass upgrading projects (Lundmark et al. 2018).

11.11 Economic Analysis of Biomass Upgrading Technologies

Building the physical infrastructure is a major financial commitment in the first stages of setting up a biomass upgrading plant. This includes constructing the enclosures, support structures, and foundations needed to house of the equipment. The purchase and installation of essential technological components such digesters, fermenters, gasifiers, pyrolysis units, reactors, and digesters are also included in capital expenses. Costs for each technology vary according to its capacity and level of complexity (Sevillano et al. 2021). Pumps, heat exchangers, condensers, and control systems are examples of additional equipment required for the effective operation of the core technologies. Pre-treatment procedures (such as size reduction and hydrolysis) are frequently necessary for efficient biomass upgrading in order to prepare the feedstock. These units require more money for specific tools and machinery (Wagle et al. 2022). The size of the biomass feedstock is decreased by technologies including shredding, grinding, and milling, which improves its suitability for further processing (Bajwa et al. 2018). Usually, this equipment requires a modest initial outlay. Reagent handling and application require the use of chemical reactors and storage tanks, which raises the overall CAPEX of processes like acid or alkaline hydrolysis (Davis et al. 2018). Certain high-temperature reactors and related heat recovery systems are needed for processes like steam explosion or torrefaction, which can be rather costly. Installing waste treatment facilities, emission control systems, and monitoring equipment is required to comply with environmental standards (Piersa et al. 2022). These mechanisms guarantee that the plant runs in accordance with the necessary environmental regulations. Purchasing protective gear, fire suppression systems, and emergency response infrastructure are necessary to guarantee worker safety.

The price of purchasing biomass feedstock, which varies greatly depending on its kind, location, and availability (Zahraee et al. 2020). Another important consideration is the cost of transportation, especially if the feedstock has to be brought from a distance. Seasonal variations in feedstock supply might affect the cost of procurement. Keeping a consistent supply often means paying more for storage (Carpenter et al. 2014). To guarantee optimum performance and lifetime, regular maintenance is essential for reactors, pre-treatment units, and other machinery. This covers the price of lubricants, spare parts, and other supplies needed for upkeep (Gutu et al. 2021). Facilities for upgrading biomass need a crew with certain expertise to operate. Pay, benefits, and training for technicians, engineers, and support personnel are all included in labor costs. Processes that upgrade biomass, especially those that include heat, like gasification and pyrolysis, require a lot of energy. The operational costs are largely influenced by the use of electricity, natural gas, or other energy sources (Jaggers and Das 2018). Significant water input is needed for many processes in order to generate steam, cool down, or carry out chemical reactions. Water procurement, treatment, and disposal costs need to be considered (Jaglas et al. 2011).

Continuous additions of acids, alkalis, enzymes, or microbial cultures are frequently needed for processes like fermentation and chemical pre-treatment. Depending on the needs of the process, the cost of these consumables may be high. Certain technologies for upgrading make use of catalysts or adsorbents, which raise operational costs because they must be periodically replaced or regenerated. Process by-products like gasification ash and anaerobic digestion residual sludge require special handling and disposal techniques that come with extra costs. By-products may be processed and sold in some circumstances to partially cover these expenses (Pérez-Botella et al. 2022).

a. Cost benefit analysis

It is imperative to perform a cost-benefit analysis on biomass upgrading technologies in order to assess their commercial feasibility and potential for widespread implementation. This analysis entails a thorough investigation of the financial and environmental advantages as well as the costs (both capital and operating) related to these technologies.

i. Capital Expenditures (CAPEX): Facilities for upgrading biomass may require large upfront costs. Reactors, digesters, separators, and other necessary equipment must be bought and installed. For example, digesters and gas collection systems cost a lot of money when building a biogas plant employing anaerobic digestion technology. Raw biomass must be pre-treated using techniques like mechanical, chemical, or thermal treatments in order to become an acceptable feedstock. These procedures increase the initial capital expenses by requiring more facilities and equipment. It will cost more to invest in monitoring and control systems to ensure adherence to safety and environmental laws (O'Connor et al. 2021).

ii. Operational Expenditures (OPEX): Biomass feedstock availability and cost can have a big impact on the overall economics. While municipal solid garbage may require collection and transportation charges, forestry byproducts and agricultural leftovers may be provided for free or at a minimal cost. Ongoing operational expenses include labor costs and routine equipment maintenance. Labor expenses may rise due to the need for competent staff in highly specialized technologies. Processes for upgrading biomass frequently need a significant energy input. For instance, pyrolysis and other heat processes require a large amount of energy, which raises operating expenses (Angulo-Mosquera et al. 2021).

b. Benefit Analysis

Revenue Streams: The sales of biofuels (such as bioethanol, biodiesel) and bioproducts (such as bioplastics, biochemicals) provide the majority of the financial gain. The market pricing of these goods can differ, which affects the possibility of making profit. Renewable energy facilities can earn Renewable Energy Certificates (RECs), which is an extra source of income. Those looking to reduce their carbon footprint can purchase these certificates. Upgrading procedures frequently produce by-products in addition to the core products, such as digestate from anaerobic digestion or biochar from pyrolysis, which can be sold as fertilizers or soil amendments (Sari et al. 2021).

Environmental Benefits: Technologies for processing biomass help reduce emissions and sequester carbon, possibly resulting in carbon credits. It is possible to monetize these credits, which would have major financial advantages. Recycling biomass waste into useful goods promotes sustainability and a circular economy by lowering the need for landfills and the associated disposal expenses (Ahmadi et al. 2020).

Social and Economic Benefits: The establishment and operation of biomass upgrading facilities boosts the local economy by generating job opportunities. Domestic biofuel production improves national energy security by lowering reliance on imported fossil fuels (Kanwal et al. et al. 2022).

i. Market Potential

The market potential for biomass upgrading technology is driven by the growing emphasis on carbon footprint reduction and the shift to renewable energy sources worldwide. The market for biofuels and bioenergy is expected to grow as nations pledge to meet their renewable energy commitments (Yang et al. 2021). The Renewable Energy Directive of the European Union and the Renewable Fuel Standard of the United States are two examples of laws and regulations that promote the use of biomass upgrading technology by offering financial support, tax breaks, and renewable energy credits (Flamos et al. 2011). Businesses are embracing sustainable practices more and more in order to satisfy legal requirements and customer expectations. Investments in sustainable products made from biomass and renewable energy are being fueled by this transition (Cantarero et al. 2020).

By producing valuable products from municipal solid trash, forestry byproducts, and agricultural residues, biomass upgrading provides a sustainable solution to waste management problems. Landfill use and related environmental effects are decreased as a result (Pérez-Almada et al. 2023). The ideas of biomass upgrading are consistent with the movement towards a circular economy, which places an emphasis on waste reduction and resource efficiency. As industries work to adopt circular practices, this alignment increases market potential (Jain et al. 2022).

Technological developments in pre-treatment, conversion, and process optimization are improving the economic feasibility and efficiency of biomass upgrading. These advancements make the technology more enticing to investors and end users. The creation of scalable technologies that may be applied at many scales, from small-scale community projects to large-scale industrial facilities, is increasing the market potential (Ab Rasid et al. 2021).

In areas like Asia, Africa, and Latin America where biomass resources are plentiful, upgrading biomass offers prospects for rural development, economic expansion, and energy security. Technologies for upgrading biomass are being adopted by these regions more frequently (Benti et al. 2021).

ii. Revenue Streams

As renewable fuel substitutes for gasoline and diesel, the production of bioethanol and biodiesel serves as the main source of income for biomass upgrading. Requirements to blend these biofuels with fossil fuels have created a sizable market for them. Second-generation biofuels made from biomass sources other than food are becoming more popular and have the potential to generate more money because of their premium pricing and regulatory backing (Valdivia et al. 2016). Many biochemicals, such as organic acids (such as lactic acid and succinic acid), solvents, and specialized compounds, can be produced via biomass upgrading processes. These goods offer significant income potential because they are used in the food and beverage, cosmetics, and pharmaceutical industries. The market for biochemicals made from biomass is becoming more specialized as consumers' need for sustainable and eco-friendly chemicals rises (Teo et al. 2020). Bioplastics provide a sizable revenue stream in response to the growing customer demand for environmentally friendly packaging options. They are preferred for their biodegradability and reduced environmental impact and are utilized in consumer items, packaging, and agriculture. The market for bioplastics is anticipated to expand quickly due to laws against single-use plastics and consumer demand for environmentally friendly goods (Moshood et al. 2021). Anaerobic digestion produces biogas, which can be utilized as fuel for vehicles as well as for heating and power generation. It offers a consistent source of income, particularly in areas where renewable energy initiatives are supported. The conversion of biogas into biomethane, a renewable natural gas replacement, opens up new revenue streams, especially for the transportation and home heating industries (Obileke et al. 2021). Biochar is created through the pyrolysis of biomass and can be sold as a filtration media, soil amendment, or carbon sequestration tool, increasing the revenue stream. Digestate, the leftover material from anaerobic digestion, can be utilized as fertilizer, generating additional income. Technologies for upgrading biomass help reduce greenhouse gas emissions and have the potential to generate carbon credits. In carbon markets, these credits can be traded to generate extra income (Thengane et al. 2021).

11.12 Case Studies and Real-World Applications

Case studies and real-world applications of techno-economic analysis (TEA) in biomass waste upgrading provide valuable insights into the practical challenges and opportunities of these technologies. One notable example is the development and commercialization of the waste-to-biofuel plant in Güssing, Austria, which employs a gasification process to convert biomass waste into synthetic gas and subsequently into biofuels. This project, operational since 2006, has demonstrated how TEA can be used to optimize process design and assess economic viability. The analysis highlighted the importance of integrating renewable energy sources and optimizing feedstock logistics to achieve cost-effectiveness and sustainability (Tschürtz et al. 2021). Another significant case is the use of hydrothermal liquefaction (HTL) at the Pacific Northwest National Laboratory (PNNL), where TEA was critical in scaling up the technology for converting wet biomass into bio-crude oil. The TEA revealed that HTL could achieve competitive economic performance by leveraging its ability to process high-moisture feedstocks that are otherwise challenging for conventional technologies, thus demonstrating HTL's potential for large-scale implementation (Klein et al. 2022).

In the United States, the BioEnergy Technologies Office (BETO) has conducted several TEAs of biomass upgrading technologies, focusing on the conversion of agricultural residues into advanced biofuels. For instance, the analysis of lignocellulosic biomass-to-ethanol technologies demonstrated that significant cost reductions could be achieved through improvements in enzymatic hydrolysis and fermentation processes. The TEA also highlighted the economic benefits of integrating biogas production from agricultural residues as a supplementary energy source to reduce overall operational costs (Davis et al. 2020). Similarly, in Sweden, the Växjö municipality's biomass upgrading project utilizes combined heat and power (CHP) plants to convert wood chips and other biomass waste into electricity and district heating. The TEA conducted for this project underscored the economic advantages of co-firing biomass with conventional fossil fuels and leveraging government incentives for renewable energy, which collectively enhanced the project's financial viability and environmental performance (Ekdahl et al. 2019).

In Brazil, the development of the "Biomass to Bioenergy" project involved the use of anaerobic digestion to convert agricultural waste into biogas. TEA of this project emphasized the economic feasibility of biogas production when combined with nutrient recovery and integrated into local agricultural practices. The analysis highlighted how integrating waste management with energy production not only provided a sustainable energy source but also contributed to soil fertility through the recycling of digestate, thus demonstrating the synergistic benefits of biomass upgrading in agricultural contexts (Tavares et al. 2021). Furthermore, the TEA of the Biomass Power Plant in India, which uses rice husks as feedstock, showed how low-cost biomass resources could be effectively utilized for power generation. The analysis pointed out the importance of optimizing the logistics and preprocessing of rice husks to minimize costs and enhance economic returns, thereby providing a model for similar applications in rice-producing regions globally (Kumar et al. 2019).

These case studies illustrate the diverse applications of TEA in biomass waste upgrading and underscore the importance of tailoring analyses to specific technologies, feedstocks, and regional contexts. They demonstrate how TEA can be used to identify cost-saving opportunities, optimize process design, and evaluate the economic and environmental impacts of biomass conversion technologies. By providing detailed insights into real-world implementations, these case studies contribute to a better understanding of the practical challenges and benefits associated with biomass waste upgrading, guiding future research and development efforts in this field.

11.13 Challenges and Barriers

Techno-economic analysis (TEA) of biomass waste upgrading faces several significant challenges and barriers that hinder the development and deployment of these technologies. One of the primary challenges is the high variability and unpredictability of biomass feedstocks, which can affect the consistency and efficiency of upgrading processes. Biomass feedstocks differ widely in their composition, quality, and availability, leading to variability in process performance and economic outcomes (Miller et al. 2019). This variability complicates the establishment of reliable and scalable models for techno-economic evaluation, making it difficult to predict costs and efficiencies accurately. Another major barrier is the high capital and operational costs associated with biomass upgrading technologies. Advanced technologies such as gasification, pyrolysis, and hydrothermal liquefaction often require substantial initial investments in equipment and infrastructure, which can be a significant financial hurdle, especially for small-scale and start-up operations (Kreutzer et al. 2020). Additionally, the operational costs, including energy consumption and maintenance, can be considerable, impacting the overall economic feasibility of these technologies.

The integration of biomass waste upgrading technologies into existing waste management and energy systems presents further complications. Traditional waste management systems and energy infrastructure are often not designed to accommodate or integrate new biomass upgrading processes, leading to challenges in scaling up and optimizing these technologies (McKendry 2021). The lack of infrastructure and logistical support for biomass collection, transportation, and preprocessing can also add to the complexity and cost of implementing upgrading technologies. Moreover, there are significant uncertainties related to policy and regulatory environments. Biomass upgrading technologies often rely on government incentives, subsidies, and supportive regulatory frameworks to be economically viable. However, these policies can be inconsistent, change frequently, or vary widely between regions, creating uncertainty and risk for investors and operators (Kumar et al. 2022). This regulatory uncertainty can deter investment and complicate long-term planning and financial analysis.

Another challenge is the limited availability of comprehensive and accurate data for techno-economic modeling. Reliable data on biomass feedstock characteristics, process efficiencies, and economic performance is crucial for conducting accurate TEA. However, such data is often scarce or incomplete, particularly for emerging or less mature technologies. This lack of data can lead to inaccuracies in economic assessments and hinder the ability to make informed decisions (Steinberg et al. 2020). The complexity of integrating environmental and social impacts into techno-economic analysis adds another layer of difficulty. Traditional TEA models primarily focus on economic factors, but there is a growing need to incorporate environmental and social considerations, such as lifecycle emissions, resource use, and community impacts. Developing comprehensive models that accurately reflect these broader impacts while maintaining economic feasibility is a challenging task (González et al. 2021). Lastly, the technological maturity and scalability of biomass upgrading technologies are often uneven. Many promising technologies are still in the experimental or pilot phase, lacking the necessary scale-up experience and commercialization success to demonstrate consistent economic viability. This technological uncertainty can lead to difficulties in accurately assessing long-term performance and cost-effectiveness (Sriram et al. 2023).

11.14 Future Trends and Innovations

The integration of digital tools such as big data analytics, artificial intelligence (AI), and machine learning is anticipated to significantly refine the accuracy and predictive capabilities of TEA models. For example, AI-driven optimization algorithms can enhance process efficiency by predicting maintenance needs and identifying cost-saving opportunities, while machine learning models can analyze large datasets to uncover patterns and insights that inform economic evaluations (Wang et al. 2020). Concurrently, the Internet of Things (IoT) will enable real-time monitoring and control of biomass upgrading processes through advanced sensors and connected devices. This continuous feedback loop will facilitate dynamic adjustments and improved operational management, thereby reducing costs and enhancing process performance (Hassall et al. 2019).

Material science innovations are poised to play a crucial role in advancing biomass upgrading technologies. The development of novel catalysts and reaction pathways can lead to more efficient conversion processes, reducing both energy consumption and operational expenses. For instance, researchers are exploring advanced catalysts that operate at lower temperatures, which could result in significant energy savings and lower production costs (Zhao et al. 2021). Additionally, breakthroughs in reaction technologies, such as catalytic pyrolysis and hydrothermal liquefaction, are improving biomass conversion efficiency and product quality, making these processes more economically viable (Li et al. 2022). Sustainability will increasingly be integrated into techno-economic analysis frameworks, reflecting a shift towards circular economy principles and comprehensive lifecycle assessments. Incorporating environmental impact metrics, such as greenhouse gas emissions and resource depletion, alongside economic factors will provide a more holistic view of biomass upgrading's viability. The development of advanced lifecycle assessment (LCA) tools will support this integration, enabling more accurate evaluations of the environmental and economic trade-offs associated with different upgrading technologies (Norton et al. 2020).

The rise of decentralized and modular biomass upgrading systems will also influence TEA. These smaller-scale, distributed technologies offer flexibility and lower initial costs compared to large, centralized facilities. Evaluating the economic implications of decentralized systems, including their potential for local energy production and waste management, will be crucial for adapting TEA methodologies to these new technologies (Mou et al. 2023). The ability to scale these systems based on feedstock availability and demand will further impact their economic viability. Policy and regulatory changes will shape the future of techno-economic analysis, with new environmental regulations and incentives affecting the economics of biomass upgrading. For instance, carbon pricing mechanisms, subsidies for renewable energy, and tax credits for waste-to-energy projects will influence the economic feasibility of upgrading technologies. TEA models will need to adapt to these evolving regulations to provide accurate assessments of economic and financial performance (Zhao et al. 2024). Understanding the interplay between policy incentives and market dynamics will be essential for stakeholders to navigate the regulatory landscape effectively. Collaboration among academia, industry, and government will drive innovation in techno-economic analysis. Research partnerships and public–private initiatives will accelerate the development and deployment of new technologies and methodologies, ensuring that TEA remains relevant and effective. Interdisciplinary approaches that integrate engineering, economics, and environmental science will enrich TEA models, providing more comprehensive and robust analyses (Smith et al. 2021).

11.15 Conclusions

To conclude, the chapter provides a comprehensive information on techno-economic analysis of biomass waste upgrading. In the current scenario, biomass driven industry has emerged as a prominent in sectors like food, feed, fuel and fine chemicals. The biomass derived from agriculture, horticulture, forests and allied sectors is renewable and offers a large-scale employment for the rural areas and also stand as a potential alternative solution to the crisis of diminishing fossil resources. In comparison to conventional fossil fuels and other renewable technology, biomass fuel is more sustainable in terms of technology, economics and environment. The market dynamics, demand analysis and stake-holder analysis showed the biomass upgrading technology offers a new vista in rural employment and entrepreneurship.

References

Ab Rasid NS, Shamjuddin A, Rahman AZA, Amin NAS (2021) Recent advances in green pretreatment methods of lignocellulosic biomass for enhanced biofuel production. J Clean Prod 321:129038

Abdel-Shafy HI, Mansour MS (2018) Solid waste issue: sources, composition, disposal, recycling, and valorization. Egypt J Pet 27(4):1275–1290

Abdulla NS, Fernandez MJF, Balanay MP (2024) Upcycling biomass waste into luminescent solid-state carbon dots. Next Mater 3:100155. https://doi.org/10.1016/j.nxmate.2024.100155

Abelha P, Kiel J (2020) Techno-economic assessment of biomass upgrading by washing and torrefaction. Biomass Bioenerg 142:105751

Abushammala H, Mao J (2020) A review on the partial and complete dissolution and fractionation of wood and lignocelluloses using imidazolium ionic liquids. Polymers 12(1):195

Achakulwisut P, Erickson P (2021) Trends in fossil fuel extraction. SEI: Stockholm, Sweden

Adeoye AO, Lawal OS, Quadri RO, Malomo D, Aliyu MT, Dang GE, Hikon BN (2023) Sustainable energy via thermochemical and biochemical conversion of biomass wastes for biofuel production. In: Transportation energy and dynamics. Springer Nature Singapore, Singapore, pp 245–306

Ahmad W, Nisar J, Anwar F, Muhammad F (2023) Future prospects of biomass waste as renewable source of energy in Pakistan: a mini review. Bioresour Technol Rep 24:101658. https://doi.org/10.1016/j.biteb.2023.101658

Ahmadi L, Kannangara M, Bensebaa F (2020) Cost-effectiveness of small-scale biomass supply chain and bioenergy production systems in carbon credit markets: a life cycle perspective. Sustain Energy Technol Assess 37:100627

Ali F, Dawood A, Hussain A, Alnasir MH, Akhan MA, Butt TM, Janju NK, Hamid A (2024) Fueling the future: biomass applications for green and sustainable energy. Discov Sustain 5:156. https://doi.org/10.1007/s43621-024-00309-z

Alipour F, Lak SZ, Rahimpour MR (2024) Economic challenges of moving beyond fossil fuels. Ref Module Earth Syst Environ Sci. https://doi.org/10.1016/B978-0-323-93940-9.00249-8

Ameh VI, Ayeleru OO, Nomngongo PN, Ramatsa IM (2024) Bio-oil production from waste plant seeds biomass as pyrolytic lignocellulosic feedstock and its improvement for energy potential: A review. Waste Manage Bull 2(2):32–48. https://doi.org/10.1016/j.wmb.2024.03.002

Amenaghawon AN, Anyalewechi CL, Okieimen CO, Kusuma HS (2021) Biomass pyrolysis technologies for value-added products: a state-of-the-art review. Environ Dev Sustain 23:14324–14378. https://doi.org/10.1007/s10668-021-01276-5

Angulo-Mosquera LS, Alvarado-Alvarado AA, Rivas-Arrieta MJ, Cattaneo CR, Rene ER, García-Depraect O (2021) Production of solid biofuels from organic waste in developing countries: a review from sustainability and economic feasibility

Antar M, Lyu D, Nazari M, Shah A, Zhou X, Smith DL (2021) Biomass for a sustainable bioeconomy: an overview of world biomass production and utilization. Renew Sustain Energy Rev 139:110691. https://doi.org/10.1016/j.rser.2020.110691

Anuardo RG, Espuny M, Costa ACF, Oliveira OJ (2022) Toward a cleaner and more sustainable world: a framework to develop and improve waste management through organizations, governments and academia. Heliyon 8(4)

Arslan B, Vasudeva G, Hirsch EB (2024) Public–private and private–private collaboration as pathways for socially beneficial innovation: evidence from antimicrobial drug-development tasks. Acad Manag J 67(2):554–582

Assunção LR, Mendes PA, Matos S, Borschiver S (2021) Technology roadmap of renewable natural gas: Identifying trends for research and development to improve biogas upgrading technology management. Appl Energy 292:116849

Awasthi MK, Sarsaiya S, Patel A, Juneja A, Singh RP, Yan B, Taherzadeh MJ (2020) Refining biomass residues for sustainable energy and bio-products: an assessment of technology, its importance, and strategic applications in circular bio-economy. Renew Sustain Energy Rev 127:109876

Babu S, Rathore SS, Singh R, Kumar S, Singh VK, Yadav SK, Yadav V, Raj R, Yadav D, Shekhawat K, Wani OA (2022) Exploring agricultural waste biomass for energy, food and feed production and pollution mitigation: a review. Biores Technol 360:127566. https://doi.org/10.1016/j.biortech.2022.127566

Bajwa DS, Peterson T, Sharma N, Shojaeiarani J, Bajwa SG (2018) A review of densified solid biomass for energy production. Renew Sustain Energy Rev 96:296–305

Balanay R, Halog A (2024) Bioenergy updates and prospects for decarbonization in the ASEAN region: a review on logistical concerns and potential solutions. Wiley Interdisc Rev Energy Environ 13(1):e499

Banerjee N (2023a) Biomass to energy—an analysis of current technologies, prospects, and challenges. BioEnergy Res 16(2):683–716

Banerjee N (2023b) Biomass to energy—an analysis of current technologies, prospects, and challenges. Bioenerg Res 16:683–716. https://doi.org/10.1007/s12155-022-10500-7

Banik R, Das P (2020) A review on architecture, performance and reliability of hybrid power system. J Inst Eng (India): Ser B 101:527–539

Bernesson S, Nilsson D, Hansson P-A (2006) A limited LCA comparing large- and small-scale production of ethanol for heavy engines under Swedish conditions. Biomass Bioenerg 30(1):46–57

Benteng W, Lin R, Bose A, Huerta JD, Kang X, Deng C, Murphy JD (2023) Economic and environmental viability of biofuel production from organic wastes: a pathway towards competitive carbon neutrality. Energy 285:129322

Benti NE, Gurmesa GS, Argaw T, Aneseyee AB, Gunta S, Kassahun GB, Asfaw AA (2021) The current status, challenges and prospects of using biomass energy in Ethiopia. Biotechnol Biofuels 14:1–24

Boretti A (2013) Renewable hydrogen to recycle CO_2 to methanol. Int J Hydrogen Energy 38:1806–1812. https://doi.org/10.1016/j.ijhydene.2012.11.097

Bulman A, Cordes KY, Mehranvar L, Merrill E, Fiedler Y (2021) Guide on incentives for responsible investment in agriculture and food systems. Food Agriculture Org.

Cantarero MMV (2020) Of renewable energy, energy democracy, and sustainable development: a roadmap to accelerate the energy transition in developing countries. Energy Res Soc Sci 70:101716

Carpenter D, Westover TL, Czernik S, Jablonski W (2014) Biomass feedstocks for renewable fuel production: a review of the impacts of feedstock and pretreatment on the yield and product distribution of fast pyrolysis bio-oils and vapors. Green Chem 16(2):384–406

Chai Y, Tong Z, Al-Dhabi NA, Zhou L, Chen A, Zhu S, Tang W, Ji Z, Zeng J (2024) Upgrading lignocellulosic biomass to high-value products through the pretreatment driven by bio-based ethylene glycol solvent. Chem Eng J 496:153797. https://doi.org/10.1016/j.cej.2024.153797

Chen X, Ding H (2023) Assessment of green economic development through the impacts of fossil fuels resources and education in China. Resour Policy 85(Part B):103910. https://doi.org/10.1016/j.resourpol.2023.103910

Cherubini F, Strømman AH (2011) Life cycle assessment of bioenergy systems: state of the art and future challenges. Bioresour Technol 102(2):437–451. https://doi.org/10.1016/j.biortech.2010.08.010

Chiquetto JB, Leichsenring AR, dos Santos GM (2022) Socioeconomic conditions and fossil fuel CO_2 in the Metropolitan Area of Rio de Janeiro. Urban Climate 43:101176. https://doi.org/10.1016/j.uclim.2022.101176

Cho EJ, Trinh LTP, Song Y, Lee YG, Bae HJ (2020) Bioconversion of biomass waste into high value chemicals. Biores Technol 298:122386

Choi PJ, Lee J, Jang A (2024) Interconnection between renewable energy technologies and water treatment processes. Water Res 261:122037. https://doi.org/10.1016/j.watres.2024.122037

Christley E, Ljungberg H, Ackom E, Nerini FF (2021) Sustainable energy for slums? using the sustainable development goals to guide energy access efforts in a Kenyan informal settlement. Energy Res Soc Sci 79:102176

Chu C, Wu K, Luo B, Cao Q, Zhang H (2023) Hydrogen storage by liquid organic hydrogen carriers: catalyst, renewable carrier, and technology—a review. Carbon Resour Conver 6(4):334–351. https://doi.org/10.1016/j.crcon.2023.03.007

Chum HL, Overend RP (2001) Biomass and renewable fuels. Fuel Process Tech 71(1–3):187–195

Clauser NM, González G, Mendieta CM, Kruyeniski J, Area MC, Vallejos ME (2021) Biomass waste as sustainable raw material for energy and fuels. Sustainability 13(2):794

Cloetingh S, Sternai P, Koptev A, Ehlers TA, Gerya T, Kovács I, Oerlemans J, Beekman F et al (2023) Coupled surface to deep Earth processes: perspectives from TOPO-EUROPE with an emphasis on climate- and energy-related societal challenges. Globl Planet Change 226:104140

Colantoni A, Villarini M, Monarca D, Carlini M, Mosconi EM, Bocci E, Rajabi HS (2021) Economic analysis and risk assessment of biomass gasification CHP systems of different sizes through Monte Carlo simulation. Energy Rep 7:1954–1961

Corsi A, Pagani RN, Kovaleski JL (2020) Technology transfer for sustainable development: social impacts depicted and some other answers to a few questions. J Clean Prod 245:118522

Corvello V, Cimino A, Felicetti AM (2023) Building start-up acceleration capability: a dynamic capability framework for collaboration with start-ups. J Open Innov Technol Market Complex 9(3):100104

Danielson N, McKay S, Bloom P, Dunn J, Jakel N, Bauer T, Shanks B (2020) Industrial biotechnology—an industry at an inflection point. Ind Biotechnol 16(6):321–332

David EE (1982) Strategies in fossil fuel technology: multiple options for unpredicted futures. In: Yuan SW (ed) Energy, resources and environment, Pergamon, pp 22–31. https://doi.org/10.1016/B978-0-08-029396-7.50010-0

Davis RE, Grundl NJ, Tao L, Biddy MJ, Tan EC, Beckham GT, Roni MS (2018) Process design and economics for the conversion of lignocellulosic biomass to hydrocarbon fuels and coproducts: 2018 biochemical design case update; biochemical deconstruction and conversion of biomass to fuels and products via integrated biorefinery pathways (No. NREL/TP-5100-71949). National Renewable Energy Lab. (NREL), Golden, CO (United States)

Davis R, Foust T, Beucke R (2020) Techno-economic analysis of biomass-to-ethanol conversion technologies. Biomass Bioenerg 137:105–118

de Jong S, Hoefnagels R, Wetterlund E, Pettersson K, Faaij A, Junginger M (2017) Cost optimization of biofuel production—the impact of scale, integration, transport and supply chain configurations. Appl Energy 195:1055–1070

Demirbas A (2008) Biomethanol production from organic waste materials. Energy Sour, Part a: Recov, Utili, Environ Effects 30:565–572

Di Gruttola F, Borello D (2021) Analysis of the EU secondary biomass availability and conversion processes to produce advanced biofuels: use of existing databases for assessing a metric evaluation for the 2025 Perspective. Sustainability 13(14):7882

dos Santos AS, Pantoja LDA (2023) Microbial conversion of biomass. In: Thomas S, Hosur M, Pasquini D, Jose Chirayil C (eds) Handbook of biomass. Springer, Singapore. https://doi.org/10.1007/978-981-19-6772-6_39-1

Ekdahl J, Jönsson A, Holm J (2019) Economic and environmental analysis of biomass CHP plants: a case study from Växjö, Sweden. Renew Energy 136:567–578

Ethaib S, Omar R, Mazlina MS, Radiah AD, Zubaidi SL (2020) Toward sustainable processes of pretreatment technologies of lignocellulosic biomass for enzymatic production of biofuels and chemicals: a review. BioResources 15(4):10063

Falcone PM, Sica E (2019) Assessing the opportunities and challenges of green finance in Italy: an analysis of the biomass production sector. Sustainability 11(2):517

Fantozzi F, Bartocci P, D'Alessandro B, Arampatzis S, Manos B (2014) Public-private partnerships value in bioenergy projects: economic feasibility analysis based on two case studies. Biomass Bioenerg 66:387–397. https://doi.org/10.1016/j.biombioe.2014.04.006

Farghali M, Osman AI, Umetsu K Rooney DW. (2022) Integration of biogas systems into a carbon zero and hydrogen economy: a review. Environ Chem Lett 20:2853–2927. https://doi.org/10.1007/s10311-022-01468-z

Farkaš A (2022). New emerging sectors. In: Economics, ecology, and policy for the bioeconomy. Routledge, pp 66–107

Fengsheng C, YunQian Z, Muhammad S, Ching-Chi H (2022) Financing for energy efficiency solutions to mitigate opportunity cost of coal consumption: an empirical analysis of Chinese industries. Environ Sci Pollut Res 29(2):2448–2465

Finnveden G, Hauschild MZ, Ekvall T, Guinée J, Heijungs R, Hellweg S, Koehler A, Pennington D, Suh S (2009) Recent developments in life cycle assessment. J Environ Manag 91(1):1–21. https://doi.org/10.1016/j.jenvman.2009.06.018

Flamos A, Georgallis PG, Doukas H, Psarras J (2011) Using biomass to achieve European Union energy targets—a review of biomass status, potential, and supporting policies. Int J Green Energy 8(4):411–428

Funk K, Milford J, Simpkins T (2020) Chapter 19—waste not, want not: analyzing the economic and environmental viability of waste-to-energy technology for site-specific optimization of renewable energy options. In: Dahiya A (ed) Bioenergy, 2nd edn, Academic Press, pp 385–423. https://doi.org/10.1016/B978-0-12-815497-7.00019-1

Gahleitner G (2013) Hydrogen from renewable electricity: an international review of power-to-gas pilot plants for stationary applications. Int J Hydrog Energy 38:2039–2061

Gandomi S, Barzegar M, Zolghadri S, Rahimpour MR (2024) The future of fossil fuels supply and impacts on sustainability. Ref Module Earth Syst Environ Sci. https://doi.org/10.1016/B978-0-323-93940-9.00270-X

Gasparatos A, Doll CN, Esteban M, Ahmed A, Olang TA (2017) Renewable energy and biodiversity: implications for transitioning to a green economy. Renew Sustain Energy Rev 70:161–184

Geddafa T, Melka Y, Sime G (2023) Cost-benefit analysis and financial viability of household biogas plant investment in South Ethiopia. Sustain Energy Res 10(1):20

Geow CH, Tan MC, Yeap SP, Chin NL (2021) A review on extraction techniques and its future applications in industry. Eur J Lipid Sci Technol 123(4):2000302

González J, Gómez M, Aznar J (2021) Integrating environmental and social impacts into techno-economic analysis of renewable energy technologies. Renew Sustain Energy Rev 140:110–124

Govorushko S (2013) Environmental problems of extraction, transportation, and use of fossil fuels. Fossil fuels: sources, environmental concerns and waste management practices 1–84

Graedel TE, Howard-Grenville JA (2005) Fossil fuel extraction and processing. Green Ind Facil Perspect Approaches Tools 129–146

Gutu L, Basitere M, Harding T, Ikumi D, Njoya M, Gaszynski C (2021) Multi-integrated systems for treatment of abattoir wastewater: a review. Water 13(18):2462

Haile A, Gelebo GG, Tesfaye T, Mengie W, Mebrate MA, Abuhay A, Limeneh DY (2021) Pulp and paper mill wastes: utilizations and prospects for high value-added biomaterials. Bioresour Bioprocess 8:1–22

Han W, Chen WY (2024) Survival of biomass and waste power generation: a global overview. Sci Total Environ 940:173593. https://doi.org/10.1016/j.scitotenv.2024.173593

Hansen S, Mirkouei A, Diaz LA (2020) A comprehensive state-of-technology review for upgrading bio-oil to renewable or blended hydrocarbon fuels. Renew Sustain Energy Rev 118:109548

Harlan T, Baka J (2024) Stacked energyscapes: conceptualizing fossil fuel and renewable energy entanglements in low-carbon transitions. Energy Res Soc Sci 115:103648. https://doi.org/10.1016/j.erss.2024.103648

Hassan Q, Viktor P, Al-Musawi TJ, Ali BM, Algburi S, Alzoubi HM, Al-Jiboory AK, Sameen AZ, Salman HM, Jaszczur M (2024) The renewable energy role in the global energy transformations. Renew Energy Focus 48:100545. https://doi.org/10.1016/j.ref.2024.100545

Hassall M, Coates S, De Silva S (2019) The role of IoT in process optimization and efficiency. J Ind Eng 75(2):89–105

Hoang AT, Nizetic S, Ong HC, Chong CT, Atabani AE (2021) Acid-based lignocellulosic biomass biorefinery for bioenergy production: advantages, application constraints, and perspectives. J Environ Manage 296:113194
Holloway S (2001) Storage of fossil fuel-derived carbon dioxide beneath the surface of the earth. Annu Rev Energy Env 26(1):145–166
Holm-Nielsen JB, Al Seadi T, Oleskowicz-Popiel P (2009) The future of anaerobic digestion and biogas utilization. Biores Technol 100(22):5478–5484
Hossain MS, Das BK, Das A, Roy TK (2024) Investigating the techno-economic and environmental feasibility of biogas-based power generation potential using food waste in Bangladesh. Renew Energy 121017. https://doi.org/10.1016/j.renene.2024.121017
Hsu DH (2007). Experienced entrepreneurial founders, organizational capital, and venture capital funding. Res Policy 36(5):722–741. https://doi.org/10.1016/j.gloplacha.2023.104140
Huang Y, Wang Q, Xu J (2023) A Stackelberg-based biomass power trading game framework in hybrid-wind/solar/biomass system: from technological, economic, environmental and social perspectives. J Clean Prod 403:136806. https://doi.org/10.1016/j.jclepro.2023.136806
Humpenöder F, Popp A, Bodirsky BL, Weindl I, Biewald A, Lotze-Campen H, Stevanovic M (2018) Large-scale bioenergy production: how to resolve sustainability trade-offs? Environ Res Lett 13(2):024011
Jagals P, Rietveld L, Cameron J, Hunter P, Pond K (2011) Estimating costs of small scale water supply interventions. Valuing Water, Valuing Livelihoods: Guidance Soc Cost-Benefit Anal Drink Water Interv Spec Ref Small Commun Water Supplies 149–166
Jagger P, Das I (2018) Implementation and scale-up of a biomass pellet and improved cookstove enterprise in Rwanda. Energy Sustain Dev 46:32–41
Jain A, Sarsaiya S, Awasthi MK, Singh R, Rajput R, Mishra UC, Shi J (2022) Bioenergy and bio-products from bio-waste and its associated modern circular economy: Current research trends, challenges, and future outlooks. Fuel 307:121859
Jiang K, Xing R, Luo Z, Huang W, Yi F, Men Y, Zhao N, Chang Z, Zhao J, Pan B, Shen G (2024) Pollutant emissions from biomass burning: a review on emission characteristics, environmental impacts, and research perspectives. Particuology 85:296–309
Julie Chandra CS, Sasi S, Bindu Sharmila TK, Varghese JR (2023) Modern biomass conversion technologies. In: Handbook of biomass. Springer Nature Singapore, Singapore, pp 1–31
Kalak T (2023) Potential use of industrial biomass waste as a sustainable energy source in the future. Energies 16(4):1783. https://doi.org/10.3390/en16041783
Kamaruzaman N, Manaf NA, Milani D, Abbas A (2023) Assessing the current state of biomass gasification technology in advancing circular economies: a holistic analysis from techno-economic-policy perspective in Malaysia and beyond. Chem Eng Res des 199:593–619. https://doi.org/10.1016/j.cherd.2023.10.023
Kanwal S, Mehran MT, Hassan M, Anwar M, Naqvi SR, Khoja AH (2022) An integrated future approach for the energy security of Pakistan: replacement of fossil fuels with syngas for better environment and socio-economic development. Renew Sustain Energy Rev 156:111978
Kataya G, Cornu D, Bechelany M, Hijazi A, Issa M (2023a) Biomass waste conversion technologies and its application for sustainable environmental development—a review. Agronomy 13(11):2833. https://doi.org/10.3390/agronomy13112833
Kataya G, Cornu D, Bechelany M, Hijazi A, Issa M (2023b) Biomass waste conversion technologies and its application for sustainable environmental development—a review. Agronomy 13(11):2833
Katiyar P, Srivastava SK, Kushwaha D (2020) Bioethanol extraction and its production from agricultural residues for sustainable development. Waste Energy: Prospects Appl 143–163
Kerzner H (2022) Project management metrics, KPIs, and dashboards: a guide to measuring and monitoring project performance. John wiley & sons
Khan MAH, Bonifacio S, Clowes J, Foulds A, Holland R, Matthews JC, Shallcross DE (2021) Investigation of biofuel as a potential renewable energy source. Atmosphere 12(10):1289

Kieliszek M, Piwowarek K, Kot AM, Pobiega K (2020) The aspects of microbial biomass use in the utilization of selected waste from the agro-food industry. Open Life Sci 22; 15(1):787–796. https://doi.org/10.1515/biol-2020-0099

Klein A, Keane S, Zhang H (2022) Scaling up hydrothermal liquefaction technology: techno-economic analysis and performance metrics. Energy Convers Manage 256:115–126

Koasidis K, Nikas A, Karamaneas A, Saulo M, Tsipouridis I, Campagnolo L, Doukas H (2022) Climate and sustainability co-governance in Kenya: a multi-criteria analysis of stakeholders' perceptions and consensus. Energy Sustain Dev 68:457–471

Kosenius AK, Ollikainen M (2013) Valuation of environmental and societal trade-offs of renewable energy sources. Energy Policy 62:1148–1156

Kreutzer M, Lattimore B, Williams P (2020) Economic barriers to the adoption of biomass conversion technologies. Biomass Bioenerg 141:105–117

Kumar A, Reddy R, Singh R (2019) Techno-economic analysis of biomass power generation from rice husk: a case study from India. Renew Energy 130:717–728

Kumar A, Singh R, Reddy R (2022) Policy and regulatory challenges in biomass waste management: a comprehensive review. Energy Policy 161:112–126

Kwatra S, Kumar A, Sharma S, Sharma P (2021) Stakeholder participation in prioritizing sustainability issues at regional level using analytic hierarchy process (AHP) technique: a case study of Goa, India. Environ Sustain Indic 11:100116

Laca A, Laca A, Díaz M (2019) Hydrolysis: from cellulose and hemicellulose to simple sugars. In: Second and third generation of feedstocks. Elsevier, pp 213–240

Lall D, Vishal V (2024) Geophysical indicators and methods for producing oceanic gas hydrates. In: Rahimpour MR, Makarem MA, Meshksar M (eds) Advances in natural gas: formation, processing, and applications, vol 3: Natural gas hydrates. Elsevier, pp 307–323. https://doi.org/10.1016/B978-0-443-19219-7.00015-1

Latif AA, Harun S, Sajab MS, Markom M, Jahim JM (2018) Ammonia-based pretreatment for ligno-cellulosic biomass conversion—an overview. J Eng Sci Technol 13(6):1595–1620

Li C, Aston JE, Lacey JA, Thompson VS, Thompson DN (2016) Impact of feedstock quality and variation on biochemical and thermochemical conversion. Renew Sustain Energy Rev 65:525–536

Liu H, Long J, Zhang K, Li M, Zhao D, Song D, Zhang W (2024) Agricultural biomass/waste-based materials could be a potential adsorption-type remediation contributor to environmental pollution induced by pesticides—a critical review. Sci Total Environ 946:174180. https://doi.org/10.1016/j.scitotenv.2024.174180

Li H, Zhang S, Wang Y (2022) Advances in catalytic pyrolysis for biomass conversion: a review. Renew Energy 182:1221–1237

Lu WM, Kweh QL. Ting IWK, Ren C (2023) How does stakeholder engagement through environmental, social, and governance affect eco-efficiency and profitability efficiency? Zooming into Apple Inc. counterparts. Bus Strategy Environ 32(1):587–601

Lumpkin GT, Bacq S (2019) Civic wealth creation: a new view of stakeholder engagement and societal impact. Acad Manag Perspect 33(4):383–404

Lundmark R, Forsell N, Leduc S, Lundgren J, Ouraich I, Pettersson K, Wetterlund E (2018) Large-scale implementation of biorefineries: new value chains, products and efficient biomass feedstock utilisation

Makepa DC, Chihobo CH, Ruziwa WR, Musademba D (2023) A systematic review of the techno-economic assessment and biomass supply chain uncertainties of biofuels production from fast pyrolysis of lignocellulosic biomass. Fuel Commun 14:100086. https://doi.org/10.1016/j.jfueco.2023.100086

Manyi-Loh CE, Lues R (2023) Anaerobic digestion of lignocellulosic biomass: substrate characteristics (challenge) and innovation. Fermentation 9(8):755

Marting Vidaurre NA, Vargas-Carpintero R, Wagner M, Lask J, Lewandowski I (2020) Social aspects in the assessment of biobased value chains. Sustainability 12(23):9843

McKendry P (2021) The role of biomass in sustainable energy systems: challenges and opportunities. Biofuels, Bioprod Biorefin 15(3):649–665
Miller C, Wright C, Smith R (2019) Variability in biomass feedstocks and its impact on process economics. J Clean Prod 225:224–233
Mohammed HI, Garba K, Ahmed SI, Abubakar LG (2023) Recent advances on strategies for upgrading biomass pyrolysis vapour to value-added bio-oils for bioenergy and chemicals. Sustain Energy Technol Assess 55:102984. https://doi.org/10.1016/j.seta.2022.102984
Mohan C, Robinson J, Vodwal L, Kumari N (2024) Sustainable development Goals for addressing environmental challenges. In: Green chemistry approaches to environmental sustainability, pp 357–374. Elsevier
Mola-Yudego B, Dimitriou I, Gagnon B, Schweinle J, Kulišić B (2024) Priorities for the sustainability criteria of biomass supply chains for energy. J Clean Prod 434:140075
Moshood TD, Nawanir G, Mahmud F, Mohamad F, Ahmad MH, Abdul Ghani A (2021) Expanding policy for biodegradable plastic products and market dynamics of bio-based plastics: challenges and opportunities. Sustainability 13(11):6170
Mou M (2024) Innovation potential of wood-based biochar for soil improvement in forestry. Master's thesis, Itä-Suomen yliopisto
Mou Z, Yang Y, Liu X (2023) Modular and decentralized biomass upgrading technologies: economic and environmental perspectives. Bioenergy Res 16(4):1525–1538
Mujtaba M, Fraceto LF, Fazeli M, Mukherjee S, Savassa SM, de Medeiros GA, Vilaplana F (2023) Lignocellulosic biomass from agricultural waste to the circular economy: a review with focus on biofuels, biocomposites and bioplastics. J Clean Prod 402:136815
Neri A, Cagno E, Lepri M, Trianni A (2021) A triple bottom line balanced set of key performance indicators to measure the sustainability performance of industrial supply chains. Sustain Prod Consum 26:648–691
Norton T, McCormick M, Shaw A (2020) Enhancing lifecycle assessment methodologies for biomass waste conversion technologies. Environ Sci Technol 54(11):6921–6932
Novakovic J, Kontogianni N, Barampouti EM, Mai S, Moustakas K, Malamis D, Loizidou M (2021) Towards upscaling the valorization of wheat straw residues: alkaline pretreatment using sodium hydroxide, enzymatic hydrolysis and biogas production. Environ Sci Pollut Res 28:24486–24498
Nunes LJ (2024) Exploring the present and future of biomass recovery units: technological innovation, policy incentives and economic challenges. Biofuels 15(4):375–387
Nunes LJR (2023) Exploring the present and future of biomass recovery units: technological innovation, policy incentives and economic challenges. Biofuels 15(4):375–387. https://doi.org/10.1080/17597269.2023.2250973
Nurdiawati A, Urban F (2021) Towards deep decarbonisation of energy-intensive industries: a review of current status, technologies and policies. Energies 14(9):2408
Obileke K, Nwokolo N, Makaka G, Mukumba P, Onyeaka H (2021) Anaerobic digestion: technology for biogas production as a source of renewable energy—a review. Energy Environ 32(2):191–225
O'Connor S, Ehimen E, Pillai SC, Black A, Tormey D, Bartlett J (2021) Biogas production from small-scale anaerobic digestion plants on European farms. Renew Sustain Energy Rev 139:110580
Odejobi OJ, Ajala OO, Osuolale FN (2024) Review on potential of using agricultural, municipal solid and industrial wastes as substrates for biogas production in Nigeria. Biomass Convers Biorefinery 14(2):1567–1579
Okafor CC, Nzekwe CA, Ajaero CC, Ibekwe JC, Otunomo FA (2022) Biomass utilization for energy production in Nigeria: a review. Clean Energy Syst 3:100043
Omer AM (2021) Energy services: concepts, applications and historical background. In: Borge-Diez D, Rosales-Asensio E (eds) Energy services and management, energy services fundamentals and financing. Academic Press, pp 3–33. https://doi.org/10.1016/B978-0-12-820592-1.00001-4
Omidkar A, Alagumalai A, Li Z, Song H (2024) Machine learning assisted techno-economic and life cycle assessment of organic solid waste upgrading under natural gas. Appl Energy 355:122321. https://doi.org/10.1016/j.apenergy.2023.122321

Østergaard PA, Duic N, Noorollahi Y, Kalogirou SA (2012) Recent advances in renewable energy technology for the energy transition. Renew Energy 179:877–884. https://doi.org/10.1016/j.renene.2021.07.111

Pan YR, Wang X, Ren ZJ, Hu C, Liu J, Butler D (2019) Characterization of implementation limits and identification of optimization strategies for sustainable water resource recovery through life cycle impact analysis. Environ Int 133:105266

Panoutsou C, Singh A (2020) A value chain approach to improve biomass policy formation. GCB Bioenergy 12(7):464–475

Park J, Kang S, Kim S, Kim H, Cho H, Lee JH (2024) Enhancing the economic viability and reliability of renewables-based electricity supply through Power-to-Gas-to-Power with green hydrogen. Energy Convers Manage 310:118485. https://doi.org/10.1016/j.enconman.2024.118485

Pérez-Almada D, Galán-Martín Á, del Mar Contreras M, Castro E (2023) Integrated techno-economic and environmental assessment of biorefineries: review and future research directions. Sustain Energy Fuels 7(17):4031–4050

Pérez-Botella E, Valencia S, Rey F (2022) Zeolites in adsorption processes: state of the art and future prospects. Chem Rev 122(24):17647–17695

Pérez-Martínez PJ, Miranda RM, Andrade MF, Kumar P (2020) Air quality and fossil fuel driven transportation in the Metropolitan Area of São Paulo. Transp Res Interdisc Perspect 5:100137. https://doi.org/10.1016/j.trip.2020.100137

Piersa P, Unyay H, Szufa S, Lewandowska W, Modrzewski R, Ślężak R, Ledakowicz S (2022) An extensive review and comparison of modern biomass torrefaction reactors vs. biomass pyrolysis—Part 1. Energies 15(6):2227

Prasad MD, Ansari S (2023) Legal and ethical accountability of fossil fuel corporations: need for a new social contract. J Clean Prod 423:138830. https://doi.org/10.1016/j.jclepro.2023.138830

Qazi A, Hussain F, Rahim NA, Hardaker G, Alghazzawi D, Shaban K, Haruna K (2019) Towards sustainable energy: a systematic review of renewable energy sources, technologies, and public opinions. IEEE Access 7:63837–63851

Raclavská H, Růžičková J, Švédová B, Kucbel M, Šafář M, Raclavský K, Abel ELDS (2022) Biodegradable waste streams. Biodegrad Waste Manage Circ Econ Chall Oppor 69–153

Rajabi MS, Moradi R (2023) Fossil fuels reservoirs and extraction. In: Reference module in earth systems and environmental sciences. Elsevier. https://doi.org/10.1016/B978-0-323-93940-9.00044-X

Rambaree K, Sundström A, Wang Z, Wright SAI (2021) Qualitative stakeholder analysis for a Swedish regional biogas development: a thematic network approach. Sustainability 13(14):8003

Ramchuran SO, O'Brien F, Dube N, Ramdas V (2023) An overview of green processes and technologies, biobased chemicals and products for industrial applications. Curr Opin Green Sustain Chem 41:100832

Ramirez-Contreras NE, Faaij AP (2018) A review of key international biomass and bioenergy sustainability frameworks and certification systems and their application and implications in Colombia. Renew Sustain Energy Rev 96:460–478

Ritchie R, Rosado P (2017) "Fossil fuels" Published online at OurWorldInData.org. Retrieved from: https://ourworldindata.org/fossil-fuels [Online Resource]

Robinson BL (2021) Giving all energy stakeholders a voice: developing a theoretical framework for the uptake and sustained use of improved/clean biomass energy technologies. Doctoral dissertation, University of Nottingham

Rocha-Meneses L, Luna-delRisco M, González CA, Moncada SV, Moreno A, Sierra-Del Rio J, Castillo-Meza LE (2023) An overview of the socio-economic, technological, and environmental opportunities and challenges for renewable energy generation from residual biomass: a case study of biogas production in Colombia. Energies 16(16):5901. https://doi.org/10.3390/en16165901

Rodionova MV, Poudyal RS, Tiwari I, Voloshin RA, Zharmukhamedov SK, Nam HG, Zayadan BK, Bruce BD, Hou HJM, Allakhverdiev SI (2017) Biofuel production: challenges and opportunities. Int J Hydrog Energy 42(12):8450–8461

Rogers JG, Brammer JG (2012) Estimation of the production cost of fast pyrolysis bio-oil. Biomass Bioenerg 36:208–217

Rosenboom JG, Langer R, Traverso G (2022) Bioplastics for a circular economy. Nat Rev Mater 7(2):117–137

Russo I, Confente I, Scarpi D, Hazen BT (2019) From trash to treasure: the impact of consumer perception of bio-waste products in closed-loop supply chains. J Clean Prod 218:966–974

Saidmamatov O, Rudenko I, Baier U, Khodjaniyazov E (2021) Challenges and solutions for biogas production from agriculture waste in the Aral Sea Basin. Processes 9(2):199

Sarathe S, Baredar PV, Dwivedi G, Tapdiya S, Gaurav A (2022) Review of various types of renewable-powered desalination technologies with economic analysis. Mater Today: Proceed 56:326–335. https://doi.org/10.1016/j.matpr.2022.01.175

Sari YW, Kartikasari K, Setyaningsih I, Lestari D (2021) Techno-economic assessment of microalgae for biofuel, chemical, and bioplastic. In: Microalgae, pp 409–432. Academic Press; Sci Total Environ 795:148816

Sayed ET, Wilberforce T, Elsaid K, Kamal M, Rabaia H, Abdelkareem MA, Chae K, Olabi AG (2020) A critical review on environmental impacts of renewable energy systems and mitigation strategies: Wind, hydro, biomass and geothermal. Sci Total Environ 766:144505. https://doi.org/10.1016/j.scitotenv.2020.144505

Sen S, Ganguly S (2017) Opportunities, barriers and issues with renewable energy development—a discussion. Renew Sustain Energy Rev 69:1170–1181

Sevillano CA, Pesantes AA, Peña Carpio E, Martínez EJ, Gómez X (2021) Anaerobic digestion for producing renewable energy—the evolution of this technology in a new uncertain scenario. Entropy 23(2):145

Shabangu S, Woolf D, Fisher EM, Angenent LT, Lehmann J (2014) Techno-economic assessment of biomass slow pyrolysis into different biochar and methanol concepts. Fuel 117:742–748. https://doi.org/10.1016/j.fuel.2013.08.053

Shah AA, Seehar TH, Sharma K, Toor SS (2022) Chapter 7—biomass pretreatment technologies. In: Sunil K, Maity KG, Bhowmick TK (eds) Hydrocarbon biorefinery. Elsevier, pp 203–228. https://doi.org/10.1016/B978-0-12-823306-1.00014-5

Shaikh RR, Thenmozhi S (2023) Organocatalytic upgrading of biomass-derived furaldehydes: An emerging chemical technology for biorefineries. Synth Commun 53(23):1963–1980. https://doi.org/10.1080/00397911.2023.2267141

Sharma HK, Xu C, Qin W (2019) Biological pretreatment of lignocellulosic biomass for biofuels and bioproducts: an overview. Waste Biomass Valoriz 10:235–251

Sherwood J (2020) The significance of biomass in a circular economy. Biores Technol 300:122755. https://doi.org/10.1016/j.biortech.2020.122755

Sikkema R, Proskurina S, Banja M, Vakkilainen E (2021) How can solid biomass contribute to the EU's renewable energy targets in 2020, 2030 and what are the GHG drivers and safeguards in energy-and forestry sectors? Renew Energy 165:758–772

Singh NB, Kumar A, Rai S (2014) Potential production of bioenergy from biomass in an Indian perspective. Renew Sustain Energy Rev 39:65–78

Singh A, Christensen T, Panoutsou C (2021) Policy review for biomass value chains in the European bioeconomy. Glob Trans 3:13–42

Singh MK, Malek J, Sharma HK, Kumar R (2024) Converting the threats of fossil fuel-based energy generation into opportunities for renewable energy development in India. Renew Energy 224:120153. https://doi.org/10.1016/j.renene.2024.120153

Skogstad G (2020) Mixed feedback dynamics and the USA renewable fuel standard: the roles of policy design and administrative agency. Policy Sci 53:349–369

Smith A, Johnson R, Liu C (2021) Interdisciplinary approaches to techno-economic analysis in biomass waste management. Renew Sustain Energy Rev 142:110–124

Smith P, Davis SJ, Creutzig F, Fuss S, Minx J, Gabrielle B, Rogelj J (2016) Biophysical and economic limits to negative CO_2 emissions. Nat Clim Change 6(1):42–50
Solarte-Toro JC, Laghezza M, Fiore S, Berruti F, Moustakas K, Alzate CAC (2022) Review of the impact of socio-economic conditions on the development and implementation of biorefineries. Fuel 328:125169
Spinelli R, Visser R, Björheden R, Röser D (2019) Recovering energy biomass in conventional forest operations: a review of integrated harvesting systems. Curr For Reports 5:90–100
Sriram R, Prasad N, Muthusamy M (2023) Scaling up biomass waste upgrading technologies: technological maturity and economic feasibility. Energy Convers Manage 268:115–128
Steinberg M, Zhang Y, Morris A (2020) Data challenges in techno-economic analysis of emerging biomass conversion technologies. Renew Energy 148:732–745
Sterling EJ, Betley E, Sigouin A, Gomez A, Toomey A, Cullman G, Porzecanski AL (2017) Assessing the evidence for stakeholder engagement in biodiversity conservation. Biol Cons 209:159–171
Strand J (2007) Technology treaties and fossil-fuels extraction. Energy J 28(4):129–142
Sun PT, Peng H (2023) Valorisation of biomass waste for sustainable bioenergy and biofuel production. Bioengineering 10(5):619. https://doi.org/10.3390/bioengineering10050619
Tanmay JD, Ahmed I, Osman DC, David B, Rooney W (2022) Methanol fuel production, utilization, and techno-economy: a review. Environ Chem Lett 20:3525–3554. https://doi.org/10.1007/s10311-022-01485-y
Task IB (2021) Progress in commercialization of biojet/Sustainable Aviation Fuels (SAF).
Tavares G, Andrade P, Costa C (2021) Economic and environmental benefits of biogas production from agricultural waste: a case study in Brazil. Waste Manage 120:206–216
Teo HL, Wahab RA (2020) Towards an eco-friendly deconstruction of agro-industrial biomass and preparation of renewable cellulose nanomaterials: a review. Int J Biol Macromol 161:1414–1430
Terrapon-Pfaff J, Dienst C, König J, Ortiz W (2014) A cross-sectional review: impacts and sustainability of small-scale renewable energy projects in developing countries. Renew Sustain Energy Rev 40:1–10
Tetteh EK, Sijadu NG, Rathilal S (2024) An overview of non-carbonaceous and renewable-powered technologies for green hydrogen production in South Africa: keywords occurrence analysis. Energ Strat Rev 54:101486. https://doi.org/10.1016/j.esr.2024.101486
Thengane SK, Kung K, Hunt J, Gilani HR, Lim CJ, Sokhansanj S, Sanchez DL (2021) Market prospects for biochar production and application in California. Biofuels, Bioprod Biorefin 15(6):1802–1819
Tomlin AS (2021) Air quality and climate impacts of biomass use as an energy source: a review. Energy Fuels 35(18):14213–14240
Toukola S, Ahola T (2022) Digital tools for stakeholder participation in urban development projects. Project Leadersh Soc 3:100053
Trimble JL, Van Hook RI, Folger AG (1984) Biomass for energy: the environmental issues. Biomass 6:3–13
Tschürtz C, Biedenkopf K, Kopp M (2021) The Güssing biomass-to-energy plant: a comprehensive techno-economic analysis. Energy 225:120–134
Tshikovhi A, Motaung TE (2023) Technologies and Innovations for biomass energy production. Sustainability 15(16):12121. https://doi.org/10.3390/su151612121
Ullah S, Lin B (2024) Green energy dynamics: analyzing the environmental impacts of renewable, hydro, and nuclear energy consumption in Pakistan. Renew Energy 232:121025. https://doi.org/10.1016/j.renene.2024.121025
Vadenbo C, Tonini D, Astrup TF (2017) Environmental multiobjective optimization of the use of biomass resources for energy. Environ Sci Technol 51(6):3575–3583
Valdivia M, Galan JL, Laffarga J, Ramos JL (2016) Biofuels 2020: biorefineries based on lignocellulosic materials. Microb Biotechnol 9(5):585–594

Vassileva AG (2022) Green Public-Private Partnerships (PPPs) as an instrument for sustainable development. J World Econ Transf Transit 2(05):22(1–18). https://doi.org/10.52459/jowett25221122

Venugopal V (2022) Green processing of seafood waste biomass towards blue economy. Curr Res Environ Sustain 4:100164. https://doi.org/10.1016/j.crsust.2022.100164

Vian T, Kohler JC (2016) Medicines transparency alliance (MeTA): pathways to transparency, accountability and access: cross-case analysis and review of phase II

Vuppaladadiyam AK, Vuppaladadiyam SV, Sahoo A, Murugavelh S, Anthony et al (2023) Bio-oil and biochar from the pyrolytic conversion of biomass: a current and future perspective on the trade-off between economic, environmental, and technical indicators. Sci Total Environ 857:159155. https://doi.org/10.1016/j.scitotenv.2022.159155

Wagle A, Angove MJ, Mahara A, Wagle A, Mainali B, Martins M, Paudel SR (2022) Multi-stage pre-treatment of lignocellulosic biomass for multi-product biorefinery: a review. Sustain Energy Technol Assess 49:101702

Wang J, Chen Q, Zhao M (2020) Machine Learning and Big Data Analytics in Biomass Conversion Processes. J Clean Prod 267:122–134

Wiedmann T, Lenzen M (2018) Environmental and social footprints of international trade. Nat Geosci 11(5):314–321

Yan J, Oyedeji O, Leal JH, Donohoe BS, Semelsberger TA, Li C, Tanjore D (2020) Characterizing variability in lignocellulosic biomass: a review. ACS Sustain Chem Eng 8(22):8059–8085

Yana S, Nizar M, Mulyati D (2022) Biomass waste as a renewable energy in developing bio-based economies in Indonesia: a review. Renew Sustain Energy Rev 160:112268

Yang Q, Zhou H, Bartocci P, Fantozzi F, Mašek O, Agblevor FA, McElroy MB (2021) Prospective contributions of biomass pyrolysis to China's 2050 carbon reduction and renewable energy goals. Nat Commun 12(1):1–12

Yogalakshmi KN, Usman M, Kavitha S, Sachdeva S, Thakur S, Kumar A, Rajesh BJ (2023) Lignocellulosic biorefinery technologies: a perception into recent advances in biomass fractionation, biorefineries, economic hurdles and market outlook. Fermentation 9(3):238

Yrjälä K, Ramakrishnan M, Salo E (2022) Agricultural waste streams as resource in circular economy for biochar production towards carbon neutrality. Curr Opin Environ Sci Health 26:100339

Yusoff MNAM, Zulkifli NWM, Sukiman NL, Chyuan OH, Hassan MH, Hasnul MH, Zakaria MZ (2021) Sustainability of palm biodiesel in transportation: a review on biofuel standard, policy and international collaboration between Malaysia and Colombia. Bioenergy Res 14(1):43–60

Zabed HM, Akter S, Yun J, Zhang G, Awad FN, Qi X, Sahu JN (2019) Recent advances in biological pretreatment of microalgae and lignocellulosic biomass for biofuel production. Renew Sustain Energy Rev 105:105–128

Zahraee SM, Shiwakoti N, Stasinopoulos P (2020) Biomass supply chain environmental and socio-economic analysis: 40-Years comprehensive review of methods, decision issues, sustainability challenges, and the way forward. Biomass Bioenerg 142:105777

Zhang P, Hao D (2023) Enterprise financial management and fossil fuel energy efficiency for green economic growth. Resour Policy 84:103763. https://doi.org/10.1016/j.resourpol.2023.103763

Zhao L, Zhou S, Li W (2021) Low-temperature catalysis for biomass conversion: recent advances and future directions. Catal Sci Technol 11(5):1634–1649

Zhao P, Zhang X, Huang Y (2024) Policy and regulatory impacts on the economics of biomass waste upgrading technologies. Energy Policy 162:112–125

Zohir S, Rahman KM, Hossain MI, Anjum A (2016) Review of SME credit-related policies of Bangladesh Bank. Nathan Associates (London) Limited

Chapter 12
Biomass Waste Valorization: Technological, Economic and Environmental Challenges

Ikrame Ayouch, Ilias Barrak, Soumia Aboulhrouz, Ilham Jioui, and Mohamed Zahouily

Abstract This chapter explores the multifaceted realm of biomass waste valorization, highlighting significant technological, economic, and environmental challenges and future prospects. It delves into the latest advancements in biomass conversion technologies, including enzymatic hydrolysis, gasification, and the integration of biorefineries. Economic considerations discuss the viability of these technologies against traditional fossil-based processes, emphasizing cost-effectiveness and market demands. Environmental challenges focus on the sustainable management of by-products and emissions, ensuring minimal impact on local ecosystems. The chapter concludes by envisioning transformative scenarios that could revolutionize biomass valorization, integrating it within a circular economy to foster sustainable development.

I. Ayouch (✉) · I. Jioui · M. Zahouily
Fert-Explorer, College for Sustainable Agriculture and Environmental Science (CSAES), Mohammed VI Polytechnic University, Ben Guerir, Morocco
e-mail: ikrame.ayouche@um6p.ma

I. Jioui
e-mail: ILham.JIOUI@um6p.ma

M. Zahouily
e-mail: Mohamed.ZAHOUILY@um6p.ma

I. Barrak
HYDROJEEL-INNOVX, Mohammed VI Polytechnic University, Ben Guerir, Morocco
e-mail: Ilias.BARRAK@innovx.ma

S. Aboulhrouz
Materials Science, Energy and Nanoengineering (MSN) Department, Mohammed VI Polytechnic University, Ben Guerir, Morocco
e-mail: Soumia.ABOULHROUZ@um6p.ma

M. Zahouily
Laboratory of Materials, Catalysis and Natural Resources Valorization, URAC 24, Faculty of Science and Technology, Hassan II University, Casablanca, Morocco

A. Hind (ed.), *Biomass Waste: Processing, Valorization, Industrial Applications*,
https://doi.org/10.1007/978-981-95-0991-1_12

12.1 Introduction

In the quest for sustainable development, the valorization of biomass waste stands as a cornerstone strategy, addressing both environmental concerns and the pressing need for renewable resources. As the world grapples with the dual challenges of waste management and energy scarcity, the transformation of biomass waste into valuable products presents a promising pathway. However, this field is fraught with technical, economic, and environmental challenges that must be navigated to unlock its full potential.

This chapter delves into the multifaceted challenges that impede the progress of biomass waste valorization. It explores the technological hurdles that limit the efficiency of conversion processes, the economic barriers that affect the viability of biomass-based products, and the environmental frameworks that shape the operational landscape. Moreover, it casts a forward-looking view on the future perspectives in this field, highlighting innovative approaches and emerging technologies that hold the promise of overcoming these obstacles. The discussion begins with an examination of the current technologies used in biomass conversion, including thermochemical, biochemical, and physicochemical processes. Each method presents unique advantages and limitations, which are critical to understanding the broader implications for industrial scalability and environmental impact. The economic challenges are also significant, with factors such as high initial investments, fluctuating raw material costs, and market competitiveness posing substantial hurdles to widespread adoption. Looking ahead, the chapter identifies cutting-edge research and novel technologies that could potentially revolutionize the field. Advances in genetic engineering, nanotechnology, and process integration promise to enhance efficiency and reduce the costs of biomass valorization. Furthermore, the development of a circular economy and stronger synergy between industries could foster more robust and sustainable valorization strategies.

12.2 Biomass Valorization

Biomass valorization has advanced significantly in recent years, becoming an increasingly critical area in the transition towards sustainable energy and materials production. The field of biomass valorization is evolving, marked by significant technological advances such as gasification, enzymatic hydrolysis, and anaerobic digestion, which have become more efficient and environmentally friendly. These technologies are pivotal in converting biomass into biofuels, biogas, and biochemicals, maximizing the utility of raw materials and reducing waste. Recent studies have focused on optimizing these processes to enhance yield and reduce energy inputs, ensuring a more sustainable approach to biomass conversion (Abu-Omar et al. 2021; Singhvi and Gokhale 2019; Vasilakou et al. 2023).

Another aspect of biomass valorization is the concept of integrated biorefineries which has been increasingly implemented. These facilities co-produce a variety of bio-based products, such as bioplastics, bioethanol, and organic acids, and energy, including electricity and heat, utilizing various components of biomass including cellulose, hemicellulose, and lignin to minimize waste and enhance economic viability. This integration allows for the comprehensive use of biomass feedstocks, converting them into fuels (biodiesel and bio-oil), chemicals (such as solvents and lubricants), composites, nanocomposites and fibers, thus supporting the principles of the circular economy, which aim to maximize resource efficiency and promote the reuse of waste products as new inputs (Clark and Deswarte 2015).

Current efforts in biomass valorization are increasingly focused on sustainability. This includes the development of resource-efficient processes such as thermochemical conversion methods that utilize lower temperatures and advanced catalytic systems (Chowdhury et al. 2020; Rani et al. 2022), and biochemical processes like anaerobic digestion, alcoholic fermentation and photobiological hydrogen production (Fig. 12.1), which reduce greenhouse gas emissions. Sustainable production practices, including the optimization of water use, the reduction of chemical inputs, and the enhancement of waste recycling within the process, not only help in mitigating the environmental impact but also improve the overall acceptance and feasibility of biomass technologies (Nzeteu et al. 2022). Researchers are continually exploring innovative approaches to enhance the energy balance and reduce the carbon footprint of biomass processing, such as integrating renewable energy sources into biomass processing facilities and improving the efficiency of biomass conversion technologies. In addition, economically, biomass valorization is becoming more viable due to technological advancements and scaling up of operations. The market for biomass-derived products is expanding, driven by increasing consumer demand for sustainable products and supportive regulatory frameworks. This economic shift is supported by a growing body of research focused on making biomass processes more cost-effective and competitive with traditional fossil-based processes (Ahmed et al. 2021).

12.3 Challenges in Biomass Waste Valorization

12.3.1 Technological Challenges in Biomass Waste Valorization

12.3.1.1 Pretreatment Inefficiencies

Pretreatment is a critical step in the biomass valorization process, where raw biomass is prepared for subsequent biochemical conversions. The main goal of pretreatment is to enhance the accessibility of cellulose and hemicellulose for hydrolysis by removing or modifying lignin and reducing the crystallinity of cellulose. Despite its importance, pretreatment presents several technological challenges that can impact

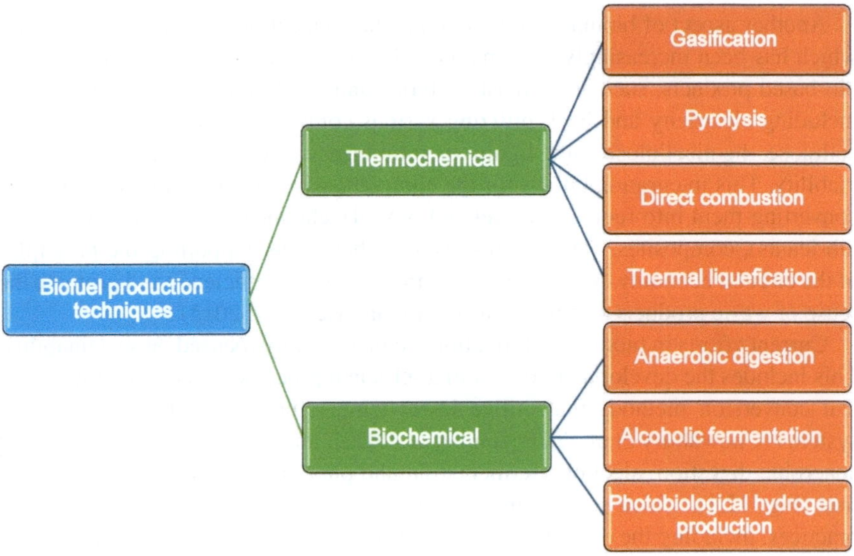

Fig. 12.1 Technology involved in the extraction of biofuels from microalgae biomass (Rani et al. 2022)

the efficiency and cost-effectiveness of the entire biomass valorization pipeline. For instance, in the process of physical pretreatment, biomass is either broken down into smaller particles through mechanical comminution or its surface area is expanded without necessarily reducing its size. This expansion of surface area is often achieved through mechanical refining, where the biomass undergoes shear-induced "fibrillation." Such physical manipulation enhances the accessibility of the biomass surface for cellulases, thus improving the efficiency of hydrolysis. Physical pretreatment is a crucial step, either preceding or following chemical or biochemical treatments. However, detailed knowledge on how physical pretreatment affects the chemical composition or the structure of the cell wall remains limited. Despite this, it is known that physical pretreatment induces modifications in the cell wall that are similar to those seen in mechanical wood pulping technologies like thermomechanical pulping, grinding, and pressure grinding. Observations indicate that certain physical pretreatment methods can modify the lignin structure. A significant limitation of physical pretreatment is its inability to remove lignin effectively, which restricts enzyme access to cellulose. Furthermore, this method of pretreatment is energy-intensive, costly at scale, and associated with substantial environmental and safety challenges. Research suggests that the need for delignification in feedstocks could contribute to the high energy demands of physical pretreatment methods, thereby impacting the overall energy efficiency of biorefineries (Aslanzadeh et al. 2014).

Moreover, chemical pretreatment was largely studied and involves acid pretreatment, alkaline pretreatment, organo-solvents, ionic liquids (ILs), and deep eutectic solvents (DES) methods. Acid pretreatment involves the treatment of biomass with various acids such as sulfuric, nitric, hydrochloric, phosphoric, formic, maleic and oxalic acids. These acids are used either in dilute form (0.1–10%) at elevated temperatures (100–250 °C) or in concentrated form (30–70%) at lower temperatures (less than 100 °C) (Awogbemi and Kallon 2022; Lorenci Woiciechowski et al. 2020). This method is highly effective in breaking down hemicellulose, improving enzymatic digestibility and achieving high glucose recovery rates, making it viable for commercial use (Nazari et al. 2018). However, acid pre-treatment presents several challenges. It can lead to the formation of inhibitory compounds such as furans and phenolics, which can interfere with subsequent processes. There is also a risk of cellulose degradation, which can reduce the overall yield of the final product (Palmqvist and Hahn-Hägerdal 2000). The use of concentrated acids can cause corrosion of equipment and require acid recovery measures, adding to the complexity of the process (Alvira et al. 2010). These disadvantages can be mitigated by careful process optimization and control. Adjusting factors such as temperature, acid concentration, biomass to acid ratio and pretreatment duration can improve the efficiency of acid pretreatment. For example, increasing temperature and acid concentration can improve hemicellulose hydrolysis, but can also increase the production of inhibitors. Therefore, optimizing the pretreatment time is crucial to maximize hemicellulose degradation while minimizing cellulose degradation (Chiaramonti et al. 2012). On the other hand, alkaline pretreatment is typically performed under gentle conditions using mild chemicals that are recoverable and reusable. These chemicals, which include the hydroxides of sodium, potassium, calcium, and ammonium, are non-hazardous, non-corrosive, non-toxic, and relatively affordable (Badiei et al. 2014; Mafa et al. 2020). This not only reduces the capital and operational costs associated with the process but also enhances its environmental sustainability (Mafa et al. 2020). Nevertheless, there are several drawbacks to the alkaline pretreatment method. It often requires longer reaction times and necessitates a post-treatment step to neutralize the resulting slurry, commonly referred to as black liquor. There is also the risk of forming non-recoverable salts that can deposit within the biomass, and the process is generally less effective for high-lignin content materials such as wood (Kumar et al. 2020; Oriez et al. 2020).

Furthermore, Ionic liquid (IL) pretreatment is an emerging, eco-friendly technique for extracting cellulose and lignin from various lignocellulosic biomass sources. This method utilizes green solvents composed of cations and anions and operates under relatively mild conditions (Mahmood et al. 2017b). Ionic liquids are distinguished by several properties: they are non-volatile, non-flammable, chemically and thermally stable, and can be easily recovered and reused. They also have a melting point below 100 °C (Mahmood et al. 2017a). Additional benefits of this method include its high selectivity for specific biomass components, reduced generation of inhibitory compounds, and lower energy requirements due to its operation at relatively low temperatures. However, there are significant drawbacks to using ionic liquids for biomass pretreatment. The overall process can be costly due to the high price of ionic liquids. Additionally, the recovery of these liquids is crucial, which can complicate the process. Another concern is the potential toxicity of ionic liquids to microorganisms and enzymes used in subsequent stages of biomass processing (Joshi and Manjare 2024).

Other pretreatment methods include biological pretreatment which uses microorganisms or enzymes to modify or degrade lignin and hemicellulose. While more environmentally benign and requiring milder processing conditions, biological pretreatments are typically slower and can be less effective compared to chemical methods. Moreover, techniques such as steam explosion or torrefaction involve the use of high temperatures and pressures to disrupt the lignocellulosic structure. Steam explosion is particularly effective at breaking down the complex structure of biomass but requires significant energy inputs and can lead to the degradation or loss of sugars through the formation of inhibitory compounds such as furfural and hydroxymethylfurfural (Joshi and Manjare 2024).

While pretreatment is indispensable for efficient biomass valorization, its inefficiencies and high costs pose significant challenges. Advances in pretreatment technologies are focusing on developing more energy-efficient, less chemically intensive methods that can reduce costs and improve the overall sustainability of biomass valorization. Innovations such as combining pretreatment steps or developing novel catalysts and solvents that minimize byproduct formation and energy use are among the strategies being explored to address these challenges.

12.3.1.2 Lignin Utilization

Lignin valorization presents numerous challenges due to its complex and heterogeneous nature. Efficiently converting lignin into valuable products is difficult because of its intricate aromatic polymer structure, which varies significantly across different plant species and even within different tissues of the same plant. This complexity makes it resistant to many chemical and biological processes that are effective on other biomass components like cellulose and hemicellulose.

Depolymerization techniques for lignin valorization involve advanced chemical and enzymatic methods that aim to break down the complex polymer into smaller, more manageable molecules like phenolic monomers and aromatic aldehydes. Chemical methods, such as oxidative depolymerization using oxygen or ozone, and enzymatic approaches utilizing laccases and peroxidases, require precise control over reaction conditions to be effective. These conditions include optimal temperature and pH, specific solvent systems, and appropriate catalyst concentrations, which must be carefully managed to maximize enzyme activity and reaction efficiency. Challenges in these techniques include the high specificity of enzymatic actions which limits their application across different types of lignin, the need for high temperatures in chemical methods which can lead to the degradation of desired products, and the overall complexity in maintaining constant optimal reaction conditions. Furthermore, the scalability of these processes to industrial levels presents significant economic and technical hurdles, necessitating innovations in process control and efficiency to make lignin valorization commercially viable (Joshi and Manjare 2024; Mahmood et al. 2015).

Biocatalytic methods for lignin valorization harness the capabilities of specialized enzymes and microbes to selectively break down the complex lignin structure into simpler, valuable compounds. Enzymes such as laccases and peroxidases are pivotal, as they can catalyze the breakdown of lignin at specific sites (Singh et al. 2023), while microbes like white-rot fungi (e.g., Phanerochaete chrysosporium) and bacteria (e.g., Rhodococcus jostii) have evolved mechanisms to metabolize various lignin-derived compounds (Grgas et al. 2023). Tailoring these biological agents to efficiently process the specific types of lignin found in different biomass sources often involves intricate genetic and microbial engineering. For instance, modifying the genes in bacteria to enhance their ability to produce lignin-modifying enzymes or engineering fungal strains to increase their tolerance to the byproducts of lignin degradation are common approaches (Grgas et al. 2023; Lee et al. 2019). However, these biocatalytic methods face significant challenges, such as the inherent variability of lignin's chemical structure which can reduce the efficacy of degradation processes across different lignin sources. Moreover, the slow rates of lignin conversion by biological means pose a scalability issue for industrial applications. The need for precise environmental conditions to maintain enzyme activity and microbial health further complicates the deployment of these methods on a commercial scale. These challenges necessitate ongoing research into more robust and versatile biocatalysts and improved bioprocesses that can operate effectively under a broader range of conditions.

12.3.1.3 Scaling Up Limitations

Scaling up biomass valorization technologies from laboratory or pilot scales to commercial levels presents multiple challenges that can significantly impact their viability and efficiency. These include issues with inconsistent feedstock properties, such as variability in biomass composition and moisture content, which can affect the efficiency and output of conversion processes. Larger system inefficiencies also emerge, such as the need for more robust and durable equipment capable of handling higher throughputs and the complexity of managing larger volumes of inputs and outputs. Moreover, increased logistical and operational costs become prominent as the scale of operations expands, encompassing larger physical footprints, higher energy requirements, and more complex supply chain management. Studies discussing these issues highlight that scale-up challenges in biomass valorization often stem from the difficulty in replicating small-scale process efficiencies at a larger scale due to the increased complexity of the systems involved. For example, studies have shown that the heterogeneity of lignocellulosic biomass can lead to unpredictable behavior in larger reactors, requiring advanced control strategies to maintain process stability and efficiency. Additionally, the economic aspects, such as the initial capital investment and ongoing operational costs, can be prohibitive without significant technological advancements or policy support (Suhartini et al. 2024; Xu et al. 2022).

12.3.2 *Economic Challenges in Biomass Waste Valorization*

In the transition from traditional fossil fuels to renewable biomass sources, the economic challenges play a critical role in determining the scale and feasibility of biomass valorization. The two main economic challenges faced in this sector are ensuring cost-effectiveness and economic viability, and navigating the complexities of market demand and commodity prices.

12.3.2.1 Cost-Effectiveness and Economic Viability

Cost-effectiveness and economic viability are crucial for the successful commercialization of biomass valorization technologies, yet they face multiple economic hurdles. High initial capital investment is required for setting up biomass processing facilities, which includes costs for land acquisition, construction, and specialized equipment necessary for processing biomass. This investment is risky due to the nascent stage of many biomass technologies that may not yet have proven economic returns at a commercial scale. Operational costs also contribute significantly, encompassing energy consumption, equipment maintenance, and logistics for the collection, storage, and transportation of biomass. Unlike fossil fuels, biomass resources are usually dispersed, necessitating substantial energy and resources for their aggregation and processing. Furthermore, the technological efficiency in converting biomass into valuable products impacts economic viability; many existing technologies exhibit lower yields and generate more waste compared to fossil fuel processes, and though improving efficiency through research and development can reduce per-unit costs, it demands continuous technological innovation (Ahmed et al. 2021; Scarlat et al. 2015). Additionally, compliance with environmental regulations and sustainability standards incurs further expenses, including investments in pollution control, waste management, and obtaining sustainability certifications, all adding to the overall costs of biomass valorization projects.

12.3.2.2 Market Demand and Commodity Prices

The economic success of biomass valorization is intricately linked to market demand and commodity prices, which are influenced by various factors including competition with fossil fuels, government policies, global economic conditions, and technological adoption and consumer preferences. Biomass-derived products often compete with cheaper, more established fossil fuel equivalents; fluctuations in oil prices can further make biomass less attractive economically. Government policies such as renewable energy subsidies and carbon taxes can significantly impact the demand for biomass products. Additionally, global economic conditions like recessions or booms can influence commodity prices and the viability of biomass investments. Finally, the rate at which new technologies are adopted and consumer preferences for eco-friendly products can drive or hinder market demand for biomass products. These complexities require strategic navigation to align biomass valorization with economic and market realities, ensuring its sustainability and growth (Ahmed et al. 2021; Králík et al. 2023).

12.3.3 Environmental Challenges in Biomass Waste Valorization

The adoption and expansion of biomass valorization technologies bring significant environmental challenges that must be addressed to ensure sustainable development. These challenges primarily involve the impact on local environments and ecosystems, and the management of emissions and by-products. Both areas have been extensively studied, revealing the complexity and urgency of mitigating potential negative effects.

12.3.3.1 Impact on Local Environments and Ecosystems

Biomass cultivation and harvesting can have profound effects on local environments and ecosystems. Large-scale biomass production often requires extensive land use, which can lead to habitat destruction, biodiversity loss, and changes in land use patterns. For instance, converting forests or grasslands to biomass crop lands can disrupt local flora and fauna, altering ecosystems and potentially leading to species displacement or loss. Moreover, intensive agricultural practices for biomass production can lead to soil degradation, reduced soil fertility, and water resource depletion, further impacting local ecosystems and biodiversity (DAUBER et al. 2010; Immerzeel et al. 2014; Ribeiro et al. 2018).

12.3.3.2 Emissions and By-Products Management

The processing of biomass into energy or other valuable products involves various chemical and physical transformations, which can generate emissions and by-products that pose environmental risks. These include greenhouse gases (GHGs) such as CO_2, CH_4, and N_2O, particulate matter, and volatile organic compounds (VOCs), all of which contribute to air pollution and climate change if not properly managed. Additionally, water pollutants from biomass processing facilities, such as organic matter, nutrients, and chemicals, can contaminate local water bodies, affecting aquatic life and water quality. Managing these emissions and by-products effectively is crucial to minimizing the environmental footprint of biomass valorization (Rasheed et al. 2021; Sun et al. 2023).

12.4 Future Perspectives and Opportunities in Biomass Waste Valorization

As the global focus on sustainable development intensifies, biomass waste recovery is at the forefront of environmental and economic innovation. The future of this field holds the promise of transformative breakthroughs and significant advances, thanks to ongoing research and evolving technologies. This section explores potential research directions, long-term prospects and transformative scenarios that could redefine the use of biomass in the coming decades.

12.4.1 Potential Breakthroughs and Research Directions

Research in biomass waste valorization is on the brink of significant advancements, with key areas poised for breakthroughs that promise to enhance both the efficiency and scalability of these processes. Innovations in enzymatic and microbial technologies are leading the way, with genetic engineering and synthetic biology shaping the development of more effective and robust biocatalysts. These biocatalysts are crucial for breaking down complex biomass into high-value products, and ongoing research aims to make these processes more resistant to inhibitors and adaptable to various biomass types. In the realm of thermochemical conversion, advancements such as improved catalysts and process optimizations in gasification, pyrolysis, and hydrothermal liquefaction are expected to reduce energy requirements and increase yield, making these options more viable at a commercial scale. Additionally, the integration of nanotechnology in biomass processing could revolutionize the efficiency of these systems, allowing for the precise and selective extraction and conversion of biomass components. Furthermore, carbon capture and utilization (CCU) techniques are being integrated into biomass valorization to tackle climate change effectively, offering a way to produce bioenergy while simultaneously sequestering carbon.

12.4.2 Long-Term Outlook and Transformative Scenarios

The long-term outlook for biomass valorization is decidedly optimistic, heralding a future marked by transformative scenarios that could redefine waste management and energy production. One notable advancement is the potential shift towards decentralized biomass processing, which involves the development of smaller, localized systems that enhance efficiency and significantly cut down logistics costs associated with the large-scale collection and processing of biomass. This shift not only promises to bolster rural development by creating local opportunities but also aims to alleviate urban waste management burdens by processing biomass closer to its origin. Furthermore, the integration of biomass valorization into the circular economy presents a compelling vision, where waste biomass from various industries could be repurposed as raw materials for others, thereby minimizing waste and maximizing resource efficiency. This model of sustainability could transform biomass valorization into a pivotal component of a greener economy. Additionally, policy-driven market expansion is anticipated, driven by an increasing governmental focus on sustainable practices. Future legislative and policy frameworks could enhance subsidies for bio-based products, impose stricter regulations on waste disposal, and provide greater support for the research and development of innovative biomass technologies. Another significant aspect of this optimistic future is the development of holistic waste management solutions. By integrating biomass valorization with other renewable energy technologies and recycling efforts, a comprehensive approach to waste management could be established, supporting both environmental sustainability and economic growth. These scenarios collectively forecast a dynamic evolution in the biomass valorization sector, highlighting its critical role in fostering sustainable development and environmental stewardship.

12.5 Conclusion

Biomass waste valorization stands at a pivotal juncture, with the potential to significantly contribute to global sustainability goals. Despite facing technological inefficiencies, economic hurdles, and environmental concerns, ongoing advancements suggest a promising future. Embracing decentralized processing, aligning with circular economy principles, and leveraging supportive policies can accelerate the adoption of biomass valorization technologies. As we move forward, the integration of innovative approaches and holistic waste management strategies will be crucial in overcoming existing challenges and unlocking the full potential of biomass as a sustainable resource.

References

Abu-Omar MM, Barta K, Beckham GT, Luterbacher JS, Ralph J, Rinaldi R, Román-Leshkov Y, Samec JSM, Sels BF, Wang F (2021) Guidelines for performing lignin-first biorefining. Energy Environ Sci 14:262–292. https://doi.org/10.1039/d0ee02870c

Ahmed I, Zia MA, Afzal H, Ahmed S, Ahmad M, Akram Z, Sher F, Iqbal HMN (2021) Socio-economic and environmental impacts of biomass valorisation: a strategic drive for sustainable bioeconomy. Sustain 13:1–32. https://doi.org/10.3390/su13084200

Alvira P, Tomás-Pejó E, Ballesteros M, Negro MJ (2010) Pretreatment technologies for an efficient bioethanol production process based on enzymatic hydrolysis: a review. Bioresour Technol 101:4851–4861. https://doi.org/10.1016/j.biortech.2009.11.093

Aslanzadeh S, Ishola MM, Richards T, Taherzadeh MJ (2014) Chapter 1—An overview of existing individual unit operations. In: Qureshi N, Hodge DB, Vertès AABT-B (eds). Elsevier, Amsterdam, pp 3–36. https://doi.org/10.1016/B978-0-444-59498-3.00001-4

Awogbemi O, Von Kallon DV (2022) Pretreatment techniques for agricultural waste. Case Stud Chem Environ Eng 6:100229. https://doi.org/10.1016/j.cscee.2022.100229

Badiei M, Asim N, Jahim JM, Sopian K (2014) Comparison of chemical pretreatment methods for cellulosic biomass. APCBEE Procedia 9:170–174. https://doi.org/10.1016/j.apcbee.2014.01.030

Chiaramonti D, Prussi M, Ferrero S, Oriani L, Ottonello P, Torre P, Cherchi F (2012) Review of pretreatment processes for lignocellulosic ethanol production, and development of an innovative method. Biomass Bioenergy 46:25–35. https://doi.org/10.1016/j.biombioe.2012.04.020

Chowdhury T, Chowdhury H, Hossain N, Ahmed A, Hossen MS, Chowdhury P, Thirugnanasambandam M, Saidur R (2020) Latest advancements on livestock waste management and biogas production: Bangladesh's perspective. J Clean Prod 272:122818. https://doi.org/10.1016/j.jclepro.2020.122818

Clark J, Deswarte F (2015) The biorefinery concept. In: Introduction to chemicals from biomass, pp 1–29. https://doi.org/10.1002/9781118714478.ch1

Dauber J, Jones MB, Stout JC (2010) The impact of biomass crop cultivation on temperate biodiversity. GCB Bioenergy 2:289–309. https://doi.org/10.1111/j.1757-1707.2010.01058.x

Grgas D, Rukavina M, Bešlo D, Štefanac T, Crnek V, Šikić T, Habuda-Stanić M, Landeka Dragičević T (2023) The bacterial degradation of lignin—a review. Water (Switzerland) 15:1–17. https://doi.org/10.3390/w15071272

Immerzeel DJ, Verweij PA, van der Hilst F, Faaij APC (2014) Biodiversity impacts of bioenergy crop production: a state-of-the-art review. GCB Bioenergy 6:183–209. https://doi.org/10.1111/gcbb.12067

Joshi M, Manjare S (2024) Chemical approaches for the biomass valorisation: a comprehensive review of pretreatment strategies. Environ Sci Pollut Res 31:48928–48954. https://doi.org/10.1007/s11356-024-34473-6

Králík T, Knápek J, Vávrová K, Outrata D, Horák M, Janota L, Jablonská Š (2023) Biomass price as a key factor for the further development of biogas and biomethane use—methodology and policy implications. Sustain Energy Technol Assess 60:103492. https://doi.org/10.1016/j.seta.2023.103492

Kumar B, Bhardwaj N, Agrawal K, Chaturvedi V, Verma P (2020) Current perspective on pretreatment technologies using lignocellulosic biomass: an emerging biorefinery concept. Fuel Process Technol 199:106244. https://doi.org/10.1016/j.fuproc.2019.106244

Lee S, Kang M, Bae J-H, Sohn J-H, Sung BH (2019) Bacterial valorization of lignin: strains, enzymes, conversion pathways, biosensors, and perspectives. Front Bioeng Biotechnol 7:209. https://doi.org/10.3389/fbioe.2019.00209

Lorenci Woiciechowski A, Dalmas Neto CJ, Porto de Souza Vandenberghe L, de Carvalho Neto DP, Novak Sydney AC, Letti LAJ, Karp SG, Zevallos Torres LA, Soccol CR (2020) Lignocellulosic biomass: acid and alkaline pretreatments and their effects on biomass recalcitrance—conventional processing and recent advances. Bioresour Technol 304:122848. https://doi.org/10.1016/j.biortech.2020.122848

Mafa MS, Malgas S, Bhattacharya A, Rashamuse K, Pletschke BI (2020) The effects of alkaline pretreatment on agricultural biomasses (corn cob and sweet sorghum bagasse) and their hydrolysis by a termite-derived enzyme cocktail. Agronomy. https://doi.org/10.3390/agronomy10081211

Mahmood H, Moniruzzaman M, Yusup S, Muhammad N, Iqbal T, Akil HM (2017a) Ionic liquids pretreatment for fabrication of agro-residue/thermoplastic starch based composites: a comparative study with other pretreatment technologies. J Clean Prod 161:257–266. https://doi.org/10.1016/j.jclepro.2017.05.110

Mahmood H, Moniruzzaman M, Yusup S, Welton T (2017b) Ionic liquids assisted processing of renewable resources for the fabrication of biodegradable composite materials. Green Chem 19:2051–2075. https://doi.org/10.1039/C7GC00318H

Mahmood N, Yuan Z, Schmidt J, Xu C (Charles) (2015) Hydrolytic depolymerization of hydrolysis lignin: Effects of catalysts and solvents. Bioresour Technol 190:416–419. https://doi.org/10.1016/j.biortech.2015.04.074

Nazari L, Sarathy S, Santoro D, Ho D, Ray MB, Xu C (Charles) (2018). Chapter 3—Recent advances in energy recovery from wastewater sludge. In: Rosendahl LBT-DTL. for EA (eds). Woodhead Publishing, pp 67–100. https://doi.org/10.1016/B978-0-08-101029-7.00011-4

Nzeteu C, Coelho F, Davis E, Trego A, O'Flaherty V (2022) Current trends in biological valorization of waste-derived biomass: the critical role of VFAs to fuel A biorefinery. Fermentation 8. https://doi.org/10.3390/fermentation8090445

Oriez V, Peydecastaing J, Pontalier P-Y (2020) Lignocellulosic biomass mild alkaline fractionation and resulting extract purification processes: conditions, yields, and purities. Clean Technol. https://doi.org/10.3390/cleantechnol2010007

Palmqvist E, Hahn-Hägerdal B (2000) Fermentation of lignocellulosic hydrolysates. I: inhibition and detoxification. Bioresour Technol 74:17–24. https://doi.org/10.1016/S0960-8524(99)00160-1

Rani L, Srivastav AL, Kaushal J, Grewal AS, Dubey AK, Kumar A, Narang SK (2022) Chapter 8—Recent advances in the production of renewable biofuels using microalgae. In: Dubey AK, Narang SK, Srivastav AL, Kumar A, García-Díaz VBT-AI for RES (eds) Woodhead publishing series in energy. Woodhead Publishing, pp 173–187. https://doi.org/10.1016/B978-0-323-90396-7.00012-2

Rasheed T, Anwar MT, Ahmad N, Sher F, Khan SUD, Ahmad A, Khan R, Wazeer I (2021) Valorisation and emerging perspective of biomass based waste-to-energy technologies and their socio-environmental impact: a review. J Environ Manage 287:112257. https://doi.org/10.1016/j.jenvman.2021.112257

Ribeiro JMC, Godina R, de Matias JCO, Nunes LJR (2018) Future perspectives of biomass torrefaction: review of the current state-of-the-art and research development. Sustain 10:1–17. https://doi.org/10.3390/su10072323

Scarlat N, Dallemand J-F, Monforti-Ferrario F, Nita V (2015) The role of biomass and bioenergy in a future bioeconomy: policies and facts. Environ Dev 15:3–34. https://doi.org/10.1016/j.envdev.2015.03.006

Singh AK, Iqbal HMN, Cardullo N, Muccilli V, Fernández-Lucas J, Schmidt JE, Jesionowski T, Bilal M (2023) Structural insights, biocatalytic characteristics, and application prospects of lignin-modifying enzymes for sustainable biotechnology. Int J Biol Macromol 242. https://doi.org/10.1016/j.ijbiomac.2023.124968

Singhvi MS, Gokhale DV (2019) Lignocellulosic biomass: hurdles and challenges in its valorization. Appl Microbiol Biotechnol 103:9305–9320. https://doi.org/10.1007/s00253-019-10212-7

Suhartini S, Sunyoto NMS, Setyawan HY, Rohma NA, Elviliana, Pratiwi DA, Fitria, Mardawati E, Kasbawati, Melville L (2024) Biomass valorization for bioenergy production. In: Lubis MAR, Lee SH, Mardawati E, Rahimah S, Antov P, Andoyo R, Krišťák Ľ, Nurhadi B (eds) Biomass conversion and sustainable biorefinery. Green Energy and Technology. Springer Nature Singapore, Singapore, pp 67–104. https://doi.org/10.1007/978-981-99-7769-7_4

Sun Y, Wang S, Yang Q, Li J, Wang L, Zhang S, Yang H, Chen H (2023) Environmental impact assessment of VOC emissions from biomass gasification power generation system based on life cycle analysis. Fuel 335:126905. https://doi.org/10.1016/j.fuel.2022.126905

Vasilakou K, Nimmegeers P, Thomassen G, Billen P, Van Passel S (2023) Assessing the future of second-generation bioethanol by 2030—a techno-economic assessment integrating technology learning curves. Appl Energy 344:121263. https://doi.org/10.1016/j.apenergy.2023.121263

Xu D, Tong A, Fan L-S (2022) State of scale-up development in chemical looping technology for biomass conversions: a review and perspectives. Waste Biomass Valoriz 13:1363–1383. https://doi.org/10.1007/s12649-021-01563-2

Chapter 13
Innovations in Biomass Waste Processing: Emerging Trends and Global Perspective

Karuppusamy Manickaraj and Sivasubramanian Palanisamy

Abstract Amid global demands for sustainable waste management and renewable energy, innovations in waste biomass are emerging as an important solution. This paper explores the key trends and ideas at work on a global scale. Advanced conversion technologies such as pyrolysis, gasification, and anaerobic digestion are key to converting biomass waste into valuable products such as biofuels, biogas, and biochemicals, reducing reliance on fossil fuels Process by integrating artificial intelligence (AI) and the Internet of Things (IoT). Increased efficiency Biochemical processes using enzymes and microorganisms further perform various biomass functions and enable the production of biofuels and bioplastics. Decentralized modular systems facilitate waste management in communities and industrial farms. The circular economy increasingly recognizes that biomass waste is a resource, leading to efforts to improve resource efficiency and reduce environmental impact Global collaboration for standardization its technology develops and promotes sustainable biomass management around the world. Governments and industry are prioritizing sustainability goals, promoting market expansion and technological innovation in biomass waste production. Developing economies use biomass resources for energy security and economic growth. Public awareness of the benefits of biomass waste is increasing, highlighting its role in reducing landfill and promoting a cleaner environment.

Keywords Biomass · Waste management · Integration of AI · Internet of Things (IoT) · Decentralized and modular systems · Global collaboration

K. Manickaraj
Department of Mechanical Engineering, CMS College of Engineering and Technology, Coimbatore 641032, Tamilnadu, India

S. Palanisamy (✉)
Department of Mechanical Engineering, P T R College of Engineering and Technology, Madurai – Tirumangalam Road, Madurai 625008, Tamilnadu, India
e-mail: sivaresearch948@gmail.com

A. Hind (ed.), *Biomass Waste: Processing, Valorization, Industrial Applications*,
https://doi.org/10.1007/978-981-95-0991-1_13

13.1 Introduction

As the world grapples with the pressing challenges of sustainable waste management and the pursuit of renewable energy sources, innovative approaches to biomass waste processing have emerged as critical solutions. Biomass waste, derived from agricultural residues, forestry by-products, and organic waste, represents a significant and underutilized resource with the potential to contribute substantially to global sustainability goals (Tshikovhi and Motaung 2023). This paper delves into the key trends and perspectives shaping the field of biomass waste processing on a global scale, highlighting advanced conversion technologies and their transformative impact. Technologies like gasification, anaerobic digestion, and pyrolysis are essential to the advances in this industry. By transforming biomass waste into useful products like biofuels, biogas, and biochemicals, these technologies play a critical role in lowering reliance on fossil fuels and advancing energy security. Through real-time monitoring and predictive maintenance, the integration of Artificial Intelligence (AI) and the Internet of Things (IoT) into these operations improves efficiency, assuring optimal functioning and minimizing downtime (Castaldi et al. 2017).

Biochemical processes leveraging enzymes and microorganisms have opened new avenues for biomass utilization, enabling the production of biofuels and bioplastics. These innovations not only diversify the applications of biomass but also contribute to the reduction of environmental pollution. Additionally, decentralized and modular systems are emerging as scalable solutions for biomass waste processing, catering to community, industrial, and agricultural sectors. The shift towards a circular economy is gaining momentum, with biomass waste being increasingly recognized as a valuable resource (Jain et al. 2022). In line with the goals of global sustainability, this paradigm shift motivates initiatives to maximise resource efficiency and minimise environmental damage. Global cooperation is accelerating the creation of technology and standards, advancing sustainable biomass management techniques everywhere. Governments and industries are playing a pivotal role in prioritizing sustainability goals, which in turn stimulates market expansion and technological innovation in biomass waste processing. In developing economies, the utilization of biomass resources is becoming a cornerstone for achieving energy security and economic development. Moreover, growing public awareness of the benefits of biomass waste processing is crucial in reducing landfill waste and promoting cleaner environments (Jayasinghe and Hawboldt 2012).This introduction sets the stage for a comprehensive exploration of the transformative potential of biomass waste processing in achieving sustainable development goals and transitioning towards a circular economy. By examining the latest trends, technologies, and collaborative efforts, this paper aims to provide a holistic understanding of how biomass waste processing can drive sustainable growth and environmental stewardship globally. As the global community faces increasingly severe environmental challenges, the urgency to develop sustainable waste management solutions and renewable energy sources has never been more apparent. Among the myriad approaches being explored, biomass waste processing stands out due to its dual benefits of waste reduction and energy production (Gumisiriza et al. 2017). Biomass waste is a huge and mostly unexplored resource that consists of organic waste from homes and businesses, forestry byproducts, and agricultural leftovers. The advancement of global sustainability objectives and the resolution of the climate issue depend heavily on the efficient management and conversion of this biomass waste.

Biomass disposal is more than waste disposal; Waste is turned into useful compounds that help mix the world's energy and matter. Converting biomass into biofuels, biogas, and biochemicals can reduce reliance on fossil fuels, reduce greenhouse gas emissions, and contribute to a circular economy by recycling waste use and recycle, this process reduces the environmental footprint and increases resource efficiency. Advanced conversion technologies are needed for biomass waste processing. Anaerobic digestion, gasification, and pyrolysis are the three processes for converting biomass into usable products (Gurusamy et al. 2024). Through pyrolysis, biomass is thermal decomposed in the absence of oxygen to produce biochar, syngas, and bio-oil. Biochar can increase soil fertility and store carbon, bio-oil can be processed into biofuel, and syngas can be used to generate electricity.

Gasification is the partial oxidation of biomass at low temperatures to produce syngas. Syngus is a mixture of methane, carbon monoxide, and hydrogen that can be used as fuel and chemical feedstocks or as electrical blends (Garland et al. 1997). Feedstock can be used as fertilizer with nutrient a valuable, and biogas can be used as a sustainable energy source. The integration of artificial intelligence (AI) and the Internet of Things (IoT) in biomass waste treatment technology is a key advance that makes the process more reliable and efficient AI algorithms can analyze data more internally to determine maintenance needs, optimize installation and improve overall system performance. IoT devices enable real-time monitoring of critical parameters such as temperature, pressure, and fuel level, enabling better control and faster response to any adverse event (Cho et al. 2020).

Enzymes and microbes are used in biochemical processes, which provide more routes for turning biomass into useful compounds. In order to lessen reliance on petrochemical resources and reduce environmental pollution, these methods are especially important for the production of biofuels and bioplastics. Complex biomass molecules may be broken down by enzymes into simpler sugars, which can subsequently be fermented to produce biofuels like butanol or ethanol (Usmani et al. 2021). The specificity and efficiency of enzymes make these processes highly effective for biomass conversion. Certain microorganisms can directly convert biomass into bioproducts through metabolic pathways. For instance, some bacteria can produce bioplastics like polyhydroxyalkanoates (PHAs) directly from organic waste streams.

The development of decentralized and modular biomass waste processing systems is a promising trend that enhances the scalability and accessibility of these technologies. These systems can be implemented at various scales, from small community-based operations to larger industrial and agricultural setups (McPhail et al. 2012). Decentralized systems reduce the need for extensive transportation of biomass, lowering associated costs and emissions. Modular systems allow for incremental scaling, where additional units can be added as needed to increase capacity. The idea of a circular economy is becoming more and more popular on a global scale, and biomass waste is starting to be seen as more than just a byproduct. A circular economy minimises the exploitation of virgin resources and lessens its impact on the environment by continuously recycling and reusing waste materials. Processing biomass waste is essential to realising this goal as it promotes sustainability and resource efficiency (Vertes et al. 2011).

International cooperation plays an important role in providing standards, technologies and best practices for biomass waste management. International cooperation and knowledge sharing are essential to overcome economic, commercial and regulatory challenges. Cooperation can accelerate the adoption of new technologies and support common policies that support biomass management worldwide. Government and industry play an important role in promoting the adoption and advancement of biomass waste technology (Vlosky and Smithhart 2011). Policies that support renewable energy production, waste reduction and sustainable practices are essential to support economic growth and new technologies. Financial incentives such as subsidies, tax credits, and subsidies can encourage investment in biomass construction. On the other hand, industry can use biomass waste as a cost-effective and convenient raw material for electricity and equipment production. Integrating biomass into the industry can increase efficiency, reduce waste costs, and help achieve business goals. Biomass waste for economic development has the opportunity to solve energy security issues, economic development, and environmental sustainability issues. Many developing countries have large amounts of biomass that can be used for energy production and industry. The use of renewable energy technologies can reduce dependence on foreign fossil fuels, create local jobs and improve rural life (Gaudino et al. 2019). Raising public awareness of the benefits of biomass waste treatment is important to increase its use. Public support can drive policy change, encourage community participation and promote sustainable practices. The environmental benefits of biomass processing, such as less waste going to landfill, reducing carbon emissions and improving air quality, are gaining momentum as communities seek cleaner, healthier environments.

By examining the latest trends, technologies, and collaborative efforts, we seek to highlight how biomass waste processing can drive sustainable growth and environmental stewardship globally. The following sections will delve deeper into the various aspects of biomass waste processing, offering detailed insights and case studies that illustrate its impact and potential (Ambaye et al. 2021).

13.2 Advanced Conversion Technologies

The use of sophisticated conversion technology is essential in converting biomass waste into useful goods like biofuels, biogas, and biochemicals. These technologies are crucial for encouraging sustainable waste management techniques and lowering dependency on fossil fuels. Recycling biomass waste into useful products like biofuels, biogas, and biochemicals is essential to lowering dependency on

Table 13.1 Key aspects to consider

Advanced conversion technologies		
Pyrolysis	Gasification	Anaerobic digestion

fossil fuels and advancing environmentally friendly trash disposal methods. These processes are all vital to the sustainable management of biomass waste (Srivastava et al. 2021). The table shows the some key aspects and considerations of the advanced technology of conversion of biomass (Table 13.1).

13.2.1 Pyrolysis Process

Pyrolysis is the process of combustion of biomass in the absence of oxygen to produce charcoal, syngas, and biooil. This method is particularly suitable for biomass and agricultural residues. Improving the quality of the final products, reducing emissions and increasing efficiency are the main objectives of the development of pyrolysis technology (Lin et al. 2013).

13.2.1.1 Process Overview

Pyrolysis is a complex thermochemical process that converts organic matter into useful products by thermal decomposition in the absence of oxygen is classified organic chemicals. In this thermal process are classified into simple chemicals. The absence of oxygen is important because it inhibits the temperature of the decomposition products and assures the formation of different products. The three main types of pyrolysis products are biochar, syngas, and bio-oil. Black and viscous bio-oil is a complex combination of hydrocarbons, acids and phenols that can be further processed into fuels or chemicals. Syngus is a combination of volatile gases composed mostly of carbon dioxide, hydrogen and carbon monoxide. It can be used as a feedstock for fuel and chemicals or as an energy source (Ochieng et al. 2022). Strong solid biochar with high carbon content has benefits in soil fertility and carbon storage, contributes to soil fertility and contributes to climate change mitigation by storing carbon Pyrolysis capacity and it can produce many valuable products, and the ability to process different kinds of biomass into feedstock, this is an important technology in conventional waste management and renewable energy.

13.2.1.2 Products and Their Applications

Bio-oil

Bio-oil is a liquid product obtained by pyrolysis of biomass. It is a complex mixture of oxygenated organic compounds and water that can be further refined into various biofuels and chemicals. Due to its high energy content, bio-oil is suitable for heating and can be burned directly in boilers and furnaces. In power generation, it can be used in diesel engines and turbines after upgrading. In addition, bio-oil serves as a valuable raw material for the production of biodiesel, where it undergoes further chemical processing to produce diesel-like fuels, which contributes to the supply of renewable energy and reduces dependence on fossil fuels (Manickaraj et al. 2019).

Syngas

Syngas, or synthesis gas, is a versatile mixture of hydrogen, carbon monoxide, and methane, produced through gasification of biomass. This gaseous product can be used directly for electricity generation in gas turbines or internal combustion engines, providing a renewable energy source. Beyond power generation, syngas serves as an essential feedstock for producing synthetic fuels and chemicals through processes like Fischer–Tropsch synthesis (Clauser et al. 2021). These synthetic fuels can replace conventional petroleum-based products, offering a cleaner alternative. Additionally, syngas can be converted into hydrogen, playing a crucial role in hydrogen economy initiatives and advancing sustainable energy solutions.

Biochar

Biochar is a carbon-rich solid residue obtained by pyrolysis of biomass. This material has significant potential as a soil amendment, increasing soil fertility, water retention and microbial activity. By improving soil health, biochar can increase agricultural productivity and sustainability. In addition, biochar's high carbon content allows it to act as a stable form of carbon sequestration, helping to mitigate climate change by keeping carbon in the soil for longer periods of time. In addition to agricultural applications, biochar can be used in water filtration, waste management and as an additive in building materials, highlighting its diverse uses (Demirbaş 2001).

13.2.1.3 Technological Advances

Advancements in pyrolysis technology focus on the following considerations and these are explained below.

Improving Efficiency

Advancements in pyrolysis technology aim to enhance the thermal efficiency of the process, thereby maximizing energy recovery and product yield. This involves optimizing reactor design, heat transfer mechanisms, and feedstock preparation to ensure uniform heating and efficient decomposition of biomass. Innovations such as the use of fluidized bed reactors, microwave-assisted pyrolysis, and catalytic pyrolysis can significantly increase the efficiency of the process. Additionally, integrating heat recovery systems and co-processing with other waste streams can further improve overall energy efficiency, making pyrolysis a more viable and sustainable option for biomass conversion on a commercial scale (Karuppusamy et al. 2023).

Reducing Emissions

Developing and putting into practice strategies to collect and reduce pollutants including tar, particulate matter, and volatile organic compounds (VOCs) is necessary to reduce emissions from pyrolysis. One of the innovations is the removal of particulate matter from exhaust gases through the use of sophisticated filtering devices like scrubbers and electrostatic precipitators. During the pyrolysis process, catalysts can be used to encourage cleaner combustion and lessen tar production (Pandey et al. 2021). Furthermore, optimizing operating conditions, such as temperature and residence time, can minimize the production of unwanted by-products. These advancements are crucial for complying with environmental regulations and improving the sustainability of pyrolysis technologies.

Enhancing Product Quality

Enhancing the quality of pyrolysis products focuses on refining processes to produce bio-oil and syngas with fewer contaminants and higher value. This includes the development of catalytic pyrolysis, where specific catalysts are used to upgrade bio-oil by reducing oxygen content and increasing stability. Innovations in feedstock preprocessing, such as torrefaction and pelletization, can also improve product quality by providing uniform and high-energy–density inputs. Advanced separation and purification techniques, such as fractional distillation and membrane separation, are employed to isolate high-value chemicals from bio-oil and improve syngas composition (Manickaraj et al. 2024a). These advancements ensure that pyrolysis products meet the standards required for various industrial applications.

13.2.2 Gasification

Biomass may be gasified to create synthesis gas, or syngas, which can be used to generate heat, power, or biofuels. It includes heating biomass to a high point and partially oxidising it. The goals of gasification innovations are to maximise the generation of syngas, reduce the amount of tar formation, and boost total energy efficiency (Ahen and Amankwah-Amoah 2021).

13.2.2.1 Process Overview

Gasification is a transformational thermochemical process that effectively converts biomass into syngas through partial oxidation. Operating at high temperatures, typically between 800 and 1200 °C, gasification differs fundamentally from pyrolysis by incorporating a controlled amount of oxygen or steam into the reaction medium. This partial oxidation is decisive for the decomposition of biomass into its elementary components without complete combustion. During gasification, biomass—such as wood chips, agricultural residues or municipal solid waste is exposed to a high-temperature environment in a gasifier, where it reacts with a limited supply of oxygen or steam (Awasthi et al. 2019). The controlled introduction of these reactants enables partial oxidation of the biomass, leading to the formation of syngas, a mixture primarily composed of hydrogen (H_2), carbon monoxide (CO) and methane (CH_4). The presence of oxygen or steam not only facilitates the conversion process, but also increases the efficiency and quality of the resulting syngas by shifting the chemical reactions towards the formation of these valuable gases, rather than leaving a large amount of unconverted carbon.

13.2.2.2 Products and Their Applications

Syngas

Methane (CH_4), carbon monoxide (CO) and hydrogen (H_2) make up most of syngas, also known as synthesis gas, which is a versatile byproduct of the gasification process. The numerous uses of this combination of gases make it extremely valuable. Syngas can be used in internal combustion engines or gas turbines to produce electricity. This method offers a flexible and efficient energy source that can be included in current energy networks (Nithyanandhan et al. 2022). Syngas is a greener alternative to conventional fossil fuels in heating applications where it can be used in combined heat and power (CHP) systems or industrial burners. Besides being used directly as energy, syngas is an essential starting point for the synthesis of many chemicals and fuels. Syngas may be transformed into liquid fuels like petrol or diesel using procedures like the Fischer–Tropsch synthesis. It may also be used to make methanol, which is an essential component of many chemical goods including plastics, fertilisers, and medications. Syngas is a versatile and useful resource in the production of energy and chemicals since it can be customised in composition through the gasification process (Bocci et al. 2014).

Biochar

Biochar is another significant byproduct of gasification, sharing similarities with the biochar produced in pyrolysis. It is a stable, carbon-rich material that forms from the thermal decomposition of biomass. Biochar's high carbon content makes it particularly effective for use in agriculture, where it can improve soil health by enhancing nutrient retention, increasing water holding capacity, and promoting beneficial microbial activity (Adeoye et al. 2023). By incorporating biochar into soil, farmers can boost crop yields and improve soil structure, leading to more sustainable agricultural practices. Additionally, biochar plays a crucial role in carbon sequestration. When added to soil, it can lock away carbon for long periods, thus mitigating greenhouse gas emissions and contributing to climate change mitigation efforts. This carbon sequestration capability is significant in the context of global efforts to reduce atmospheric CO_2 levels and combat climate change. The dual benefits of enhancing soil fertility and capturing carbon make biochar a valuable byproduct in sustainable waste management and environmental stewardship (Kataya et al. 2023).

13.2.2.3 Technological Advances

Technological advances in gasification are driving significant improvements in efficiency and effectiveness, with several key areas of focus aimed at optimizing the overall process and maximizing the benefits of this technology.

Optimize Syngas Production

One of the primary goals of recent innovations in gasification is to enhance the composition and yield of syngas. By improving the process conditions—such as temperature, pressure, and the ratio of oxygen or steam—engineers and researchers aim to increase the concentration of valuable gases like hydrogen (H_2), carbon monoxide (CO), and methane (CH_4) in the syngas. Advanced catalytic materials and reaction chamber designs are being developed to facilitate more complete and efficient gasification reactions. These improvements can lead to a higher energy content in the syngas, making it more effective for use in power generation, heating, and as a precursor for chemical synthesis. Furthermore, optimizing syngas production involves fine-tuning the gasifier's operation to ensure a consistent and high-quality output, which is crucial for integrating syngas into various industrial applications (Jha et al. 2022).

Minimize Tar Formation

Tar formation is a significant challenge in the gasification process. Tars are complex hydrocarbons that can condense into a sticky, viscous substance, potentially causing operational issues such as clogging of equipment and reducing overall efficiency. Innovations are focused on minimizing tar production through several strategies. These include developing advanced gasifier designs that improve the thermal and chemical conditions to reduce tar formation, and integrating tar removal systems such as scrubbers or filters to clean the syngas before it is used (Awasthi et al. 2020). Researchers are also exploring the use of catalysts that can break down tar into simpler compounds during the gasification process, thus enhancing the overall efficiency and reliability of the system.

Increase Energy Efficiency

Enhancing energy efficiency is another critical focus of technological advancements in gasification. Innovations are being pursued to recover and utilize a greater proportion of the energy contained in the biomass feedstock. This involves optimizing system designs to improve heat recovery and integration with other processes (Bhatia et al. 2020). For example, advanced heat exchangers and combined heat and power (CHP) systems are being developed to capture and use excess heat generated during gasification. These systems can convert the heat into additional electricity or provide heating for industrial processes, thereby maximizing the overall energy yield from the biomass. Furthermore, improvements in system control and automation help in maintaining optimal operating conditions, reducing energy losses, and increasing the overall efficiency of the gasification process. Overall, these technological advances in gasification are aimed at making the process more efficient, cost-effective, and environmentally friendly, thereby enhancing its role in sustainable energy production and waste management.

13.2.2.4 Products and Their Applications

Biogas

Biogas is a valuable by-product of anaerobic digestion, mainly composed of methane (CH_4) and carbon dioxide (CO_2), with small amounts of other gases such as hydrogen sulfide (H_2S) and ammonia (NH_3). The composition of biogas makes it a versatile and renewable energy source with several key applications (Ali et al. 2020):

- Electricity generation: Biogas can be used to generate electricity by burning it in gas engines or turbines. This application is particularly beneficial in rural or remote areas where access to the power grid may be limited. The electricity produced can be used on site or fed into the local grid, contributing to the diversification of energy sources and reducing dependence on fossil fuels.

- Heating: In addition to generating electricity, biogas can be used for heating. It can be burned in industrial burners or residential heating systems to provide thermal energy for a variety of applications, including space heating and process heating in industrial facilities. This use of biogas can lower heating costs and lower greenhouse gas emissions compared to conventional fossil fuels.
- Upgraded to biomethane: Biogas can be upgraded to biomethane through a process that removes impurities such as carbon dioxide, hydrogen sulfide and water vapor, resulting in a product with a higher methane content. Biomethane is a high-energy fuel that can be used as a replacement for natural gas in vehicles and offers a cleaner alternative to gasoline and diesel. In addition, biomethane can be injected into existing natural gas distribution networks and blended with conventional natural gas to increase the sustainability of energy supplies.

Digestate

Digestate is the nutrient-rich material that remains after the anaerobic digestion process is complete. It consists of organic matter that has not been converted into biogas and contains valuable nutrients such as nitrogen, phosphorus and potassium. Applications of digestate are numerous and beneficial for the environment (Siwal et al. 2021):

- Fertilizer: Digestate can be used as a high quality fertilizer in agriculture. Its rich nutrient content helps improve soil fertility, promotes plant growth and increases crop yields. Unlike synthetic fertilizers, digestate also improves soil structure and water retention, contributing to more sustainable farming practices.
- Soil conditioner: In addition to being used as a fertilizer, digestate serves as an effective soil conditioner. It helps improve soil texture and structure, increases microbial activity and increases soil organic matter content. This can lead to better soil health, reduced erosion and increased resistance to drought and other environmental stresses.
- Composting: Digestate can be further processed by composting to create stabilized, high-quality compost. This compost can be used in landscaping, gardening and horticulture to improve soil quality and promote plant growth.

By efficiently using biogas and digestate, anaerobic digestion not only contributes to the production of renewable energy, but also supports sustainable agricultural practices and waste management. The ability to generate clean energy while producing valuable soil amendments underscores the multifaceted benefits of this technology.

13.2.2.5 Technological Progress

Advances in anaerobic digestion are driving significant improvements in efficiency, biogas quality and feedstock versatility, increasing the overall performance and applicability of the technology. These innovations are essential to optimize the process and maximize the benefits of anaerobic digestion in renewable energy production and waste management (Manickaraj et al. 2024).

Improving Digestive Efficiency

One of the key areas of progress in anaerobic digestion is increasing the efficiency of digestion. This includes the use of various pretreatment methods to improve biomass decomposition and increase biogas yield. Pretreatment techniques such as thermal, chemical or mechanical treatments are used to disrupt the structure of the biomass, making it more accessible to microorganisms. By heating biomass to high temperatures during the heat pretreatment process, complex organic components are broken down and their digestibility is improved (Motghare et al. 2016). To further facilitate microbial access, chemical pretreatment dissolves lignin and other resistant components using acids, bases, or oxidizing agents. Biomass is ground or crushed as part of a mechanical pretreatment process to reduce particle size and increase the surface area available for microbial activity. Researchers and engineers can improve waste management and biogas output by increasing the efficiency of the digestion process through the optimization of various pretreatments.

Expanding Feedstock Diversity

Another important area of technical improvement is to increase the variety of feedstocks that can be digested effectively. Conventional anaerobic digestion systems tend to concentrate on organic materials that are readily broken down, including food scraps and agricultural waste. But in order to fully use anaerobic digestion, methods for processing a larger range of biomass feedstocks including those with high lignin concentration, such as woody materials and some industrial byproducts need to be developed (Manickaraj et al. 2022). Innovations in enzyme technology, microbial strain development, and process optimization are helping to improve the digestion of more complex and resistant materials. By expanding feedstock diversity, anaerobic digestion systems can better accommodate different types of organic waste and biomass, leading to more flexible and sustainable waste management solutions. All things considered, these technical developments are improving anaerobic digestion's performance and flexibility, making it a more effective and flexible technique for managing waste and producing renewable energy. Anaerobic digestion has enhanced feedstock diversity, biogas quality, and digestion efficiency, making it more suitable to address the increasing need for sustainable energy and environmental management solutions.

13.2.3 *Comparative Analysis and Summary*

The Table 13.2 shows the feature of Comparative Table for advanced Technologies of bio mass.

13.3 Integration of AI and IoT

The integration of Artificial Intelligence (AI) and Internet of Things (IoT) technologies in biomass waste processing enhances operational efficiency, predictive maintenance, and process optimization. These technologies enable real-time monitoring, data analytics, and decision-making, contributing to sustainable waste management practices (Duarah et al. 2022). The Fig. 13.1 shows the important consideration in an Integration of AI and IoT in biomass waste processing.

Table 13.2 Comparative table of technologies

Feature	Pyrolysis	Gasification	Anaerobic digestion
Operating temperature	300–800 °C	800–1,200 °C	Ambient—60 °C
Main products	Bio-oil, syngas, biochar	Syngas, biochar	Biogas, digestate
Feedstock suitability	Woody biomass, agri-residues	Various biomass types	Organic waste streams
Energy efficiency	Moderate	High	Moderate
Product quality focus	Bio-oil enhancement	Syngas optimization	Biogas quality improvement
Technological maturity	Developing	Established in large-scale	Widely adopted

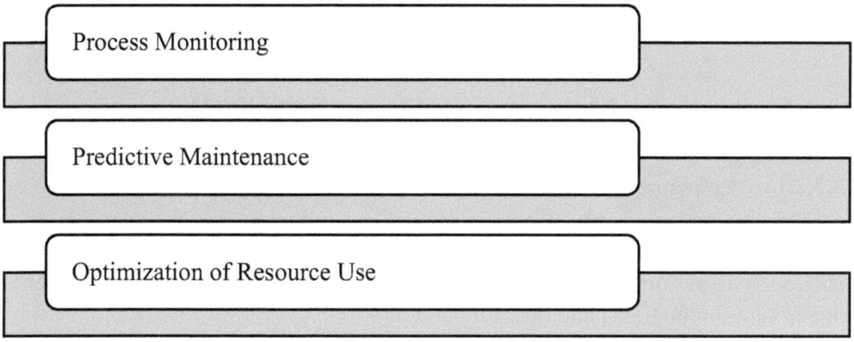

Fig. 13.1 Integration of AI and IoT in biomass waste processing

13.3.1 Process Monitoring (Jatoi et al. 2021)

13.3.1.1 Real-Time Monitoring

AI-powered systems, coupled with Internet of Things (IoT) sensors, continuously gather and analyze data from various points within biomass processing facilities. These sensors measure crucial parameters like temperature, moisture content, gas composition, and flow rates, providing a comprehensive view of the operational conditions.

13.3.1.2 Data Analysis and Visualization

AI algorithms process this data to detect patterns, anomalies, and trends. Advanced visualization tools present this information in real-time dashboards, allowing operators to monitor the system's performance immediately.

13.3.1.3 Dynamic Adjustments

Based on the analyzed data, AI systems can make automatic adjustments to process parameters. For instance, if the temperature or moisture content deviates from the optimal range, the system can tweak the settings to maintain ideal conditions, thereby improving the quality and consistency of the biomass product.

13.3.1.4 Enhanced Quality Control

By continuously monitoring and adjusting processes, AI helps in maintaining high standards of product quality. This includes ensuring the right balance of components in the final product and adhering to regulatory standards.

13.3.2 Predictive Maintenance (Gautam et al. 2020)

13.3.2.1 Performance Data Collection

AI systems collect and analyze performance data from various equipment components, such as motors, pumps, and conveyors. This data includes vibration patterns, temperature fluctuations, and operational hours.

13.3.2.2 Predictive Analytics

Advanced machine learning models analyze historical and real-time data to predict when equipment is likely to fail or require maintenance. These models can identify subtle signs of wear and tear that might not be apparent through traditional monitoring methods.

13.3.2.3 Maintenance Scheduling

Based on these predictions, maintenance schedules are optimized to perform necessary repairs or replacements before actual failures occur. This approach minimizes unexpected downtime and extends the lifespan of equipment.

13.3.2.4 Cost Savings and Efficiency

Predictive maintenance reduces the need for costly emergency repairs and minimizes production interruptions. It also helps in optimizing resource allocation, as maintenance tasks are carried out based on actual need rather than routine schedules.

13.3.3 Optimization of Resource Use (Heinimö and Junginger 2009)

13.3.3.1 Efficient Resource Management

- AI-driven optimization algorithms integrate real-time data from various sensors and process control systems to manage resource use efficiently. This includes adjusting parameters such as feed rates, combustion conditions, and energy input levels.
- AI algorithms can optimize processes to maximize energy recovery from biomass. For example, they can adjust the combustion process to enhance heat recovery, or optimize the use of by-products in energy generation.
- By continuously analyzing process efficiency and material usage, AI systems identify opportunities to reduce waste. This includes optimizing the separation of valuable components from waste streams and minimizing off-spec production.
- Effective resource optimization contributes to sustainability goals by reducing the environmental impact of biomass processing. This includes lowering emissions, minimizing waste, and improving overall resource efficiency.

In summary, AI and IoT technologies revolutionize biomass processing by providing real-time insights, predicting maintenance needs, and optimizing resource use. These advancements lead to improved process efficiency, reduced downtime, and enhanced product quality, aligning with both operational and environmental goals.

13.4 Biochemical Processes

Enzymes and microbes are used in biochemical processes to transform biomass into useful compounds such as sugars, biofuels, and bioplastics. Expanding the spectrum of biomass feedstocks and attaining high conversion efficiencies need these techniques (Kushwaha et al. 2022) (Fig. 13.2).

Biochemical processes play a pivotal role in transforming biomass into a variety of high-value products such as sugars, biofuels, bioplastics, and other chemicals. These processes harness the natural capabilities of enzymes and microorganisms to achieve high conversion efficiencies and broaden the scope of usable biomass feedstocks. Here's a closer look at the key components and advancements within biochemical processes.

13.4.1 Enzymatic Hydrolysis

13.4.1.1 Principles and Process

Enzymatic hydrolysis uses special enzymes to break down complex polyphases such as cellulose and hemicellulose in biomass into simple sugars Cellulose is converted to glucose, while hemicellulose yields a mixture of sugars such as xylose and arabinose (Antar et al. 2021). These simple sugars are then fermented into various biofuels and chemicals. For instance, glucose can be fermented into ethanol, while a combination of sugars might be used to produce butanol.

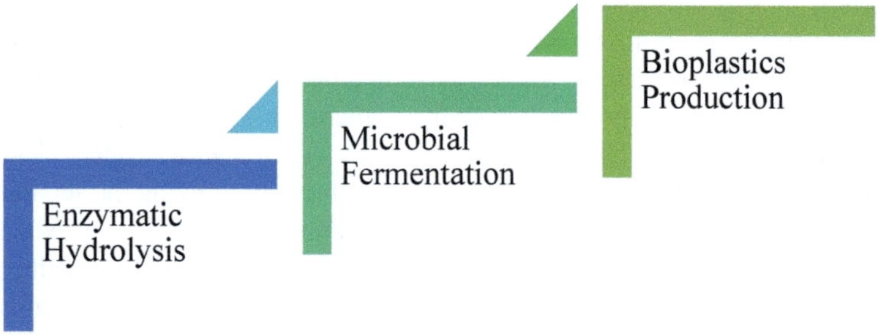

Fig. 13.2 Biochemical processes

13.4.1.2 Advancements and Innovations

Research is continually focused on developing more efficient enzymes that can accelerate the hydrolysis process. This includes engineering enzymes with enhanced activity and stability under varying operational conditions. Enzyme production and application can be costly, so efforts are directed at reducing costs through more efficient production methods, enzyme recycling, and using less expensive feedstocks (Dam et al. 2008). Enhancements in enzyme specificity can lead to more effective breakdown of specific types of biomass, enabling better utilization of diverse feedstocks and improving overall conversion rates.

13.4.2 Microbial Fermentation

13.4.2.1 Principles and Process

Microbial fermentation uses microorganisms such as yeast and bacteria to convert fermentable sugars derived from biomass into biofuels and biochemicals. Common biofuels include ethanol and biodiesel, while valuable chemicals such as lactic acid and succinic acid are also produced. These microorganisms utilize their metabolic pathways to convert sugars into desired products (Scarlat et al. 2015). For example, yeast converts glucose into ethanol, while certain bacteria produce lactic acid from sugars.

13.4.2.2 Advancements and Innovations

Significant progress is being made in genetically engineering microorganisms to enhance their metabolic pathways. These engineered strains are designed to achieve higher product yields, exhibit increased tolerance to inhibitors present in biomass, and handle various types of feedstocks more efficiently. Biomass processing can produce inhibitory compounds that affect microbial activity. Research focuses on developing strains with improved tolerance to these inhibitors, thereby maintaining high fermentation efficiency and product yield (Sansaniwal et al. 2017). Innovations in fermentation technology, such as optimizing fermentation conditions (e.g., temperature, pH) and developing more efficient fermentation vessels, contribute to higher productivity and reduced costs.

13.4.3 Bioplastics Production

13.4.3.1 Principles and Process

Bioplastics are produced through biochemical processes that convert biomass-derived sugars or oils into plastic materials. As an environmentally friendly substitute for traditional plastics made from fossil fuels, these bioplastics are compostable or biodegradable. Two popular bioplastics are polyhydroxyalkanoates (PHA), which are made by microorganisms from sugars or lipids, and polylactic acid (PLA), which is made from fermented lactic acid (Vertes et al. 2011).

13.4.3.2 Advancements and Innovations

Research aims to improve the properties of bioplastics, such as their strength, flexibility, and durability, to match or exceed those of conventional plastics. This involves modifying production processes and using different biomass feedstocks. Developing cost-effective methods for producing bioplastics is crucial for widespread adoption. Innovations include optimizing the fermentation and polymerization processes and using more economical feedstocks. The production of bioplastics is designed to minimize environmental impact (Clauser et al. 2021). Bio plastics can reduce reliance on fossil fuels, decrease greenhouse gas emissions, and offer end-of-life options such as composting. In summary, biochemical processes are at the forefront of converting biomass into valuable products using enzymes and microorganisms. Advances in enzymatic hydrolysis, microbial fermentation, and bioplastics production are continually improving the efficiency, cost-effectiveness, and sustainability of these processes, enabling a broader range of applications and contributing to a more sustainable future (Duarah et al. 2022).

13.5 Decentralized and Modular Systems

There is a growing trend towards decentralized and modular biomass waste processing systems that can be scaled and adapted to different geographical locations and feedstock types (Fig. 13.3).

13.5.1 Community-Scale Systems

The rise of decentralized and modular biomass waste processing systems reflects a shift towards more adaptable and scalable solutions for handling biomass across various settings. Small-scale biomass processing systems, which convert locally

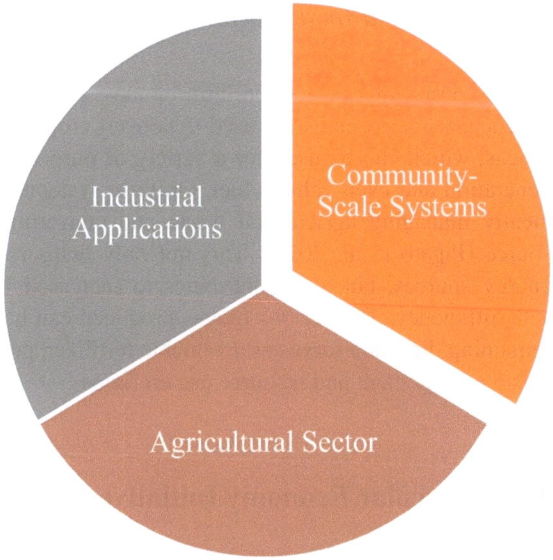

Fig. 13.3 Decentralized and modular system

obtained biomass waste—such as municipal garbage, wood chips, and agricultural residues—into useful energy sources including heat, power, and biofuels, provide a feasible solution for local communities. These systems reduce the expenses and environmental effects of transporting garbage to central processing facilities, while simultaneously promoting energy independence for nearby communities (Antar et al. 2021). By generating energy close to where the biomass is produced, communities can also achieve greater self-reliance and sustainability.

13.5.2 Industrial Applications

In industrial contexts, modular biomass processing units are becoming increasingly prevalent, particularly in sectors like agriculture, food processing, and forestry where large volumes of biomass waste are generated. These modular systems are designed to be flexible and scalable, allowing industries to adapt their waste management and energy production processes according to the specific types and quantities of biomass they handle (Zabaniotou 2018). This flexibility enables seamless integration with existing industrial operations, optimizing waste-to-energy conversion and reducing overall operational costs. The ability to tailor these systems to handle diverse feedstocks ensures that industries can maximize the utility of their biomass waste while maintaining efficiency (Worrell et al. 2001).

13.5.3 Agricultural Sector

For the agricultural sector, on-farm biomass processing systems represent a major advance. These systems are used to convert crop residues and animal manure into biogas, which can be used for a variety of purposes including heating, electricity generation and even vehicle fuel. On-farm systems increase sustainability by efficiently managing agricultural waste and converting it into a renewable energy source (Pigato et al. 2020). This not only helps reduce dependence on external energy sources, but also contributes to increased farm productivity and energy self-sufficiency. The various biogas produced can be used to power farms and the remaining digestate serves as a valuable fertilizer, promoting a circular approach to waste management and resource use on farms.

13.6 Circular Economy Initiatives

In the context of the circular economy, biomass waste is increasingly seen as a critical resource that can be harnessed to drive sustainability and efficiency. This approach seeks to redefine waste as a valuable input, transforming traditional linear models of consumption and disposal into more regenerative systems (Tigabu et al. 2015a) (Fig. 13.4).

13.6.1 Waste Valorization

Circular economy initiatives promote the valorization of biomass waste through recycling, reusing, and recovering valuable materials. For example, organic waste streams

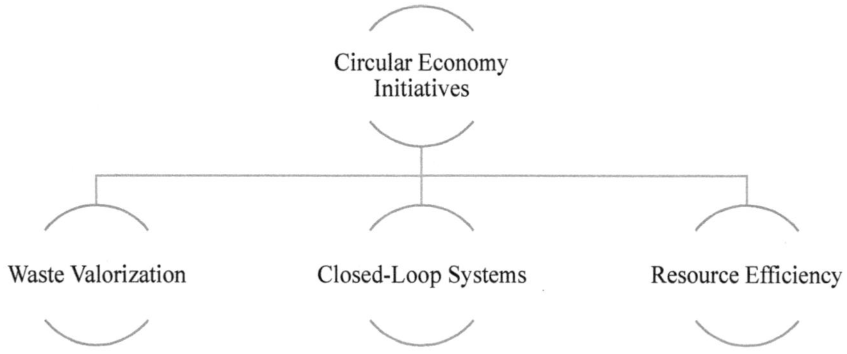

Fig. 13.4 Circular economy initiatives

are converted into bio-based products such as biofuels, biochemicals, and biocomposites (Dahunsi et al. 2020). Waste Valorization is a central concept in circular economy initiatives, focusing on maximizing the value extracted from biomass waste through processes like recycling, reusing, and recovering. For instance, organic waste streams, including agricultural residues and food waste, can be converted into bio-based products such as biofuels, biochemicals, and biocomposites. This transformation not only provides alternative resources but also reduces the burden on landfills and decreases the environmental impact associated with waste disposal (Ockwell et al. 2010). By turning waste into valuable products, we close the loop in resource use, turning potential pollutants into beneficial materials.

13.6.2 Closed-Loop Systems

Closed-loop biomass processing systems aim to create circular supply chains where waste streams are continuously recycled and reused. By-products and residues from biomass processing are utilized as feedstocks for other industrial processes or returned to the soil as organic fertilizers. Closed-Loop Systems represent a more integrated approach to managing biomass waste, where the goal is to establish circular supply chains (Vasco-Correa et al. 2018). In these systems, by-products and residues from biomass processing are systematically redirected back into the production cycle. For example, the residuals from biofuel production can be used as animal feed or organic fertilizers, thereby maintaining the value of the original biomass throughout its lifecycle. This continuous recycling and reuse reduce the need for new raw materials and help in creating a more sustainable production model. By returning organic matter to the soil, closed-loop systems also enhance soil health and support sustainable agricultural practices (Dam et al. 2008).

13.6.3 Resource Efficiency

Biomass waste processing contributes to reducing dependence on limited resources and mitigating the environmental influences associated with conventional waste disposal techniques. It supports sustainable improvement goals by using keeping herbal sources and decreasing greenhouse gas emissions. Resource efficiency is some other key benefit of incorporating biomass waste into round economic system practices. Biomass waste processing reduces dependence on confined sources and decreases the environmental effect usually related to traditional waste management methods (Golembiewski et al. 2015). By recycling biomass waste, industries and groups make contributions to the conservation of herbal resources and the discount of greenhouse fuel emissions. This no longer best supports the wider dreams of sustainable development, however, additionally aligns with global efforts to reduce ecological footprints and fight climate alternate. Through these initiatives, biomass waste processing turns into an vital part of an extra sustainable and circular financial version.

13.7 International Collaboration and Standards

International collaboration and the establishment of standards play a crucial role in advancing biomass waste processing technologies globally. These efforts foster innovation, ensure consistent practices, and support the widespread adoption of sustainable solutions. Global collaboration in research, development, and standardization of biomass waste processing technologies facilitates technology transfer and adoption across different regions are shown in Fig. 13.5 (Ben-Iwo et al. 2016).

13.7.1 Knowledge Sharing

International partnerships and collaborative research initiatives accelerate innovation in biomass waste processing technologies. Shared knowledge and best practices enable countries to learn from each other's experiences and overcome technological barriers. Knowledge sharing through international partnerships accelerates advancements in biomass waste processing technologies. Collaborative research initiatives and joint projects enable countries to pool their expertise, share successful strategies, and address common challenges (Tigabu et al. 2015b). By learning from each other's experiences, nations can overcome technological barriers more efficiently and

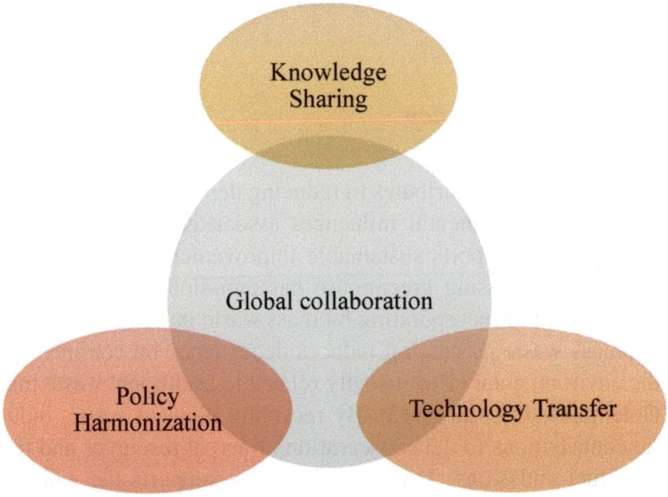

Fig. 13.5 Global collaboration in different regions

develop innovative solutions tailored to their specific needs. For example, research institutions and industry leaders from different countries can work together on projects that enhance the efficiency of biofuel production or improve waste-to-energy technologies, thus driving progress in the field on a global scale.

13.7.2 Policy Harmonization

Standardization of policies and regulations related to biomass waste management promotes investment in sustainable technologies and creates a level playing field for industry stakeholders. Harmonized standards ensure compliance with environmental, health, and safety requirements across international markets (Schmidt and Dabur 2014). Policy Harmonization ensures that the regulatory framework for biomass waste management is consistent across different regions. Standardizing policies and regulations helps promote investment in sustainable technologies by creating a predictable and fair regulatory environment. This harmonization ensures that industry stakeholders operate under the same set of rules, which can enhance market access and reduce trade barriers. Additionally, unified standards help ensure compliance with environmental, health, and safety requirements, providing assurance to both consumers and regulators that biomass processing technologies meet high standards of performance and safety.

13.7.3 Technology Transfer

International collaboration facilitates the transfer of proven biomass processing technologies to developing countries, where waste management challenges are particularly acute. Capacity building and technical assistance programs support the adoption of sustainable practices and promote economic development (Dutz and Sharma 2012). Technology Transfer is a key aspect of international collaboration, especially in facilitating the adoption of biomass processing technologies in developing countries. Many developing regions face significant waste management challenges and could benefit greatly from proven technologies and practices. Through international cooperation, these technologies can be transferred, along with the necessary training and technical assistance. Capacity building initiatives help local stakeholders understand and implement sustainable practices, leading to improved waste management, energy production, and economic development. For instance, technology transfer programs can support the establishment of small-scale biomass processing units in developing countries, providing both the tools and knowledge needed to address local waste management issues and promote sustainable growth (Hiloidhari et al. 2023).

13.8 Focus on Sustainability and Carbon Neutrality

Governments and industries worldwide are prioritizing sustainability and carbon neutrality goals, driving investments in sustainable biomass waste management solutions. Figure 13.6 shows the key considerations in the bio waste management solutions (Chen et al. 2022).

13.8.1 Climate Change Mitigation

Climate change mitigation considering the following key points, and these are explained in a brief version.

13.8.1.1 Biomass Waste Processing

The process of converting organic waste materials into valuable goods such as compost, biogas and fuels is known as biomass waste processing. By removing organic waste from incinerators and landfills, this technique dramatically reduces greenhouse gas emissions. Methane is a powerful greenhouse gas that is produced in landfills during the decomposition of organic waste. Methane emissions are avoided when these wastes are processed into biofuels or biogas, which allows beneficial use of the waste (Mohan et al. 2024).

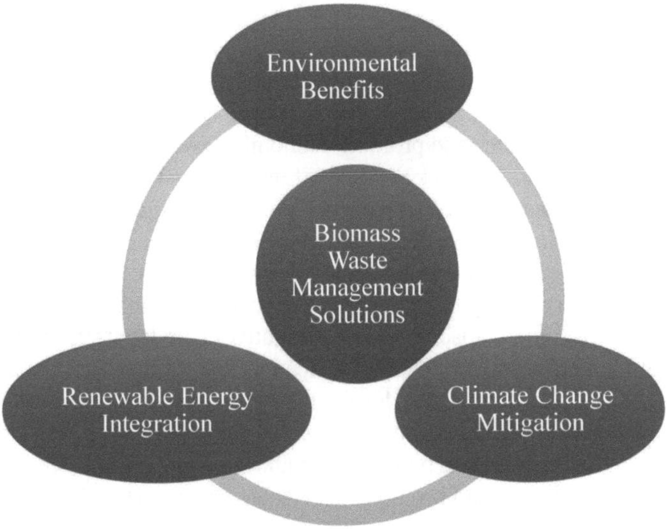

Fig. 13.6 Biomass waste management solutions

13.8.1.2 Biofuels and Decarbonization

Biofuels derived from biomass, such as ethanol and biodiesel, are renewable alternatives to fossil fuels. They help decarbonize the transport sector by reducing dependence on petroleum-based fuels. Combustion of biofuels releases carbon dioxide, but this CO_2 was previously absorbed by the biomass during its growth, creating a closed carbon cycle and reducing net greenhouse gas emissions (Tisi et al. 2023).

13.8.1.3 Biogas in Energy Production

Heat and strength can be constructed from biogas, that is produced during the anaerobic digestion of organic waste. Emissions of carbon dioxide and other pollution are reduced when biogas is used instead of fossil fuels (Tsui and Wong 2019). In addition, biogas structures can provide a sustainable and nearby source of electricity, assisting to fulfill universal emission discount goals.

13.8.2 Integration of Renewable Energy

Integrating renewable energy into bio-waste processing considering the following points and these are explained below.

13.8.2.1 Biofuels and Energy Systems

Biofuels obtained from biomass and biogas are an integral part of renewable energy systems. Their use increases energy security by providing a reliable locally sourced energy alternative to imported fossil fuels. Incorporating these biofuels into existing energy infrastructure, such as blending them with gasoline or diesel, supports a gradual transition to cleaner energy sources (Ngan et al. 2019).

13.8.2.2 Sustainable Biomass Input Raw Materials

To make sure that biomass energy production is honestly sustainable, it's miles vital to apply raw substances that do not compete with food vegetation and do not cause deforestation. Agricultural residues including straw and chaff and specialized power crops including millet and miscanthus are examples of sustainable raw materials. These materials make contributions to renewable energy diversification techniques with the aid of imparting regular and environmentally friendly strength resources (Ben-Iwo et al. 2016).

13.8.3 Environmental Benefits

The following environment benefits are considered in the bio waste management solutions (Raj et al. 2023).

13.8.3.1 Air and Water Quality Improvement

There are significant environmental advantages to sustainable biomass waste management techniques. Composting organic waste, for example, lessens the need for synthetic fertilizers, which can contaminate rivers when they run off. Furthermore, using the right biomass processing methods reduces air pollutants including particulate matter and volatile organic compounds, which are frequently emitted while burning garbage.

13.8.3.2 Reduction in Soil Contamination

By diverting waste from landfills and preventing uncontrolled burning, biomass waste processing helps to reduce soil contamination. This method ensures that potentially harmful substances in the waste do not leach into the soil or groundwater, thereby preserving soil health and preventing contamination of water resources.

13.8.3.3 Conservation of Biodiversity

Sustainable biomass processing supports conservation efforts by reducing the various environmental impact of the waste disposal. The traditional wastes disposal methods, such as landfilling and incineration, can have detrimental effects on local ecosystems. By mitigating these impacts, biomass processing helps maintain ecosystem resilience and supports biodiversity. Additionally, using biomass from agricultural residues and dedicated energy crops minimizes land-use changes that could otherwise threaten natural habitats.

13.8.3.4 Ecosystem Resilience and Resource Conservation

Biomass processing encourages behaviours that preserve and restore natural equilibrium, which strengthens ecosystem resilience. Biomass processing promotes overall ecological health and the preservation of natural resources like forests and water bodies by lowering dependency on fossil fuels and minimizing waste.

13.9 Market Expansion

The global market for biomass waste processing technologies is expanding due to supportive government policies, subsidies, and increasing awareness of environmental issues (Safferman et al. 2017) is shown in Fig. 13.7.

13.9.1 Investment Opportunities

Growing demand for renewable energy and sustainable products drives investment in biomass waste processing infrastructure and technologies. Financial incentives, various tax credits, and the feed-in tariffs to encourage private sector participation in the biomass-to-energy conversion projects.

13.9.2 Technological Innovation

Market expansion fosters technological innovation in biomass waste processing, including cost-effective solutions for biomass collection, preprocessing, and conversion. The goals of research and development are to increase market competitiveness, lower capital costs, and improve efficiency.

13.9.3 Market Dynamics

Biomass waste processing technologies cater to diverse market segments, including municipal waste management, agriculture, forestry, and industrial sectors. Market dynamics vary by region, influenced by regulatory frameworks, energy policies, and availability of biomass resources.

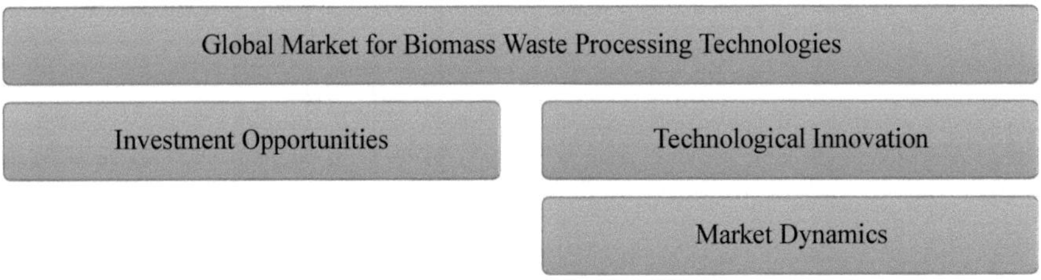

Fig. 13.7 Global market for biomass waste processing technologies

13.10 Emerging Economies

Emerging economies are increasingly turning to biomass waste processing as a viable and sustainable solution to tackle waste management issues and bolster energy security. As these countries grapple with rapid urbanization, population growth, and limited resources, biomass waste processing offers a multifaceted approach to address these challenges while promoting economic and environmental benefits (Adamu et al. 2023). Developing countries are exploring biomass waste processing as a sustainable solution to address waste management challenges and enhance energy security (Fig. 13.8).

13.10.1 Local Resource Utilization

Biomass waste processing leverages the abundance of locally available feedstocks, including agricultural residues, forestry waste, and organic municipal solid waste, to generate energy and valuable products. This approach minimizes the need for imported fossil fuels, which can be costly and environmentally damaging. By tapping into local resources, developing countries not only reduce their dependency on external energy sources but also enhance the sustainability of their energy systems. The use of biomass contributes to a more circular economy by recycling waste materials into useful products, thereby reducing landfill use and associated greenhouse gas emissions. Furthermore, processing biomass locally supports rural development by creating opportunities for local farmers and communities to participate in and benefit from the energy sector (Nižetić et al. 2019). This shift towards utilizing local resources helps in strengthening local economies and promotes energy security by making communities more self-reliant.

Fig. 13.8 Waste management challenges

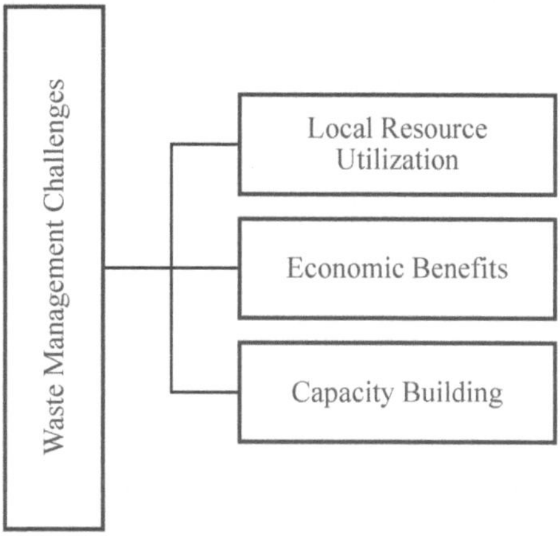

13.10.2 Economic Benefits

The establishment of biomass-to-energy projects brings substantial economic advantages. These projects generate numerous employment opportunities across various stages, including the construction, operation, and maintenance of biomass processing facilities. Effective waste management solutions provided by biomass processing also contribute to cost savings for municipalities and businesses. By fostering economic growth through these channels, biomass projects play a critical role in poverty reduction, helping to uplift communities and support sustainable economic development (Lange et al. 2021). Overall, the economic ripple effect of these projects supports a broader economic ecosystem and contributes to long-term prosperity.

13.10.3 Capacity Building

International development assistance programs are instrumental in building capacity for biomass waste processing in developing countries. These programs offer crucial support in the form of technical expertise, training, and financial resources, which are essential for establishing and maintaining effective biomass processing systems. Knowledge transfer initiatives enable local stakeholders to acquire the necessary skills and understanding to operate biomass facilities efficiently. Additionally, technology transfer provides access to advanced technologies and best practices, ensuring that developing countries can adopt and adapt innovative solutions tailored to their specific needs. This support not only promotes sustainable development but also enhances the resilience of communities to the impacts of climate change (Li et al. 2017). By strengthening local capabilities and fostering an environment of continuous learning and improvement, these international programs contribute to a more sustainable and resilient future for developing countries.

13.11 Public Awareness and Acceptance

The increasing public awareness of biomass waste processing underscores its significant environmental and economic benefits, which in turn influence policy decisions and consumer behaviors. This heightened awareness is driven by efforts to educate the public about the advantages of sustainable waste management and the positive impacts of transitioning to biomass processing technologies (Jain et al. 2022). There is growing public awareness of the environmental and economic benefits of biomass waste processing, influencing policy decisions and consumer behavior (Fig. 13.9).

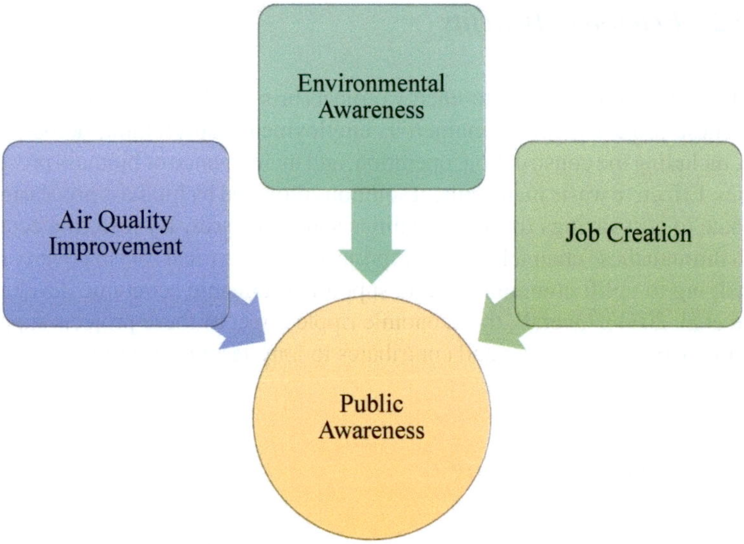

Fig. 13.9 Influence of public awareness

13.11.1 Environmental Awareness

Public education campaigns raise awareness about the environmental impacts of conventional waste disposal methods and promote the benefits of biomass waste processing. Community engagement fosters support for sustainable waste management practices and encourages participation in recycling and composting programs. Environmental Awareness is a key driver of public support for biomass waste processing. Educational campaigns and community outreach initiatives highlight the environmental drawbacks of conventional waste disposal methods, such as landfilling and open burning, which contribute to greenhouse gas emissions and soil contamination. By contrast, biomass waste processing offers a more sustainable alternative, converting organic waste into valuable resources like biofuels and biocomposites. These campaigns also emphasize the role of biomass processing in reducing waste volumes and promoting recycling and composting (Moustakas and Loizidou 2021). As communities become more informed about these benefits, they are more likely to engage in and support sustainable waste management practices, fostering a culture of environmental responsibility.

13.11.2 Air Quality Improvement

Biomass waste processing reduces air pollution by diverting organic waste from open burning and incineration. Biogas production from anaerobic digestion displaces fossil fuels in cooking, heating, and transportation, contributing to cleaner air and improved public health. Air Quality Improvement is another significant benefit of biomass waste processing that resonates with the public. Traditional waste disposal methods, such as open burning and incineration, release harmful pollutants into the air, adversely affecting air quality and public health. Biomass waste processing, particularly through anaerobic digestion, mitigates these issues by converting organic waste into biogas, which can replace fossil fuels in cooking, heating, and transportation. This shift not only reduces reliance on polluting fossil fuels but also decreases emissions of particulate matter and other air pollutants, leading to cleaner air and better health outcomes for communities (Omer 2011). By showcasing these improvements, public awareness campaigns can garner support for biomass technologies and underscore their role in enhancing air quality.

13.11.3 Job Creation

Job Creation is a crucial aspect of the economic impact of biomass waste processing that appeals to the public and policymakers alike. The renewable energy sector, including biomass processing, generates employment opportunities across various stages of the value chain, from manufacturing and construction to operation and maintenance. The renewable energy sector, including biomass waste processing, creates job opportunities in manufacturing, construction, operation, and maintenance (Yana et al. 2022). Local employment benefits communities and enhances social inclusion, particularly in rural and marginalized areas. These jobs contribute to local economies, particularly in rural and marginalized areas where economic opportunities may be limited. By creating stable, well-paying jobs, biomass processing supports economic development and social inclusion, helping to bridge employment gaps and improve livelihoods. Public recognition of these benefits can drive support for biomass initiatives and encourage investment in the renewable energy sector (Scarlat et al. 2015).

13.12 Conclusion

In conclusion, the ongoing innovations in biomass waste processing are revolutionizing global waste management practices, reflecting a broader movement toward sustainable development, renewable energy, and climate change mitigation. Advanced conversion technologies, such as gasification, pyrolysis, and hydrothermal liquefaction, are enhancing the efficiency of transforming biomass waste into valuable resources like biofuels and chemicals. These technologies, coupled with the integration of Artificial Intelligence (AI) and the Internet of Things (IoT), are optimizing processing systems through real-time monitoring and predictive analytics, leading to greater efficiency and reduced costs. Additionally, biochemical processes, including anaerobic digestion and microbial fermentation, leverage the power of microorganisms to convert organic waste into biogas and bioethanol, further underscoring the versatility of biomass waste processing. Decentralized systems are also emerging as a key trend, bringing waste processing closer to its source and thereby reducing transportation costs and emissions. The push towards circular economy initiatives is driving a shift from linear waste management to systems that emphasize recycling, material recovery, and closed-loop processes, minimizing environmental impact and enhancing resource efficiency. International collaboration is essential in advancing these innovations, facilitating knowledge exchange, and standardizing practices to tackle global challenges effectively. The focus on sustainability and carbon neutrality is further shaping the field, with efforts concentrated on reducing greenhouse gas emissions and promoting cleaner energy alternatives. As market opportunities expand, particularly in emerging economies, and public awareness of environmental issues grows, biomass waste is increasingly recognized as a valuable resource. This recognition is pivotal in fostering a transition towards a more sustainable and resilient society, where waste management practices not only address immediate challenges but also contribute to long-term environmental and economic sustainability.

References

Adamu H, Bello U, Yuguda AU, Tafida UI, Jalam AM, Sabo A, Qamar M (2023) Production processes, techno-economic and policy challenges of bioenergy production from fruit and vegetable wastes. Renew Sustain Energy Rev 186:113686

Adeoye AO, Lawal OS, Quadri RO, Malomo D, Aliyu MT, Dang GE, Hikon BN (2023) Sustainable energy via thermochemical and biochemical conversion of biomass wastes for biofuel production. In: Transportation energy and dynamics. Singapore, Springer Nature Singapore, pp 245–306

Ahen F, Amankwah-Amoah J (2021) Sustainable waste management innovations in Africa: new perspectives and research agenda for improving global health. Sustainability 13(12):6646

Ali J, Rasheed T, Afreen M, Anwar MT, Nawaz Z, Anwar H, Rizwan K (2020) Modalities for conversion of waste to energy—challenges and perspectives. Sci Total Environ 727:138610

Ambaye TG, Vaccari M, Bonilla-Petriciolet A, Prasad S, van Hullebusch ED, Rtimi S (2021) Emerging technologies for biofuel production: a critical review on recent progress, challenges and perspectives. J Environ Manage 290:112627

Antar M, Lyu D, Nazari M, Shah A, Zhou X, Smith DL (2021) Biomass for a sustainable bioeconomy: an overview of world biomass production and utilization. Renew Sustain Energy Rev 139:110691

Awasthi MK, Sarsaiya S, Wainaina S, Rajendran K, Kumar S, Quan W, Taherzadeh MJ (2019) A critical review of organic manure biorefinery models toward sustainable circular bioeconomy: technological challenges, advancements, innovations, and future perspectives. Renew Sustain Energy Rev 111:115–131

Awasthi MK, Sarsaiya S, Patel A, Juneja A, Singh RP, Yan B, Taherzadeh MJ (2020) Refining biomass residues for sustainable energy and bio-products: an assessment of technology, its importance, and strategic applications in circular bio-economy. Renew Sustain Energy Rev 127:109876

Ben-Iwo J, Manovic V, Longhurst P (2016) Biomass resources and biofuels potential for the production of transportation fuels in Nigeria. Renew Sustain Energy Rev 63:172–192
Bhatia RK, Ramadoss G, Jain AK, Dhiman RK, Bhatia SK, Bhatt AK (2020) Conversion of waste biomass into gaseous fuel: present status and challenges in India. BioEnergy Res 13:1046–1068
Bocci E, Di Carlo A, McPhail SJ, Gallucci K, Foscolo PU, Moneti M, Carlini M (2014) Biomass to fuel cells state of the art: a review of the most innovative technology solutions. Int J Hydrogen Energy 39(36):21876–21895
Castaldi M, van Deventer J, Lavoie JM, Legrand J, Nzihou A, Pontikes Y, Verstraete W (2017) Progress and prospects in the field of biomass and waste to energy and added-value materials. Waste Biomass Valoriz 8:1875–1884
Chen L, Msigwa G, Yang M, Osman AI, Fawzy S, Rooney DW, Yap PS (2022) Strategies to achieve a carbon neutral society: a review. Environ Chem Lett 20(4):2277–2310
Chithra NV, Karuppasamy R, Manickaraj K, Ramakrishnan T (2024) Effect of reinforcement addition on mechanical behavior of Al MMC-A critical review. J Environ Nanotechnol 13(2):65–79
Cho EJ, Trinh LTP, Song Y, Lee YG, Bae HJ (2020) Bioconversion of biomass waste into high value chemicals. Biores Technol 298:122386
Clauser NM, González G, Mendieta CM, Kruyeniski J, Area MC, Vallejos ME (2021) Biomass waste as sustainable raw material for energy and fuels. Sustainability 13(2):794
Dahunsi SO, Fagbiele OO, Yusuf EO (2020) Bioenergy technologies adoption in Africa: a review of past and current status. J Clean Prod 264:121683
Demirbaş A (2001) Biomass resource facilities and biomass conversion processing for fuels and chemicals. Energy Convers Manage 42(11):1357–1378
Demirbas MF, Balat M (2006) Recent advances on the production and utilization trends of bio-fuels: a global perspective. Energy Convers Manage 47(15–16):2371–2381
Dessie W, Luo X, Wang M, Feng L, Liao Y, Wang Z, Qin Z (2020) Current advances on waste biomass transformation into value-added products. Appl Microbiol Biotechnol 104(11):4757–4770
Duarah P, Haldar D, Patel AK, Dong CD, Singhania RR, Purkait MK (2022) A review on global perspectives of sustainable development in bioenergy generation. Biores Technol 348:126791
Dutz MA, Sharma S (2012) Green growth, technology and innovation. World Bank Policy Res Work Pap (5932)
Garland JL, Mackowiak CL, Strayer RF, Finger BW (1997) Integration of waste processing and biomass production systems as part of the KSC breadboard project. Adv Space Res 20(10):1821–1826
Gaudino EC, Cravotto G, Manzoli M, Tabasso S (2019) From waste biomass to chemicals and energy via microwave-assisted processes. Green Chem 21(6):1202–1235
Gautam P, Upadhyay SN, Dubey SK (2020) Bio-methanol as a renewable fuel from waste biomass: current trends and future perspective. Fuel 273:117783
Golembiewski B, Sick N, Bröring S (2015) The emerging research landscape on bioeconomy: what has been done so far and what is essential from a technology and innovation management perspective? Innov Food Sci Emerg Technol 29:308–317
Gumisiriza R, Hawumba JF, Okure M, Hensel O (2017) Biomass waste-to-energy valorisation technologies: a review case for banana processing in Uganda. Biotechnol Biofuels 10:1–29
Gurusamy M, Soundararajan S, Karuppusamy M, Ramasamy K (2024) Exploring the mechanical impact of fine powder integration from ironwood sawdust and COCO dust particles in epoxy composites. Matéria (Rio De Janeiro) 29(3):e20240216
Heinimö J, Junginger M (2009) Production and trading of biomass for energy–an overview of the global status. Biomass Bioenerg 33(9):1310–1320
Hiloidhari M, Sharno MA, Baruah DC, Bezbaruah AN (2023) Green and sustainable biomass supply chain for environmental, social and economic benefits. Biomass Bioenerg 175:106893
Hoornweg D, Thomas L, Otten L (1999) Composting and its applicability in developing countries. World Bank Work Pap Ser 8:1–46

Jain A, Sarsaiya S, Awasthi MK, Singh R, Rajput R, Mishra UC, Shi J (2022) Bioenergy and bio-products from bio-waste and its associated modern circular economy: current research trends, challenges, and future outlooks. Fuel 307:121859

Jatoi AS, Abbasi SA, Hashmi Z, Shah AK, Alam MS, Bhatti ZA, Iqbal A (2021) Recent trends and future perspectives of lignocellulose biomass for biofuel production: a comprehensive review. Biomass Convers Biorefinery 1–13

Jayasinghe P, Hawboldt K (2012) A review of bio-oils from waste biomass: focus on fish processing waste. Renew Sustain Energy Rev 16(1):798–821

Jha S, Nanda S, Acharya B, Dalai AK (2022) A review of thermochemical conversion of waste biomass to biofuels. Energies 15(17):6352

Karuppusamy M, Ramamoorthi R, Karuppasamy R, Navin M (2023) Review on fabrication and applications of jute fiber epoxy composite reinforced bio composite. J Adv Mech Sci 2(3)

Kataya G, Cornu D, Bechelany M, Hijazi A, Issa M (2023) Biomass waste conversion technologies and its application for sustainable environmental development—a review. Agronomy 13(11):2833

Kaur S, Dhillon GS, Sarma SJ, Brar SK, Misra K, Oberoi HS (2014) Waste biomass: a prospective renewable resource for development of bio-based economy/processes. Biotransform Waste Biomass High Value Biochem 3–28

Kushwaha A, Goswami S, Sultana A, Katiyar NK, Athar M, Dubey L, Kareem MA (2022) Waste biomass to biobutanol: recent trends and advancements. Waste-to-Energy Approach Towards Zero Waste 393–423

Lange L, Connor KO, Arason S, Bundgård-Jørgensen U, Canalis A, Carrez D, Vieira H (2021) Developing a sustainable and circular bio-based economy in EU: by partnering across sectors, upscaling and using new knowledge faster, and for the benefit of climate, environment & biodiversity, and people & business. Front Bioeng Biotechnol 8:619066

Li M, Luo N, Lu Y (2017) Biomass energy technological paradigm (BETP): trends in this sector. Sustainability 9(4):567

Lin CSK, Pfaltzgraff LA, Herrero-Davila L, Mubofu EB, Abderrahim S, Clark JH, Luque, R. (2013). Food waste as a valuable resource for the production of chemicals, materials and fuels. Current situation and global perspective. Energy Environ Sci 6(2):426–464

Lisbona P, Pascual S, Pérez V (2023) Waste to energy: trends and perspectives. Chem Eng J Adv 14:100494

Manickaraj K, Ramamoorthy R, Babu MS, Jeevabharath KV (2019) Bio-fiber reinforced polymer matrix composites: a Review. Turkish J Comput Math Educ (TURCOMAT) 10(2):740–748

Manickaraj K, Ramamoorthi R, Sathish S, Johnson Santhosh A (2023) A comparative study on the mechanical properties of African teff and snake grass fiber-reinforced hybrid composites: effect of bio castor seed shell/glass/SiC fillers. Int Polym Proc 38(5):551–563

Manickaraj K, Ramamoorthi R, Sathish S, Makeshkumar M (2022) Effect of hybridization of novel African teff and snake grass fibers reinforced epoxy composites with bio castor seed shell filler: experimental investigation. Polym Polym Compos 30

Manickaraj K, Ramamoorthi R, Ramakrishnan T, Karuppasamy R (2024) Enhancing solid waste sustainability with iroko wooden sawdust and African oil bean shell particle-strengthened epoxy composites. Glob Nest J 26(1)

Manickaraj K, Ramamoorthi R, Karuppasamy R, Kannan S, Vijayaprakash B (2024) Experimental investigation of steel and porous Al foam LM vehicle leaf spring by using mechanical and computer method 107–112. https://doi.org/10.1002/9781394198221.ch8

Manickaraj K, Ramamoorthi R, Karuppasamy R, Sakthivel KR, Vijayaprakash B (2024) A review of natural biofiber-reinforced polymer matrix composites 135–141. https://doi.org/10.1002/9781394198221.ch11

Manikandan S, Vickram S, Sirohi R, Subbaiya R, Krishnan RY, Karmegam N, Awasthi MK (2023) Critical review of biochemical pathways to transformation of waste and biomass into bioenergy. Biores Technol 372:128679

McPhail SJ, Cigolotti V, Moreno A, Cigolotti V (2012) Biomass and waste as sustainable resources. Fuel Cells Waste-to-Energy Chain: Distrib Gener Through Non-Conv Fuels Fuel Cells 23–44

Mohan C, Robinson J, Vodwal L, Kumari N (2024) Sustainable development goals for addressing environmental challenges. In: Green chemistry approaches to environmental sustainability. Elsevier, pp 357–374

Motghare KA, Rathod AP, Wasewar KL, Labhsetwar NK (2016) Comparative study of different waste biomass for energy application. Waste Manage 47:40–45

Moustakas K, Loizidou M (2021) Waste and biomass management and valorization. Environ Sci Pollut Res 28:24224–24229

Ngan SL, How BS, Teng SY, Promentilla MAB, Yatim P, Er AC, Lam HL (2019) Prioritization of sustainability indicators for promoting the circular economy: the case of developing countries. Renew Sustain Energy Rev 111:314–331

Nithyanandhan T, Sivaraman P, Manickaraj K, Raj NM, Pragash MS, Tharun A (2022) Tribological behaviour of aluminium 6061 reinforced with graphite and chicken bone ash by using stir casting. Int J Veh Struct Syst 14(7):849–854

Nižetić S, Djilali N, Papadopoulos A, Rodrigues JJ (2019) Smart technologies for promotion of energy efficiency, utilization of sustainable resources and waste management. J Clean Prod 231:565–591

Ochieng R, Gebremedhin A, Sarker S (2022) Integration of waste to bioenergy conversion systems: a critical review. Energies 15(7):2697

Ockwell D, Watson J, Mallett A, Haum R, MacKerron G, Verbeken AM (2010) Enhancing developing country access to eco-innovation: the case of technology transfer and climate change in a post-2012 policy framework

Okolie JA, Epelle EI, Tabat ME, Orivri U, Amenaghawon AN, Okoye PU, Gunes B (2022) Waste biomass valorization for the production of biofuels and value-added products: a comprehensive review of thermochemical, biological and integrated processes. Process Saf Environ Prot 159:323–344

Omer AM (2011) Biomass energy resources utilisation and waste management. J Agric Biotechnol Sustain Dev 3(8):149

Pandey A, Tyagi RD, Varjani S (eds) (2021) Biomass, biofuels, biochemicals: circular bioeconomy—current developments and future outlook. Elsevier

Perea-Moreno MA, Sameròn-Manzano E, Perea-Moreno AJ (2019) Biomass as renewable energy: worldwide research trends. Sustainability 11(3):863

Pigato M, Black S, Dussaux D, Mao Z, Rafaty R, Touboul S (2020) Technology transfer and innovation for low-carbon development. World Bank Publications

Posmanik R, Cantero DA, Malkani A, Sills DL, Tester JW (2017) Biomass conversion to bio-oil using sub-critical water: study of model compounds for food processing waste. J Supercrit Fluids 119:26–35

Raj T, Sompura S, Chandrasekhar K, Singh SK, Pandey S, Singh LK, Singhania RR (2023) Technology development and challenges for the transformation of municipal solid waste into sustainable energy production. Biomass Bioenerg 178:106965

Safferman SI, Smith JS, Dong Y, Saffron CM, Wallace JM, Binkley D, Lenz J (2017) Resources from wastes: benefits and complexity. J Environ Eng 143(11):03117005

Sansaniwal SK, Rosen MA, Tyagi SK (2017) Global challenges in the sustainable development of biomass gasification: an overview. Renew Sustain Energy Rev 80:23–43

Scarlat N, Dallemand JF, Monforti-Ferrario F, Nita V (2015) The role of biomass and bioenergy in a future bioeconomy: policies and facts. Environ Dev 15:3–34

Schmidt TS, Dabur S (2014) Explaining the diffusion of biogas in India: a new functional approach considering national borders and technology transfer. Environ Econ Policy Stud 16:171–199

Siwal SS, Zhang Q, Devi N, Saini AK, Saini V, Pareek B, Thakur VK (2021) Recovery processes of sustainable energy using different biomass and wastes. Renew Sustain Energy Rev 150:111483

Srivastava RK, Shetti NP, Reddy KR, Kwon EE, Nadagouda MN, Aminabhavi TM (2021) Biomass utilization and production of biofuels from carbon neutral materials. Environ Pollut 276:116731

Tekin K, Karagöz S, Bektaş S (2014) A review of hydrothermal biomass processing. Renew Sustain Energy Rev 40:673–687

Tigabu AD, Berkhout F, van Beukering P (2015a) Technology innovation systems and technology diffusion: adoption of bio-digestion in an emerging innovation system in Rwanda. Technol Forecast Soc Chang 90:318–330

Tigabu AD, Berkhout F, van Beukering P (2015b) The diffusion of a renewable energy technology and innovation system functioning: COMPARING bio-digestion in Kenya and Rwanda. Technol Forecast Soc Chang 90:331–345

Tisi YSAB, Matos FA, Carneiro MLN (2023) Development of waste-to-energy through integrated sustainable waste management: the case of ABREN WtERT Brazil towards changing status quo in Brazil. Waste Disposal Sustain Energy 5(3):295–308

Tripathi N, Hills CD, Singh RS, Atkinson CJ (2019) Biomass waste utilisation in low-carbon products: harnessing a major potential resource. NPJ Clim Atmos Sci 2(1):35

Tshikovhi A, Motaung TE (2023) Technologies and innovations for biomass energy production. Sustainability 15(16):12121

Tsui TH, Wong JW (2019) A critical review: emerging bioeconomy and waste-to-energy technologies for sustainable municipal solid waste management. Waste Disposal Sustain Energy 1:151–167

Usmani Z, Sharma M, Awasthi AK, Sivakumar N, Lukk T, Pecoraro L, Gupta VK (2021) Bioprocessing of waste biomass for sustainable product development and minimizing environmental impact. Biores Technol 322:124548

Van Dam J, Junginger M, Faaij A, Jürgens I, Best G, Fritsche U (2008) Overview of recent developments in sustainable biomass certification. Biomass Bioenerg 32(8):749–780

Vasco-Correa J, Khanal S, Manandhar A, Shah A (2018) Anaerobic digestion for bioenergy production: global status, environmental and techno-economic implications, and government policies. Biores Technol 247:1015–1026

Vertes AA, Qureshi N, Yukawa H, Blaschek HP (eds) (2011) Biomass to biofuels: strategies for global industries. John Wiley & Sons

Vlosky R, Smithhart R (2011) A brief global perspective on biomass for bioenergy and biofuels

Worrell E, Van Berkel R, Fengqi Z, Menke C, Schaeffer R, Williams RO (2001) Technology transfer of energy efficient technologies in industry: a review of trends and policy issues. Energy Policy 29(1):29–43

Yana S, Nizar M, Mulyati D (2022) Biomass waste as a renewable energy in developing bio-based economies in Indonesia: a review. Renew Sustain Energy Rev 160:112268

Zabaniotou A (2018) Redesigning a bioenergy sector in EU in the transition to circular waste-based bioeconomy—a multidisciplinary review. J Clean Prod 177:197–206

Chapter 14
Comparative Analysis of Biomass Valorization Methods: Environmental Impact Assessment

Selvaraj Dinesh Kirupha, Mayakrishnan Muthukrishnan, and Selvaganapathi Anbazhagan

Abstract With increase in human population and need of fuel increase day by day, even the most cost-effective method for the removal of waste is found to be bizarre. As a result, concerned authorities started searching for suitable solutions which may provide solutions without any performance decline. Researchers over a past two decade started searching for alternate to fossil fuels. The present study focuses on biomass valorisation and their socio-economic significance. Since all the traditional methods implied failed to meet up the standards due to their negative impacts after treatment process. Hence a better method with synergic effect in the process outcome was researched and found that through modern synthetic methods, the outcome will be more reliable handling all serious threats to the humans. These emerging techniques in near future will not only be a game-changer but also lead to us to better lives with biofuel, production of bioenergy, pharmaceuticals and bio-chemicals at cheaper price. For instance electricity generation from bio-waste using bio-augmentation technique directly depends on the quality and measure of the biomass. Integrated techniques in the industrial sector emphasize green energy production and the creation of sustainable, industry-specific products. However, current technological advancements face limitations in efficiently handling biomass and waste. Addressing these challenges requires innovative solutions that integrate advanced detection, classification, and separation methods, fostering more effective resource management and promoting sustainable practices. Proposed methods are reviewed to address the growing volume of biomass and waste at various operational levels.

S. D. Kirupha (✉)
Department of Science and Humanities, Christ the King Engineering College, Coimbatore, India
e-mail: dineshkirupha@gmail.com

M. Muthukrishnan (✉)
Department of Mechanical Engineering, Christ the King Engineering College, Coimbatore, India
e-mail: mk24in@gmail.com

S. Anbazhagan
Department of Science and Humanities, KGISL Institute of Technology, Coimbatore, India
e-mail: pathyaug@gmail.com

A. Hind (ed.), *Biomass Waste: Processing, Valorization, Industrial Applications*,
https://doi.org/10.1007/978-981-95-0991-1_14

Keywords Biomass · Circular economy · Waste valorization · Renewable · Microbial biotechnology

14.1 Introduction

Producing clean, affordable, and reliable energy is a complex challenge requiring a balance between environmental sustainability, economic feasibility, and technological innovation. The growing global demand for energy, coupled with the need to reduce greenhouse gas emissions, emphasizes the importance of developing efficient and sustainable energy systems. Addressing this challenge necessitates collaborative efforts in research, policy-making, and the adoption of renewable energy technologies to secure a sustainable energy future. To overcome this issue, biomass is considered to a valuable source of renewable energy with an ability to produce only carbon source during emission. Hence it is mentioned as a valuable source to reduce CO_2 emission. The conversion of biomass into biofuels and chemicals is increasingly being used globally to diversify energy sources and reduce climate change (Berndes et al. 2003). Biomass, a renewable and carbon–neutral energy resource, generally burns cleaner than fossil fuels (Demirbas 2005; Ahmed et al. 2021; Hall et al. 1993; Awasthi et al. 2019). Various forms of biomass, such as wood waste, agricultural plants, industrial waste, sludge, and food processing by-products, are converted into fuels and chemicals. Wood logs are used directly, while wood waste from sawmills, logging, and municipal sources is recycled as fertilizer (Hall et al. 1993). Biomass conversion methods produce energy and biofuels like biogas, bioethanol, biodiesel, and syngas (Awasthi et al. 2019; Corona et al. 2019; Demirbas 2001; Mata et al. 2010). Integrated bio-refineries, which combine multiple processes with downstream upgrades and separations, are key to maximizing feedstock efficiency and product value. Biomass is becoming an important contributor to global energy, offering a cleaner-burning alternative to fossil fuels (Welker et al. 2015).

14.2 Sources of Biomass Production

The some of the sources of biomass production are described below.

14.2.1 Agricultural Wastes

Agricultural systems produce substantial crop residues. Crops like wheat, rice, vegetables, and fruits generate biomass that significantly supports annual renewable energy production. Biomass from agriculture plays a vital role in supporting sustainable energy for industrial and domestic purposes (Welker et al. 2015). Unused agricultural biomass can be categorized into two types: (1) plant parts left in the field, and (2) loss of plant material due to harvesting techniques. In the past, not all residues were suitable for biomass use, but today, they are repurposed for various applications such as biofuels (e.g., biogas, bioethanol), energy production, and animal feed, contributing to economic growth (Phaff and Miller 2015). In horticulture, the woody endocarp tissues of drupe fruits represent a valuable source of lignin, a biopolymer constituting about 50% of the total biomass. This hard, inedible component offers a high energy density compared to cellulose, making it a key resource for biofuel production. The lignin-rich endocarps of drupe fruits are increasingly utilized in sustainable energy applications, leveraging their potential to enhance the efficiency and yield of biofuel processes while contributing to waste valorization.

14.2.2 Biomass Produced from Industry

Microorganisms are utilized for biomass production in industrial processes. Yeast, in particular, functions as a biocatalyst and is employed in applications such as bakeries, lactic acid production (for use as a starter culture in dairy manufacturing), breweries, probiotics, and feed production for aquatic and terrestrial animals. On a large scale, biomass can be produced through fermentation using cost-effective substrates, including soybean meal, sugarcane molasses, and industrial waste. Solid biomass remains the dominant choice for industrial applications, complemented by the use of liquid waste and biogas (Phaff and Miller 2015). Over the last seven years, its industrial relevance has grown, with 80% of global industrial biomass utilization sourced mainly from U.S. forests, particularly black liquor (Liu et al. 2011). In 2010, 40.6% of Sweden's industries depended on biomass, while Finland's timber sector used 56% of industrial energy in 2011, with wood-based biomass contributing 44%.

14.2.3 Biomass from Domestic Food Waste

Global economic growth is hindered by rising waste accumulation, causing energy loss, environmental harm, and health issues (Agarwal et al. 2019). The European Union seeks efficient resources for a 'Recycling Society' to minimize waste. Biological waste, particularly food waste, forms the largest share of municipal solid waste (MSW). With population growth comes higher food consumption and a projected increase in food waste generation. At the EU level, food waste management is primarily regulated by the Waste Framework Directive. Consequently, alternative and efficient methods for managing biodegradable waste at the household level are needed (Kumar et al. 2020). Common approaches like domestic composting and anaerobic digestion are widespread but often fail, leading to additional problems such as unpleasant odors and greenhouse gas emissions. Of the 4.85 million tons of MSW generated annually, approximately 25% is recycled (mainly packaging waste), while the remaining 77% is improperly disposed of. Without new, innovative solutions, waste generation is expected to keep rising, exacerbating environmental problems (Zhang et al. 2015a; Chiaramonti et al. 2013; Mata-Alvarez et al. 2000).

14.2.4 Biomass Composition

A wide variety of biomass is produced, leading to significant variations in its composition, whether generated industrially or domestically. The primary components of biomass include cellulose, hemicellulose, lignin, starch, and proteins (Chheda et al. 2007).

Cellulose

Cellulose is a key polysaccharide composed of D-glucose units linked by β-glucosidic bonds, forming microfibrils in plant cell walls and bacterial biofilms. Hemicellulose, consisting of both 5-carbon and 6-carbon sugars, has a simpler structure, lower molecular weight, and growing industrial relevance due to advances in processing (Ofori et al. 2018; Sun et al. 2019). Lignin, a complex polymer of phenylpropane units, provides structural support but resists processing due to its intricate and irregular structure. Recent studies have focused on overcoming lignin's resistance to hydrolysis, opening up new avenues for its industrial application (Lynd et al. 2002).

The chemical composition of plant biomass varies with species, typically containing lignin and carbohydrates like cellulose and hemicellulose. Biomass moisture content is a critical factor for energy applications, as high moisture levels reduce efficiency by requiring more energy for water evaporation during thermochemical processes. Dry biomass, with lower moisture, offers higher heating values, making it more suitable for biofuel production and energy generation. Additionally, biomass composition influences its suitability for different bioenergy processes, such as gasification or fermentation, where cellulose and hemicellulose play key roles in yield optimization (Zhang et al. 2015b; Kumar et al. 2018; Bridgwater and Peacocke 2000).

The Carbon content in biomass is around 46%, lower than coal, which contains 60% or more, resulting in a lower heating value. Hydrogen content is approximately 6%, and a higher hydrogen content enhances heating value. Biomass nitrogen content ranges from 0.2% to over 1%, contributing to NOx emissions during combustion. Sulfur content is generally below 0.2% but may reach 0.5–0.7% in some cases. Elevated chloride levels can cause combustion fouling, while hydrochloric acid formation poses a risk of boiler corrosion (Demirbas 2004; Jenkins et al. 1998).

Waste valorization is the process of converting waste materials into valuable products such as chemicals, fuels, and materials. This concept revolves around the idea that waste contains unused polymeric components, which can be transformed into energy and chemicals, creating valuable by-products. These by-products highlight the potential of waste residues as an important resource, urging the need for their efficient exploitation. Waste valorization plays a critical role in both artificial and biological waste management, forming the foundation for Waste-to-Energy (WtE) strategies. It not only addresses the depletion of natural resources but also contributes to the development of renewable energy sources. Additionally, waste valorization enables the creation of high-value products, such as nano-bioplastics and ethanol, which are cost-effective, sustainable, and contribute to environmental preservation. Through this approach, waste is no longer seen as a burden but as a potential resource for sustainable innovation.

Waste-to-value (WtV) and Waste-to-Energy (WtE) methods are essential for sustainable resource recovery. When combined with tools like Life Cycle Assessment (LCA) and Ecological Footprint Analysis (EFA), they help recover chemicals and energy from waste. Key techniques such as anaerobic digestion, incineration, and gasification, alongside landfill and agro-industrial residue use, are crucial for improving waste management and supporting modern waste-to-energy facilities. Detailed examples are provided in later section (Lucie et al. 2013).

14.3 Eco-Friendly Method for Biomass Valorisation

Waste valorization is the process of converting waste materials into valuable goods such as chemicals, fuels, and materials by using underutilized polymeric components in garbage. This strategy is applicable to both artificial and biological waste, supporting waste-to-energy (WtE) systems and stressing trash as a resource rather than an unwanted byproduct. It promotes renewable energy generation and environmentally friendly products like nano-bioplastics and ethanol.

WtV and WtE strategies use methods such as LCA, EI, and EFA to encourage a zero-waste approach. These methods use green technologies, modified organisms, and sustainable processing processes to generate energy and fine chemicals. Anaerobic digestion, incineration, gasification, pyrolysis, and the conversion of landfill and agro-industrial waste into bioenergy are essential methods for improving waste management and modernising facilities.

14.3.1 Fermentation Technology

"Commercial ethanol production uses fermentation processes with sugar crops (e.g., sugar beet) and starch crops (e.g., wheat)." Starch is enzymatically converted into simple sugars, which are subsequently fermented by yeast to produce alcohol. The solid residues from fermentation serve valuable purposes, such as being repurposed as bovine feed or used as bagasse to fuel boilers or gasification systems. For complex biomass like lignocellulosic materials, breaking down long-chain polysaccharides into fermentable sugars requires additional steps. These include pretreatment processes such as acid hydrolysis, which help degrade the rigid structure of lignin and cellulose, making the sugars more accessible for fermentation. This multi-step approach enhances the efficiency of biofuel production while maximizing the utility of biomass residues (Koutinas et al. 2014; Razik et al. 2019).

14.3.2 Flow Technology

The high physical and chemical complexity of biomass presents a significant challenge in developing new processes. One potential solution is breaking down biomass into simpler fractions that are easier to manage in downstream processes (Makarichi et al. 2018; Balaman et al. 2018; Bilal and Iqbal 2020). In bio-refinery systems, similar to petroleum facilities, chemicals, energy, and fuels are produced from biomass. These key biomass derivatives, often referred to as platform molecules or building blocks, are relatively simple compounds with versatile functionalities that can be used in the production of various chemical compounds. Continuous flow systems enable precise control over reaction parameters, making them particularly effective for handling highly reactive materials derived from biomass. Additionally, these systems facilitate scalability, which is crucial as many biomass conversion methods are still in the developmental stage. The advancement of flow technologies will likely aid in the commercialization of biomass processing. Since the chemical structure of biomass feedstock differs significantly from that of the final products, multiple processing steps are usually required, which can negatively impact process efficiency. Stream processing techniques help intensify chemical reactions, contributing to the simplification of biomass processing technologies (Hahn-Hägerdal et al. 2006a).

14.3.3 Microbial Digestion

The metabolic activity of microorganisms decomposes organic matter, producing carbon dioxide (CO_2) and methane (CH_4) as primary by-products. This process, known as microbial decomposition, plays a central role in anaerobic digestion and natural carbon cycling, contributing to biogas production and greenhouse gas emissions. This mixture of CO_2 and methane is referred to as biogas. Anaerobic digestion provides a means for renewable energy production and improves the treatment of agricultural residue waste. Recently, this technology has gained traction in Europe as an effective method for biodegrading solid organic fractions of municipal waste. The process occurs in specialized vessels known as anaerobic bioreactors or anaerobic digesters. A biogas facility consists of the complete system, including feedstock input, digester, biogas storage, and a digestate tank (Baños et al. 2011; Poonam Singh Nigam 2011; Brinchi et al. 2013).

14.3.4 Gasification

Gasification is a thermochemical process that partially oxidizes organic or natural materials to produce a synthesis gas (syngas) composed primarily of hydrogen, carbon monoxide, and trace amounts of methane. This process occurs at high temperatures and involves the interaction of carbonaceous materials with gasification agents such as oxygen, steam, and carbon dioxide. Within the gasifier, these components react and quickly reach an equilibrium state, optimizing the conversion of feedstock into energy-rich syngas.

The syngas produced can be used as a clean and efficient fuel for heat and electricity generation or as a raw material for synthesizing chemicals and fuels like methanol and hydrogen. Gasification is particularly advantageous due to its ability to handle diverse feedstocks, including biomass, coal, and municipal solid waste, and its potential for lower emissions compared to traditional combustion methods. This process also supports carbon capture technologies, enhancing its role in sustainable energy solutions and reducing environmental impact.. However, certain challenges must be addressed before biomass gasification can be widely commercialized. These challenges include the removal of tar, issues with biomass feedstock production and pre-treatment, and a clear understanding of biomass properties. Supercritical water gasification (SCWG), which is an advanced form of gasification, has gained significant attention globally. SCWG has the advantage of eliminating the need for drying wet biomass before the gasification process (Sansaniwal et al. 2017).

14.3.5 Microwave Technology

Biomass feedstock is typically subjected to heating under pressure, around 60–100 bar, within a closed vessel at temperatures between 180 and 200 °C for a minimum of 12 h, without the presence of oxygen. Heat transfer occurs through conduction and convection due to temperature gradients. However, conventional heating methods can lead to extended residence times and uneven surface heating (Kant Bhatia et al. 2022). To address these issues, the microwave hydrothermal carbonization method has been proposed. Over time, microwave technology has replaced traditional heating methods in processes such as carbohydrate digestion, sterilization, synthesis, and more recently, in the dry pyrolysis of biomass feedstock to produce char and gas. Microwave heating offers advantages such as reduced residence times, rapid, selective, and volumetric heating, which fosters new reaction pathways and optimal conditions for product formation. Microwave technology offers advantages due to its precision and the ability to control processes, as well as its potential for portable applications, thanks to the smaller size of the reactor (Gupta et al. 2022). The efficiency of microwave processing of human bio-waste is primarily due to the high water content, which interacts effectively with electromagnetic fields, facilitating "microwave dielectric heating." (Šelo et al. 2021).

14.4 Implemented Possibilities—A Push for Sustainability

Waste and biomass can yield a wide range of products when appropriate techniques and conditions are employed. The global trend is shifting towards a preference for green energy over fossil fuels. Green energy offers two primary benefits: the use of renewable resources, especially waste, and the reduction of pollution. In addition to their role in energy production, biomass and waste materials are increasingly recognized for their value in various industrial applications. They serve as raw materials in the production of chemicals, polymers, and bio-based additives. These materials can enhance the quality and properties of chemical products by acting as renewable feedstocks, replacing traditional petrochemical inputs. Biomass and waste are utilized in the manufacturing of bio-plastics, adhesives, and even pharmaceuticals, where they contribute to reducing environmental impact by substituting non-renewable resources. This not only supports a circular economy but also helps in lowering the overall carbon footprint of industrial processes (Velvizhi et al. 2023; Sun and Peng 2023).

Global energy consumption has consistently increased since the early nineteenth century, primarily driven by industrialization and technological advancements. This surge in energy demand has led to greater reliance on fossil fuels, such as coal, oil, and natural gas, which are non-renewable resources. The continued use of fossil fuels raises concerns about resource depletion and environmental degradation. Notably, the burning of these fuels results in higher carbon dioxide emissions, contributing significantly to climate change, air pollution, and the degradation of air quality. As the world continues to face these challenges, transitioning to renewable energy sources and implementing sustainable practices has become critical in mitigating the negative impacts of fossil fuel consumption. Consequently, alternative energy sources are necessary (Duarah et al. 2022). This has led to the exploration of renewable energy alternatives that reduce environmental harm. Biomass and organic waste represent valuable renewable energy sources collectively referred to as "biofuels." These biofuels include a range of products such as bioethanol, biodiesel, bio-butanol, bioether, biogas, biohydrogen, and syngas. The conversion of biomass into biofuels involves various processes such as fermentation, gasification, and anaerobic digestion. These biofuels can be used for electricity generation, heating, transportation, and even in the production of chemicals, offering a sustainable alternative to fossil fuels (Kaur et al. 2021; Calijuri et al. 2022; Zhou et al. 2022; Ambaye et al. 2021).

14.4.1 Bioethanol

Bioethanol production has traditionally relied on fermenting sugars derived from sugar and starch crops, utilizing yeast species such as Saccharomyces cerevisiae or Zymomonas mobilis to convert six-carbon sugars into ethanol. Common feedstocks for bioethanol include sugarcane and sugar beet, as these crops are rich in fermentable sugars. However, there has been a shift towards utilizing lignocellulosic biomass as an alternative, offering a more sustainable source of bioethanol. Lignocellulosic biomass, derived from materials like wood chips, agricultural residues, and municipal or industrial solid wastes, presents a renewable option for ethanol production (Jayakumar et al. 2023; Hahn-Hägerdal et al. 2006b; Gray et al. 2006).

For efficient bioethanol production from lignocellulosic materials, pre-treatment is necessary to break down the complex structure of cellulose and hemicellulose, which are more challenging to convert into fermentable sugars. The pre-treatment process enhances the accessibility of these components, making them easier to hydrolyze into monomeric sugars. Acid hydrolysis, often using sulfuric acid, and enzymatic hydrolysis, which utilizes enzymes like cellulase and xylanase, are two common methods for saccharification. Enzymatic hydrolysis, although more expensive, is preferred due to its higher yields and minimal by-products, offering a cleaner and more efficient conversion process (Cardona and Sánchez 2007).

Following saccharification, the released sugars are fermented by S. cerevisiae to produce ethanol. The bioethanol can then be used directly as a fuel or blended with petroleum products. Blending bioethanol with gasoline not only reduces carbon emissions during combustion but also contributes to a more sustainable energy future by decreasing dependence on fossil fuels. This process, as well as its advancements, plays a vital role in the ongoing transition towards greener, renewable energy solutions (Arifeen et al. 2007; Kaparaju et al. 2009; John et al. 2011).

14.4.2 Biodiesel

Biodiesel is produced through transesterification, a process that combines oils extracted from seeds and nuts with alcohol (usually methanol) (Singh et al. 2018; Santos-Ballardo et al. 2015). Transesterification, a crucial process in biodiesel production, can occur through catalytic or non-catalytic methods. The catalytic route, typically employing alkali catalysts, is favoured for its high efficiency and ability to reach yields of up to 98.5% under mild temperature and pressure conditions. This makes it the most commonly used method in large-scale biofuel production. In contrast, enzyme-catalyzed transesterification, though technically feasible, faces limitations such as high enzyme costs and slower reaction rates, which reduce its commercial viability. Non-catalytic transesterification, although promising for certain conditions, generally requires higher temperatures and pressures, making it less efficient than catalytic methods. This makes catalytic transesterification the preferred choice in industrial applications for its higher yield and efficiency. Non-catalytic transesterification, using supercritical methanol, offers higher yields in shorter time frames (Corma et al. 2011; Soares Dias et al. 2023).

14.4.3 *Biogas*

Derived from lignocellulosic biomass, biogas production through anaerobic digestion is recognized as a cost-effective and stable method for generating biofuels, an established method to recover significant energy from biomass sources. The degradation of cellulose and hemicellulose in lignocellulosic biomass enables anaerobic microorganisms to produce biogas (Methane) but its efficiency is limited by the presence of lignin in the biomass. To enhance the yield of methane, various pre-treatment techniques can be employed to modify the physicochemical properties of the biomass. Alkaline pre-treatment is one of the most widely used methods, as it effectively breaks down the ester bonds between lignin and carbohydrates, particularly affecting the crystalline structure of cellulose. This alteration significantly improves the efficiency of anaerobic digestion, thereby increasing methane production. The application of Sodium hydroxide (NaOH) for pretreatment is found to increase methane production from lignocellulosic biomass by up to 170% compared to untreated biomass (Keerthana Devi et al. 2022).. Factors influencing biomethane production are diverse and critical to optimizing the anaerobic digestion process. Key factors include the type of inoculum used, which affects the microbial community involved in methane production. Reactor design is another important consideration, as it influences the efficiency of gas transfer and substrate degradation. The organic loading rate (OLR), which defines the amount of organic material added to the reactor, must be carefully controlled to prevent overloading, which can inhibit methane production. Additionally, maintaining the appropriate pH and temperature is essential, as these factors significantly impact microbial activity. Finally, the carbon-to-nitrogen (C/N) ratio plays a crucial role in optimizing microbial growth and methane production, with an ideal ratio supporting the best balance of nutrient availability for microbial digestion (Deena et al. 2022; Donoso-Bravo et al. 2021; Zhao et al. 2021; Mishra et al. 2021).

Bio-hydrogen: It serves as a sustainable fuel alternative to fossil fuels, particularly for transportation. It can be produced through three main processes: bio-photolysis, photo-fermentation, and anaerobic fermentation. Combining anaerobic fermentation with photo-fermentation can enhance bio-hydrogen yield, utilizing metabolites from the former in the latter (Samrot et al. 2023; Chandran and Mohan 2023; Ntaikou et al. 2010; Chong et al. 2009).

14.5 Organic Acid and Chemicals

Biomass and waste, beyond biofuels and energy, are key for producing synthetic chemicals and organic acids. Research shows that kitchen waste can be converted into valuable organic acids via anaerobic digestion. Formic acid (FA) is often produced as a by-product during the hydrolysis of biomass. It plays a crucial role as a hydrogen donor in the hydrogenation of levulinic acid (LA), a process that leads to the production of valuable fuel additives such as γ-valerolactone. When it comes to the synthesis of lactic acid, two primary approaches are commonly used. One method for producing D-lactic and L-lactic acids is chemical synthesis, where lactonitrile is subjected to acidic hydrolysis using strong acids. This process results in the formation of solutions containing both D-lactic and L-lactic acids. The method is efficient for industrial-scale production but requires careful control of reaction conditions to ensure high yield and selectivity. Although effective, this approach can be costly and less efficient compared to microbial fermentation.

The second and more efficient method is microbial fermentation, which uses lactic acid bacteria (LAB) such as Lactobacillus and Streptococcus to convert pyruvate into lactic acid through the action of lactate dehydrogenase. This process is generally considered more cost-effective and environmentally friendly compared to chemical synthesis. Furthermore, propionic acid (PA), a monocarboxylic acid, is typically produced by propionibacteria under anaerobic conditions. However, the acid-sensitive nature of these bacteria can limit PA production. Recent advancements in microbial biotechnology, such as immobilization techniques, can enhance the acid tolerance of these bacteria, allowing them to thrive in the acidic environment created by PA production. Genetic modifications, such as knocking out the acetate kinase gene, can help reduce acetate accumulation in bacteria, thereby improving the yield of propionic acid (PA) and enhancing their resistance to acidity. These innovations are part of ongoing efforts to optimize microbial production of organic acids. By addressing metabolic bottlenecks, such as acetate buildup, researchers are developing more sustainable and efficient processes for producing valuable organic acids, including PA, through fermentation. These strategies represent significant advancements in industrial biotechnology, offering potential for higher yields and cost-effective production methods (Joshi and Manjare 2024).

Succinic acid (SA) is a versatile dicarboxylic acid that can be produced by various microbial strains, particularly bacteria. Common producers include Escherichia coli, Actinobacillus succinogenes, and Mannheimia succiniciproducens. These bacteria ferment sugars, often glucose, under anaerobic or aerobic conditions to produce succinic acid as a metabolic byproduct. The production pathway involves converting phosphoenolpyruvate (PEP) to oxaloacetate (OAA), which is then transformed into succinic acid through a series of reactions requiring electron donors like NADH. Malic acid (MA) is another dicarboxylic acid produced using various yeast strains, including genetically engineered ones like modified E. coli. Aspergillus flavus is known to produce notable amounts of malic acid when grown on glucose. However, it also has a significant drawback: the production of aflatoxins, which are potent carcinogens and pose a serious risk in food-grade applications. Due to this, the use of A. flavus in industrial processes for malic acid production is limited to non-food applications, where the risk of contamination can be better managed. Other strains, such as Aspergillus niger and Zygosaccharomyces rouxii, have also been reported to produce malic acid efficiently. Fumaric acid, also a dicarboxylic acid, can be produced by fungi such as Rhizopus arrhizus in aerobic conditions (Dugmore et al. 2017). Research has shown that controlling dissolved oxygen levels during production can optimize yields. Butyric acid, a four-carbon organic acid, can be produced from biomass and waste using various microorganisms, including genera like Clostridium and Fusobacterium. The strain Clostridium tyrobutyricum is known for producing butyric acid under anaerobic conditions and is commonly used globally. High volumetric production has been achieved during the fermentation of cane molasses with this strain. Enhancing butyric acid production efficiency can be accomplished through cell immobilization (Tuck et al. 2012). However, metabolic by-products often inhibit the fermentation process, complicating butyric acid production. Using mutant strains, such as C. tyrobutyricum ATCC 25,755, has not effectively reduced acetic acid production compared to wild-type strains. Genetically engineered immobilized Clostridium tyrobutyricum strains offer a promising solution for improving butyric acid production, especially when glucose is used as the carbon source. These strains enhance yield and efficiency by stabilizing microorganisms through immobilization, reducing contamination risks and facilitating larger-scale processes (Corma et al. 2007).

14.6 Lipids and Enzymes

Biomass and waste, particularly by-products from the food industry, are increasingly being utilized in the production of industrial enzymes. The cost-efficiency of using food waste as a raw material for enzyme production is a key factor in its growing use. This approach not only reduces production costs but also supports waste valorization by converting organic waste into valuable bioproducts. It aligns with sustainable practices by helping meet the increasing demand for industrial enzymes while reducing environmental impact. Additionally, recycling food industry waste promotes the circular economy, ensuring that by-products are utilized rather than discarded. This method reflects an eco-friendly approach to enzyme production that benefits both industry and the environment. Enzymatic hydrolysis is part of the lignocellulose pre-treatment process, which, when combined with fermentation using fungi like *Aspergillus, Melanocarpus*, and *Pleurotus*, can yield industrial enzymes. Various oxidative enzymes, such as amylase, can also be synthesized from food waste (Calijuri et al. 2022; Hijazi et al. 2022). However, the heterogeneous nature of this waste complicates downstream processing, leading to high costs for enzyme isolation and purification. To overcome these challenges, various strategies for one-step purification and enzyme immobilization are applied. These include techniques such as immobilizing enzymes at a single point, using specially designed Supports designed for selective binding, along with directed mutagenesis, are employed to improve the interaction between enzymes and their supports, thereby enhancing the affinity and efficiency of enzyme immobilization. Solid-state fermentation is also a cost-effective method for producing essential bacteria, as it supports high yields of industrial enzymes. Among the different organisms used in this process, Basidiomycetes fungi stand out for their efficiency in synthesizing industrial enzymes, primarily due to their optimal compatibility with solid-state fermentation conditions (García-Sancho et al. 2023).

Non-enzymatic proteins, such as collagen, can be extracted from biomass, with fish waste, particularly from cod, being an important source. One method involves using sodium hydroxide (NaOH) to isolate acid-soluble collagen from cod bones. This approach efficiently recycles fish waste, turning it into a valuable bioproduct. Fish-based collagen is commonly used in the food, cosmetic, and pharmaceutical industries due to its beneficial properties. Cheese whey wastewater contains various proteins that precipitate based on their denaturation properties, which can be improved through centrifugation and filtration methods, though achieving thermal stability and functionality in isolated proteins remains a challenge (Kim et al. 2024).

Biomass and food waste also serve as valuable sources for lipid production, offering a cost-effective alternative for biofuel generation. Lignocellulosic biomass and food waste are rich in organic carbon, significantly reducing the overall cost of lipid production. Oleaginous microorganisms—such as certain species of yeasts, algae, and fungi—are utilized to convert these materials into lipids through fermentation processes. However, The crystalline structure of cellulose and lignin presents significant barriers to efficient biomass conversion, necessitating pre-treatment processes. These methods, including acid, base, or enzymatic treatments, are used to break down these complex compounds and make them more accessible for fermentation. By altering the structure of cellulose and lignin, pre-treatment enhances the efficiency of subsequent processes like hydrolysis, enabling the extraction of fermentable sugars and improving overall biomass utilization for biofuel production.

Following pre-treatment, biomass can be processed using either solid-state fermentation or submerged fermentation, depending on the specific properties of the material. To convert cellulose into lipids, enzymes play critical roles in the metabolic pathways. The choice of oleaginous microorganisms also depends on the feedstock used; for instance, Trichosporon fermentas is commonly employed for fermenting rice straw, while Cryptococcus curvatus is effective for processing wheat straw (Sathish et al. 2015; Juszczyk et al. 2013).

14.7 Biological Macromolecule

Biological macromolecules like bioplastics, polyhydroxyalkanoates (PHAs), lignin, and cellulose can be derived from biomass. Among these, bioplastics, primarily composed of PHAs, are gaining recognition as a sustainable alternative to traditional petroleum-based plastics. By using biomass and food waste for PHA production, both production costs are reduced, and dependence on fossil fuels is minimized (Baiano 2014).

14.7.1 Cellulose n($C_6H_{10}O_5$)

Cellulose ($C_6H_{10}O_5$) is a key component of plant cell walls, composed of glucose chains linked by β-1,4-glycosidic bonds. It provides structural strength and is used in paper, textiles, and biofuels. Certain microorganisms, including specific algae and bacteria, can synthesize cellulose to bypass these issues of impurities. Synthesized cellulose, while structurally similar to plant cellulose, offers enhanced properties such as higher polymerization, increased purity, and better water retention, making it suitable for industrial use. Acetobacter xylinum produces cellulose synthase, incorporating uridine diphosphate glucose (UDP glucose) into the cellulose chain. The polymerization rate in bacterial cellulose production is determined by the crystallization process, which creates a 3D network as the cellulose is expelled from the bacterial cells. Common carbon sources for bacterial cellulose production include glucose, glycerol, sucrose, and maltose. These sugars act as key substrates for the bacteria, fostering efficient cellulose synthesis (Torres-León et al. 2021; Sadh et al. 2018). Research has shown that nutrient supplements, such as lignosulfonate, can improve yields, as seen in fermentation with Gluconacetobacter xylinus using glucose. Agitated cultures of Acetobacter xylinum also lead to higher cellulose concentrations, with carboxymethylcellulose enhancing the yield (Sharma et al. 2022; Pal et al. 2024).

14.8 Food/Feed Products

Acids enhance sugar extraction from food waste. For example, acid hydrolysis with low concentrations of hydrochloric acid (HCl) or sulfuric acid (H_2SO_4) treatments help break down complex carbohydrates into fermentable monomeric sugars like glucose and fructose, which are key for bioethanol production. Additionally, enzymes such as α-amylase and glucoamylase facilitate the breakdown of starch-like polymers into simpler sugars, further boosting the yield. However, cellulase efficiency can be hindered by non-productive binding to lignin in the biomass. To address this, pretreatment methods such as enzymatic hydrolysis and physical methods are applied to make lignocellulosic material more accessible. Pre-treatment also helps improve sugar yield, making the subsequent fermentation processes for bioethanol more efficient. The use of biological treatments, such as the addition of Trichoderma viride fungi, has shown promise in increasing the availability of fermentable sugars from waste biomass. Combining enzymatic treatment with acid hydrolysis significantly enhances sugar yields, and continued research into optimizing these processes is vital for reducing the costs of bioethanol production from waste biomass.

MDP To mitigate this, surfactants are added to prevent enzyme-lignin complexes from forming. Surfactants bind to the cellulase binding sites on lignin, allowing cellulase to effectively target cellulose and produce sugars. Surfactant-assisted cellulose hydrolysis has been shown to increase sugar yields from recycled materials, such as newspapers (Mattioli et al. 2024).

Common pre-treatment methods for paper products include steam explosion, bacterial pre-treatment, and ozonolysis. Steam explosion applies high-pressure steam to break down the material, facilitating easier access to cellulose. Bacterial pre-treatment involves using microorganisms to degrade lignin, while ozonolysis utilizes ozone to disrupt the lignin-carbohydrate bonds, improving the efficiency of subsequent processes. These methods enhance the digestibility and yield of cellulose for biofuel production or other uses. Weak acid pre-treatment, particularly with phosphoric acid, has shown promising results for lingo-cellulosic biomass. Fermentation of agricultural waste using fungi has gained significant attention for improving the nutritional value of animal feed. Fungi like Pleurotus ostreatus, Lentinus edodes, and Trametes versicolor are particularly effective in breaking down lignocellulosic material, which improves the digestibility and protein content of the feed. These fungi enhance nutrient availability by producing enzymes that degrade complex plant polymers, thus increasing the protein levels in agricultural by-products like wheat straw, rice straw, and soy meal. For example, Pleurotus ostreatus has been used to ferment rice and wheat straw, boosting protein levels, while Trametes versicolor improves nutrient digestibility in wheat straw. Moreover, certain edible mushrooms like Agaricus and Grifola have also shown potential in fermenting food waste, further enhancing protein content and making the feed safer for animal consumption.

In addition to improving feed quality, this process can reduce waste and offer an eco-friendly alternative to traditional animal feed production. The choice of fungal species and fermentation conditions—such as moisture content and fermentation duration—are crucial for maximizing protein yield and optimizing the process (Şahin and Kurtulbaş 2024; Vurukonda et al. 2024).

14.9 Microbial Fuel Cell (MFC) Technology

Microbial Fuel Cell (MFC) technology is a promising method for generating bioelectricity from biomass and waste materials. It utilizes electroactive bacteria, such as Shewanella putrefaciens and Geobacter sulfurreducens, which metabolize organic substrates to produce electrical energy. These bacteria facilitate the oxidation of biomass or organic waste at the anode, releasing electrons and protons. The electrons travel through an external circuit to provide electrical power, while protons move through a membrane to the cathode. At the cathode, electron reduction occurs, enabling energy capture. Recent studies have explored integrating MFCs with solid-state fermentation to boost bioelectricity production from food waste, utilizing volatile fatty acids from food waste leachate after hydrolysis and acidogenic digestion. Despite their potential, MFCs still face challenges, particularly related to low bioelectricity output due to high internal resistance in the system.

To address the challenge of low power output in microbial fuel cells (MFCs), researchers found that increasing the conductivity or ionic strength of the anolyte and catholyte can enhance performance. MFCs operating under neutral or alkaline conditions demonstrated better COD removal and higher power output. Additionally, the pH range of the anolyte, along with salt concentration and the type of biomass or food waste used, plays a significant role in generating elevated power density and effective COD removal in bioelectricity production (Rout et al. 2023).

14.10 Pulp and Paper Industry Sector

In the pulp and paper industry, biomass and waste materials play a vital role in both energy production and product enhancement. These materials are utilized in two primary ways. Firstly, they are burned to generate green energy for the production process, reducing reliance on fossil fuels and helping companies cut costs while lowering their carbon footprintand waste are incorporated as additives to improve the quality of paper products, such as enhancing durability, strength, and other desirable properties.

The shift towards energy production has become a preferred alternative due to its eco-friendly nature and cost-effectiveness. By replacing fossil fuels with renewable organic waste, the paper and pulp industry not only reduces its environmental impact but also contributes to the circular economy by reusing waste materials. This dual-use of biomass anh as energy sources and raw materials—supports the industry's transition toward more sustainable and efficient practices (Ververis et al. 2007).

14.11 Pharmaceutical Sector

In the pharmaceutical sector, biomass and waste materials are increasingly recognized as valuable sources of bioactive compounds with various therapeutic applications. These compounds, including antitumor, antioxidant, and antibacterial agents are derived from diverse biomass sources such as lignocellulosic waste, fungal biomass, and seafood processing by-products. For example, lentinan, a well-known polysaccharide with antitumor properties, has been isolated from Lentinus edodes (shiitake mushrooms) using techniques like centrifugation and anion-exchange column chromatography. Studies have shown that oral administration of lentinan is highly efficient in suppressing colon carcinoma growth in murine models. In addition to its anticancer potential, lentinan has demonstrated effectiveness against various tumors, reducing their size by up to 95% in some cases. Other biomass-derived bioactive compounds continue to be explored for their medicinal properties, contributing to the development of new treatments and improving existing therapies in the pharmaceutical industry (Bejenaru et al. 2024; Bala et al. 2023).

14.12 Conclusion

The bio-economy, which includes activities such as utilizing agricultural residues to generate biomass, plays a pivotal role in the creation of sustainable, renewable resources. However, while biomass is a renewable resource, its production requires land, water, and nutrients, which raises environmental concerns. The bio-economy, relying on land and marine feedstocks, contrasts with fossil fuel-based systems but still faces challenges, particularly in ensuring a consistent biomass supply. This demand can lead to resource depletion, such as the conversion of forests into arable land, contributing to CO_2 emissions.

A significant challenge in the transition to a bio-based economy is the shift from fossil fuel-based industries to biomass-based ones. This transition necessitates the development of bio-refineries and other markets. While the bio-economy can support environmental goals, its increasing demand for biomass can put pressure on natural resources, contributing to land-use changes that can impact food security and exacerbate water and soil pollution. Overuse of land can also deplete soil fertility, requiring more intensive inputs like fertilizers and pesticides.

To mitigate these issues, the use of degraded land, improving agricultural productivity, and utilizing waste materials for biomass production could ease resource demands. Biofuels, derived from biomass, can replace fossil fuels such as gasoline and diesel, potentially reducing CO_2 emissions and offering cost-effective alternatives for transportation. Biomass-based fuels like hydrogen and methanol also provide adaptable energy solutions.

Despite the potential environmental concerns, these challenges are not insurmountable. With careful management, the bio-economy can contribute to long-term environmental sustainability and societal benefits, fostering a cleaner, greener future while generating economic opportunities.

References

Agarwal A, Jain A, Singh A (2019) Renewable energy sources for sustainable development: a review. Energy Rep 5:925–935

Ahmed I, Zia MA, Afzal H, Ahmed S, Ahmad M, Akram Z, Sher F, Iqbal HMN (2021) Socio-economic and environmental impacts of biomass valorisation: a strategic drive for sustainable bioeconomy. Sustain 13:4200

Ambaye TG, Vaccari M, Bonilla-Petriciolet A, Prasad S, van Hullebusch ED, Rtimi S (2021) Emerging technologies for biofuel production: a critical review on recent progress, challenges and perspectives. J Environ Manage 15(290):112627

Arifeen N, Wang R, Kookos IK, Webb C, Koutinas AA (2007) Process design and optimization of novel wheat-based continuous bioethanol production system. Biotechnol Prog 23(6):1394–1403

Awasthi MK, Zhao J, Soundari PG, Kumar S, Chen H, Awasthi SK, Duan Y, Liu T, Pandey A, Zhang Z (2019) Sustainable management of solid waste. In: Sustainable resource recovery and zero waste approaches. Elsevier, Amsterdam, The Netherlands, pp 79–99

Baiano A (2014) Recovery of biomolecules from food wastes—a review. Molecules 19:14821–14842

Bala S, Garg D, Sridhar K, Inbaraj BS, Singh R, Kamma S, Tripathi M, Sharma M (2023) Transformation of agro-waste into value-added bioproducts and bioactive compounds: micro/nano formulations and application in the agri-food-pharma sector. Bioengineering (Basel). 10(2):152

Baños R, Manzano-Agugliaro F, Montoya FG, Gil C, Alcayde A, Gómez J (2011) Optimization methods applied to renewable and sustainable energy: a review. Renew Sustain Energy Rev 15(4):1753–1766

Bejenaru LE, Radu A, Segneanu AE, Biţă A, Manda CV, Mogoşanu GD, Bejenaru C (2024) Innovative strategies for upcycling agricultural residues and their various pharmaceutical applications. Plants (Basel) 13(15):2133

Berndes G, Hoogwijk M, Van Den Broek R (2003) The contribution of biomass in the future global energy supply: a review of 17 studies. Biomass Bioenergy 25:1–28

Bilal M, Iqbal HM (2020) Recent advancements in the life cycle analysis of lignocellulosic biomass. Curr Sustain Renew Energy Rep 7:100–107

Bridgwater AV, Peacocke GVC (2000) Fast pyrolysis processes for biomass. Renew Sustain Energy Rev 4:1–73

Brinchi L, Cotana F, Fortunati E, Kenny JM (2013) Production of nanocrystalline cellulose from lignocellulosic biomass: technology and applications. Carbohyd Polym 94(1):154–169

Calijuri ML, Silva TA, Magalhães IB, Pereira ASAP, Marangon BB, Assis LR, Lorentz JF (2022) Bioproducts from microalgae biomass: technology, sustainability, challenges and opportunities. Chemosphere 305:135508

Cardona CA, Sánchez OJ (2007) Fuel ethanol production: process design trends and integration opportunities. Bioresour Technol 98(12):2415–2457

Chandran EM, Mohan E (2023) Sustainable biohydrogen production from lignocellulosic biomass sources—metabolic pathways, production enhancement, and challenges. Environ Sci Pollut Res 30:102129–102157

Chheda JN, Czernik S, Dumesic JA (2007) An overview of biomass to biofuels via thermochemical processes. Energy Environ Sci 1:24–34

Chiaramonti D, Calabro V, Gigliotti G (2013) Biomass fast pyrolysis: an overview of the technology and its applications. Renew Energy 49:8–18

Chong M-L, Sabaratnam V, Shirai Y, Hassan MA (2009) Biohydrogen production from biomass and industrial wastes by dark fermentation. Int J Hydrogen Energy 34(8):3277–3287

Corma A, Iborra S, Velty A (2007) Chemical routes for the transformation of biomass into chemicals. Chem Rev 107(6):2411–2502

Corma A, de la Torre O, Renz M, Villandier N (2011) Production of high-quality diesel from biomass waste products. Angew Chem Int Ed Engl 50(10):2375–2378

Corona B, Shen L, Reike D, Carreón JR, Worrell E (2019) Towards sustainable development through the circular economy—a review and critical assessment on current circularity metrics. ResourConserv Recycl 151:104498

Deena SR, Vickram AS, Manikandan S, Subbaiya R, Karmegam N, Ravindran B, Chang SW, Awasthi MK (2022) Enhanced biogas production from food waste and activated sludge using advanced techniques—a review. Bioresour Technol 355:127234

Demirbas A (2001) Biomass resource facilities and biomass conversion processing for fuels and chemicals. Energy Convers Manag 42:1357–1378

Demirbas A (2004) Combustion characteristics of different biomass fuels. Prog Energy Combust Sci 30:219–230

Demirbas A (2005) Potential applications of renewable energy sources, biomass combustion problems in boiler power systems and combustion related environmental issues. Progress Energy Combus Sci 31:123–192

Donoso-Bravo A, Toledo-Alarcón J, Ortega V, Barría V, Lesty Y, Fontana J, Bossche HV (2021) New findings on the anaerobic co-digestion of thermally pretreated sludge and food waste: laboratory and pilot-scale studies. Water Sci Technol 84(9):2530–2540

Duarah P, Haldar D, Patel AK, Dong CD, Singhania RR, Purkait MK (2022) A review on global perspectives of sustainable development in bioenergy generation. Bioresour Technol 348:126791

Dugmore TIJ, Clark JH, Bustamante J, Houghton JA, Matharu AS (2017) Valorisation of biowastes for the production of green materials using chemical methods. Top Curr Chem (Cham). 375(2):46
García-Sancho C, Mérida-Robles JM, Cecilia-Buenestado JA, Moreno-Tost R, Maireles-Torres PJ (2023) The role of copper in the hydrogenation of furfural and Levulinic acid
Gray KA, Zhao L, Emptage M (2006) Bioethanol. Curr Opin Chem Biol 10(2):141–146
Gupta VK, Nguyen QD, Liu S, Taherzadeh MJ, Sirohi R (2022) Microbes in valorisation of biomass to value-added products. Bioresour Technol 347:126738
Hahn-Hägerdal B, Galbe M, Gorwa-Grauslund MF, Lidén G, Zacchi G (2006a) Bio-ethanol-the fuel of tomorrow from the residues of today. Trends Biotechnol 24(12):549–556
Hahn-Hägerdal B, Galbe M, Gorwa-Grauslund MF, Lidén G, Zacchi G (2006b) Bio-ethanol–the fuel of tomorrow from the residues of today. Trends Biotechnol 24(12):549–556
Hall DO, Rosillo-Calle F, Williams RH, Woods J (1993) Biomass for energy: supply prospects. In: Biomass energy supply prospect. Earthscan, London, UK, pp 593–651. ISBN 1853831557
Hijazi A, Khalaf N, Kwapinski W, Leahy JJ (2022) Catalytic valorisation of biomass levulinic acid into gamma valerolactone using formic acid as a H2 donor: a critical review. RSC Adv 12(22):13673–13694
Jayakumar M, Gindaba GT, Gebeyehu KB, Periyasamy S, Jabesa A, Baskar G, John BI, Pugazhendhi A (2023) Bioethanol production from agricultural residues as lignocellulosic biomass feedstock's waste valorization approach: a comprehensive review. Sci Total Environ 25(879):163158
Jenkins BM, Baxter LL, Miles TR, Miles TR (1998) Combustion properties of biomass. Fuel Process Technol 54:17–46
John RP, Anisha GS, Nampoothiri KM, Pandey A (2011) Micro and macroalgal biomass: a renewable source for bioethanol. Bioresour Technol 102(1):186–193
Joshi M, Manjare S (2024) Chemical approaches for the biomass valorisation: a comprehensive review of pretreatment strategies. Environ Sci Pollut Res Int 31(36):48928–48954
Juszczyk P, Tomaszewska L, Kita A, Rymowicz W (2013) Biomass production by novel strains of Yarrowia lipolytica using raw glycerol, derived from biodiesel production. Bioresour Technol 137:124–131
Kant Bhatia S, Ahuja V, Chandel N, Gurav R, Kant Bhatia R, Govarthanan M, Kumar Tyagi V, Kumar V, Pugazendhi A, Rajesh Banu J, Yang YH (2022) Advances in algal biomass pretreatment and its valorisation into biochemical and bioenergy by the microbial processes. Bioresour Technol 358:127437
Kaparaju P, Serrano M, Thomsen AB, Kongjan P, Angelidaki I (2009) Bioethanol, biohydrogen and biogas production from wheat straw in a biorefinery concept. Bioresour Technol 100(9):2562–2568
Kaur P, Thakur M, Tondan D, Bamrah GK, Misra S, Kumar P, Pandohee J, Kulshrestha S (2021) Nanomaterial conjugated lignocellulosic waste: cost-effective production of sustainable bioenergy using enzymes. Biotech 11(11):480
Keerthana Devi M, Manikandan S, Oviyapriya M, Selvaraj M, Assiri MA, Vickram S, Subbaiya R, Karmegam N, Ravindran B, Chang SW, Awasthi MK (2022) Recent advances in biogas production using agro-industrial waste: a comprehensive review outlook of techno-economic analysis. Bioresour Technol 363:127871
Kim JH, Kim M, Park G, Kim E, Song H, Jung S, Park YK, Tsang YF, Lee J, Kwon EE (2024) Chemicals and fuels from lipid-containing biomass: a comprehensive exploration. Biotechnol Adv 75:108418
Koutinas AA, Vlysidis A, Pleissner D, Kopsahelis N, Garcia LI, Kookos IK (2014) Valorization of industrial waste and by-product streams via fermentation for the production of chemicals and biopolymers. Chem Soc Rev 43(8):2587–2627
Kumar P, Singh S, Singh OV (2018) Bioconversion of lignocellulosic biomass: a review. Renew Sustain Energy Rev 81:67–79
Kumar A, Singh M, Sinha A (2020) Biomass energy: a review on its potential and challenges. Energy Rep 6:929–938

Liu Y, Chen X, Zhang Y (2011) Lactic acid production from renewable resources: a review. Biotechnol Adv 29:239–248
Lucie AP, Mario D, Emma CC, Vitaly B, James HC (2013) Food waste biomass: a resource for high-value chemicals. Green Chem 15(2):307–314
Lynd LR, Weimer PJ, Van Zyl WH, Pretorius IS (2002) Microbial cellulose utilization: fundamentals and biotechnology. Microbiol Mol Biol Rev 66:506–577
Makarichi L, Jutidamrongphan W, Techato K (2018) The evolution of waste-to-energy incineration: a review. Renew Sustain Energy Rev 91(C):812–821
Mata TM, Martins AA, Caetano NS (2010) Microalgae for biodiesel production and other applications: a review. Renew Sustain Energy Rev 14:217–232
Mata-Alvarez J, Dosta J, Romero F, Steyer JP (2000) Anaerobic digestion of organic solid waste: an overview of the current status. Biores Technol 74:3–16
Mattioli LB, Corazza I, Budriesi R, Hrelia S, Malaguti M, Caliceti C, Amoroso R, Maccallini C, Crupi P, Clodoveo ML, Muraglia M, Carocci A, Tardugno R, Barbarossa A, Corbo F (2024) From waste to health: olive mill wastewater for cardiovascular disease prevention. Nutrients 16(17):2986
Mishra A, Kumar M, Bolan NS, Kapley A, Kumar R, Singh L (2021) Multidimensional approaches of biogas production and up-gradation: opportunities and challenges. Bioresour Technol 338:125514
Ntaikou I, Antonopoulou G, Lyberatos G (2010) Biohydrogen production from biomass and wastes via dark fermentation: a review. Waste Biomass Valor 1:21–39
Nigam PS, Singh A (2011) Production of liquid biofuels from renewable resources. Prog Energy Combust Sci 37(1):52–68
Ofori J, Nduro D, Boakye P (2018) Biomass resources and biofuels potential in West Africa: a review. Renew Energy 127:557–564
Pal P, Singh AK, Srivastava RK, Rathore SS, Sahoo UK, Subudhi S, Sarangi PK, Prus P (2024) Circular bioeconomy in action: transforming food wastes into renewable food resources. Foods 13(18):3007
Phaff HJ, Miller MW (2015) Chapter 4—Yeast in biotechnology. In: The yeasts: a taxonomic study. Elsevier, Amsterdam, The Netherlands, pp 45–70
Razik AHA, Khor CS, Elkamel A (2019) A model-based approach for biomass-to-bioproducts supply chain network planning optimization. Food Bioprod Process 118:293–305
Rout PR, Goel M, Pandey DS, Briggs C, Sundramurthy VP, Halder N, Mohanty A, Mukherjee S, Varjani S (2023) Technological advancements in valorisation of industrial effluents employing hydrothermal liquefaction of biomass: strategic innovations, barriers and perspectives. Environ Pollut 316(Pt 2):120667
Sadh PK, Kumar S, Chawla P, Duhan JS (2018) Fermentation: a boon for production of bioactive compounds by processing of food industries wastes (by-products). Molecules 23:2560
Şahin S, Kurtulbaş E (2024) Green extraction and valorization of by-products from food processing. Foods 13(10):1589
Samrot AV, Rajalakshmi D, Sathiyasree M, Saigeetha S, Kasipandian K, Valli N, Jayshree N, Prakash P, Shobana N (2023) A review on biohydrogen sources, production routes, and its application as a fuel cell. Sustainability 15(16):12641
Sansaniwal SK, Rosen MA, Tyagi SK (2017) Global challenges in the sustainable development of biomass gasification: an overview. Renew Sustain Energy Rev 80(C):23–43
Santos-Ballardo DU, Font-Segura X, Ferrer AS, Barrena R, Rossi S, Valdez-Ortiz A (2015) Valorisation of biodiesel production wastes: anaerobic digestion of residual Tetraselmis suecica biomass and co-digestion with glycerol. Waste Manag Res. 33(3):250–257
Sathish A, Marlar T, Sims RC (2015) Optimization of a wet microalgal lipid extraction procedure for improved lipid recovery for biofuel and bioproduct production. Bioresour Technol 193:15–24
Šelo G, Planinić M, Tišma M, Tomas S, Koceva D, Bucić-Kojić A (2021) A comprehensive review on valorization of agro-food industrial residues by solid-state fermentation. Foods 10(5):927

Sharma V, Tsai M-L, Nargotra P, Chen C-W, Kuo C-H, Sun P-P, Dong C-D (2022) Agro-industrial food waste as a low-cost substrate for sustainable production of industrial enzymes: a critical review. Catalysts 12:1373

Singh V, Belova L, Singh B, Sharma YC (2018) Biodiesel production using a novel heterogeneous catalyst, magnesium zirconate (Mg2Zr5O12): process optimization through response surface methodology (RSM). Energy Convers Manag [internet]. 174:198–207

Soares Dias AP, Rijo B, Santos F, Galhano Dos Santos R, Frade T (2023) Overview on biofuels production in a seaweed biorefinery. Sci Total Environ 884:163714

Sun PT, Peng H (2023) Valorisation of biomass waste for sustainable bioenergy and biofuel production. Bioengineering (Basel). 10(5):619

Sun S, Sun X, Zhang Y, Wang H, Li J (2019) Interactions of cellulose and hemicellulose in lignocellulosic biomass: a review. Carbohyd Polym 214:197–205

Torres-León C, Chávez-González ML, Hernández-Almanza A, Martínez-Medina GA, Ramírez-Guzmán N, Londoño-Hernández L, Aguilar CN (2021) Recent advances on the microbiological and enzymatic processing for conversion of food wastes to valuable bioproducts. Curr Opin Food Sci 38:40–45

Tuck CO, Pérez E, Horváth IT, Sheldon RA, Poliakoff M (2012) Valorization of biomass: deriving more value from waste. Science 337(6095):695–699. https://doi.org/10.1126/science.1218930. Erratum.In:Science.2012Nov2;338(6107):604

Velvizhi G, Jacqueline PJ, Shetti NP, K L, Mohanakrishna G, Aminabhavi TM (2023) Emerging trends and advances in valorization of lignocellulosic biomass to biofuels. J Environ Manage 345:118527

Ververis C, Georghiou K, Danielidis D, Hatzinikolaou DG, Santas P, Santas R, Corleti V (2007) Cellulose, hemicelluloses, lignin and ash content of some organic materials and their suitability for use as paper pulp supplements. Bioresour Technol 98:296–301

Vurukonda SSKP, Fotopoulos V, Saeid A (2024) Production of a rich fertilizer base for plants from waste organic residues by microbial formulation technology. Microorganisms 12(3):541

Welker CM, Balasubramanian VK, Petti C, Rai KM, De Bolt S, Mendu V (2015) Engineering plant biomass lignin content and composition for biofuels and bioproducts. Energies 8:7654–7676

Yılmaz Balaman Ş, Wright DG, Scott J, Matopoulos A (2018) Network design and technology management for waste to energy production: an integrated optimization framework under the principles of circular economy. Energy143(C):911–933

Zhang Y, He Y, Li Y, Chen G (2015a) Municipal solid waste management in China: a review. Waste Manage 45:256–263

Zhang Y, Xiong Y, Xu Y, Chen H (2015b) Life cycle assessment of bio-based propionic acid production from lignocellulosic biomass. Biores Technol 198:200–207

Zhao J, Li Y, Dong R (2021) Recent progress towards in-situ biogas upgrading technologies. Sci Total Environ 15(800):149667

Zhou Y, Liu L, Li M, Hu C (2022) Algal biomass valorisation to high-value chemicals and bioproducts: recent advances, opportunities and challenges. Bioresour Technol 344(Pt B):126371

Chapter 15
Sustainable Development: Towards a Circular Economy with Biomass Waste

Ratan Chowdhury, Ambarish Bhuyan, and Gisandu K. Malunguja

Abstract The transition to a circular economy (CE) shift from the linear model by reducing waste, reusing resources, and fostering sustainable growth. Biomass waste, particularly lignocellulosic biomass from agricultural residues, forestry by-products, and municipal waste, is a key renewable resource for this shift. This chapter explores how biomass waste can reduce greenhouse gas emissions, improve waste management, and create economic opportunities, especially in developing regions. Biomass waste plays a crucial role in the CE by adhering to the principles of reducing, reusing, and recycling materials. Through technologies like anaerobic digestion, pyrolysis, and gasification, biomass waste can be converted into bioenergy, biochar, and valuable bioproducts. These closed-loop systems promote sustainable development, addressing global challenges such as climate change and resource depletion. The chapter also discusses the technological, regulatory, and economic factors influencing biomass waste utilization, emphasizing the need for supportive policies and investment. Case studies demonstrate successful applications of biomass waste in various sectors, highlighting collaboration among governments, industries, and communities. Challenges such as technological limitations and economic feasibility are explored, along with future prospects for scaling up biomass waste utilization.

Keywords Circular economy · Biomass waste · Lignocellulosic biomass · Bioenergy

---

R. Chowdhury (✉)
Department of Botany, Rangapara College (Autonomous), Rangapara, Assam, India
e-mail: ratan@rcr.ac.in

A. Bhuyan
Department of Botany, D.H.S.K. College, Dibrugarh, Assam, India

G. K. Malunguja
Department of Earth Science, Mbeya University of Science and Technology, Mbeya, Tanzania

A. Hind (ed.), *Biomass Waste: Processing, Valorization, Industrial Applications*,
https://doi.org/10.1007/978-981-95-0991-1_15

15.1 Introduction

The concept of sustainable development has gained increasing importance in global environmental policy and economic strategies. Central to achieving sustainable development is the transition from a linear to a circular economy, where waste is minimized, resources are reused, and environmental impacts are reduced. Biomass waste, a readily available resource, offers a promising pathway towards sustainability by serving as an alternative to fossil-based products and energy sources. Biomass waste refers to organic waste materials derived from agricultural, forestry, and industrial activities, which can be converted into valuable products such as biofuels, bioenergy, and bio-based chemicals (Yana et al. 2022). This not only helps to mitigate environmental degradation but also fosters economic growth, particularly in rural and industrial regions.

The circular economy aims to redefine growth by decoupling economic activity from the consumption of finite resources. In contrast to the linear model—extract, use, and dispose—the circular economy emphasizes the continuous use of resources, extending product life cycles, and regenerating natural systems. At the core of the circular economy are three principles viz; designing out waste and pollution, reducing the creation of waste at the source by designing products and processes that minimize resource use and environmental impact (Gaur et al. 2022). Extending the lifespan of products through reuse, repair, remanufacturing, and recycling. Returning nutrients to the biosphere, enhancing ecosystem services, and avoiding damage to natural resources. The integration of biomass waste into these principles allows for the recovery of energy, materials, and nutrients from organic waste streams, facilitating the transition to a regenerative, circular economy (Adami and Schiavon 2021).

15.2 Biomass Waste and Its Role in the Circular Economy

Biomass waste, particularly lignocellulosic materials, plays a crucial role in advancing the circular economy by transforming organic waste into renewable resources for energy, materials, and nutrient recycling. Lignocellulosic biomass, comprising plant-based materials that include agricultural residues, forestry waste, and industrial by-products, is one of the most abundant forms of biomass globally (Mujtaba et al. 2023). These materials can be repurposed into biofuels, bioenergy, bio-based products, and compost, contributing to sustainability and waste minimization. The ability of biomass waste to be reused and recycled in closed-loop processes ensures that nutrients and valuable materials are continuously returned to the ecosystem, reducing the need for fossil fuel-based resources and synthetic inputs (Nagarajan et al. 2021).

15.2.1 Agricultural Residues: A Renewable Resource

Agricultural residues are by-products of crop production that hold significant potential for bioenergy and material production within the circular economy. These residues include straw, husks, corn stover, and other leftover plant materials, which are often discarded or underutilized. According to recent estimates, more than 4 billion tonnes of agricultural residues are produced globally each year, with vast potential for conversion into biofuels, biochemicals, and bio-based materials (Mujtaba et al. 2023).

Lignocellulosic agricultural residues contain cellulose, hemicellulose, and lignin, which can be broken down into simpler molecules for various applications. For instance, straw can be processed through anaerobic digestion to produce biogas or fermented to generate bioethanol, which serves as a renewable transport fuel (Alrefai et al. 2017). Furthermore, agricultural residues like rice husks and corn stover can be used to produce bio-based plastics and composites, offering an alternative to petroleum-derived plastics. These bio-based products contribute to reducing plastic pollution while promoting the circular economy by replacing non-renewable materials with biodegradable alternatives.

Additionally, agricultural residues play a role in reducing soil degradation. Returning biochar or composted agricultural residues to soils can improve soil fertility, enhance water retention, and increase carbon sequestration, thus contributing to sustainable agricultural practices (Agegnehu et al. 2017). This closed-loop nutrient cycling exemplifies the principles of the circular economy by minimizing waste and enhancing resource efficiency in agriculture. Table 15.1 represents different biomass waste and their specific uses.

Table 15.1 Biomass waste sources and their utilization

Type of biomass waste	Examples	Specific uses	References
Agricultural residues	Straw, husks, corn stover, rice husks	Bioenergy (biogas, bioethanol), biochar, bioplastics	Yana et al. (2022), Alrefai et al. (2017), Agegnehu et al. (2017)
Forestry by-products	Wood chips, bark, sawdust	Bioenergy (bio-oil, syngas), biochar, wood composites	Emenike et al. (2024), Millati et al. (2019), Lee et al. (2022)
Municipal organic waste	Food scraps, yard trimmings, sewage sludge	Biogas, compost, biofertilizers, hydrochar	Manea et al. (2024), Ddiba et al. (2022), Wang et al. (2022)

15.2.2 Forestry By-Products: A Sustainable Source of Bioenergy and Biomaterials

Forestry operations generate substantial amounts of biomass waste, including wood chips, bark, sawdust, and other by-products, which can be effectively repurposed within the circular economy. In many regions, these by-products are considered a sustainable source of bioenergy, as they are renewable and have a lower carbon footprint than fossil fuels. Recent studies indicate that the global production of forestry residues exceeds 1.3 billion tonnes annually, with significant potential for energy and material recovery (Millati et al. 2019).

Wood chips and sawdust can be processed through thermal technologies such as pyrolysis and gasification to produce biochar, bio-oil, and syngas, which serve as renewable energy sources or feedstocks for bio-based chemicals (Emenike et al. 2024). Biochar, in particular, has garnered attention for its dual role in carbon sequestration and soil improvement, aligning with the circular economy's focus on environmental sustainability. When applied to soils, biochar enhances nutrient retention and promotes plant growth while sequestering carbon in a stable form, thus contributing to both agricultural productivity and climate change mitigation (Sarfraz et al. 2019).

In addition to bioenergy, forestry residues can be utilized in the production of bio-based materials such as wood composites, bio-based adhesives, and biodegradable packaging. These materials reduce dependence on petrochemical-based products, contributing to a more sustainable and circular materials economy. For example, wood chips are increasingly used in the manufacture of particleboard and fibreboard, providing durable and recyclable alternatives to conventional building materials (Lee et al. 2022).

15.2.3 Municipal Organic Waste: A Valuable Resource for Urban Circularity

Municipal organic waste, including food scraps, yard trimmings, and other organic fractions of household and industrial waste, represents a valuable resource in the circular economy. In urban areas, organic waste accounts for a significant portion of total waste streams, with estimates suggesting that over 60% of municipal waste in low- and middle-income countries is organic (Ddiba et al. 2022). Efficient management and utilization of municipal organic waste are essential for reducing landfill reliance and promoting sustainable urban development.

Anaerobic digestion and composting are two primary methods for converting municipal organic waste into valuable outputs. Anaerobic digestion of food waste and yard trimmings produces biogas, which can be used for electricity generation, heating, or as a transportation fuel. Recent advancements in AD technology have improved the efficiency of biogas production, making it an increasingly viable solution for managing urban organic waste (Ellacuriaga et al. 2021). Additionally, the digestate produced as a by-product of AD can be used as a biofertilizer, closing the nutrient loop and reducing the need for synthetic fertilizers.

Composting is another widely used method for managing municipal organic waste, turning food waste and yard trimmings into nutrient-rich compost that can be applied to soils. Composting not only diverts waste from landfills but also contributes to soil health by improving its organic matter content and nutrient availability (Manea et al. 2024). In urban areas, composting can be integrated into community-level waste management systems, supporting urban agriculture and green infrastructure projects, thus promoting circularity in cities.

Emerging technologies such as hydrothermal carbonization (HTC) are also gaining attention for their ability to process wet organic waste streams, such as food waste and sewage sludge, into valuable outputs like hydrochar and biogas. HTC mimics natural coal formation processes, converting wet biomass into carbon-rich materials under moderate temperatures and pressure. This technology offers a promising solution for managing urban organic waste efficiently, reducing both the volume of waste and the associated environmental impacts (Wang et al. 2022).

15.2.4 Biomass Waste in the Circular Economy: A Carbon–Neutral and Renewable Resource

Biomass waste, whether derived from agriculture, forestry, or municipal sources, is inherently renewable and carbon–neutral, making it an attractive resource for the circular economy. When managed properly, biomass waste can be continuously recycled and reused in closed-loop systems, minimizing waste and contributing to resource efficiency. For example, anaerobic digestion and composting return valuable nutrients to soils, reducing the need for synthetic fertilizers and enhancing soil health. Similarly, bioenergy technologies such as gasification and pyrolysis convert biomass into energy and bio-based products, reducing reliance on fossil fuels and mitigating greenhouse gas emissions (Yang et al. 2021).

The integration of biomass waste into circular economy frameworks has the potential to address global sustainability challenges, including waste management, resource scarcity, and climate change. By transforming biomass waste into renewable energy, bio-based materials, and compost, the circular economy reduces environmental impacts while creating new economic opportunities in rural and urban areas alike. Furthermore, the carbon–neutral nature of biomass waste ensures that its use aligns with global climate goals, contributing to carbon sequestration and the reduction of greenhouse gas emissions (Chen et al. 2022).

15.3 Technologies for Biomass Waste Utilization

Technological advancements are at the core of transforming biomass waste into valuable energy and bio-based products, contributing significantly to the circular economy. The diverse composition of biomass waste, including agricultural residues, forest by-products, and organic municipal waste, provides opportunities for the development of multiple technologies. Each process aims to extract value from biomass, converting waste into biofuels, bioenergy, and bio-based materials while promoting sustainability and resource efficiency. Key technologies include anaerobic digestion (AD), pyrolysis, and gasification, each with unique outputs and applications in the circular economy.

15.3.1 Anaerobic Digestion (AD)

Anaerobic digestion (AD) is a biological process where microorganisms break down organic matter in the absence of oxygen, resulting in the production of biogas and digestate. Biogas primarily consists of methane (CH_4) and carbon dioxide (CO_2), with small amounts of other gases like hydrogen sulfide (H_2S). The biogas generated through AD can be used for electricity generation, heating, or as a transport fuel after purification. This process has been widely adopted for treating organic waste streams, such as food waste, manure, and sewage sludge, due to its environmental and economic benefits.

Biogas produced from AD is considered a renewable and low-carbon energy source, contributing to energy security and climate change mitigation. For instance, when used as a transport fuel, biogas can reduce greenhouse gas emissions by up to 90% compared to diesel and gasoline (Bordelanne et al. 2011). Furthermore, AD technology provides a valuable solution for managing organic waste, reducing the environmental impacts associated with landfill disposal and methane emissions.

The by-product of AD, digestate, is a nutrient-rich material that can be applied to soils as a biofertilizer. Digestate contains essential nutrients like nitrogen, phosphorus, and potassium, which are vital for plant growth. Using digestate as a biofertilizer reduces the need for synthetic fertilizers, thereby lowering the carbon footprint of agricultural practices (Sharma et al. 2023). This nutrient recycling exemplifies the circular economy by closing the loop between waste and agricultural inputs.

AD technology has been evolving, with recent innovations focusing on improving the efficiency of biogas production. For example, co-digestion, which involves combining different types of organic waste (e.g., agricultural residues with food waste), has been shown to increase biogas yields by optimizing the balance of nutrients required by the microorganisms involved in the digestion process (Siddique and Wahid 2018). Additionally, advancements in AD reactor designs, such as high-rate digesters, improve the process's overall efficiency and scalability.

15.3.2 Pyrolysis

Pyrolysis is a thermal decomposition process that occurs in the absence of oxygen and at high temperatures, typically between 400 and 600 °C. This process breaks down biomass into three main products: biochar, bio-oil, and syngas (a mixture of carbon monoxide, hydrogen, and methane). Each of these products has distinct applications, contributing to the versatility of pyrolysis as a biomass waste utilization technology.

Biochar is a carbon-rich solid that remains after the pyrolysis of biomass. Biochar has gained attention for its ability to sequester carbon, providing a stable form of carbon storage in soils for hundreds to thousands of years (Lorenz and Lal 2014). When applied to soils, biochar improves soil health by enhancing water retention, nutrient availability, and microbial activity. This makes it a valuable tool for improving agricultural productivity while simultaneously mitigating climate change through carbon sequestration. Research has shown that biochar can reduce soil emissions of methane and nitrous oxide, two potent greenhouse gases, further enhancing its environmental benefits (Kammann et al. 2017).

Bio-oil, another product of pyrolysis, is a liquid fuel that can be used for heat and power generation or refined into higher-value chemicals and fuels. However, bio-oil has limitations due to its high acidity and instability. Recent advancements in catalytic upgrading techniques are addressing these challenges by improving the quality and stability of bio-oil, making it more suitable for use as a transportation fuel or chemical feedstock (Yang et al. 2015).

Syngas, produced during pyrolysis, is a gaseous mixture that can be used for electricity generation, heating, or as a precursor for synthesizing other chemicals. The energy content of syngas depends on the composition of the biomass feedstock and the specific conditions of the pyrolysis process. Syngas offers flexibility in energy production, as it can be used directly in gas engines or further processed into synthetic fuels through the Fischer–Tropsch synthesis, a well-established method for producing liquid fuels (Dieterich et al. 2020).

15.3.3 Gasification

Gasification is a thermochemical process that converts biomass into syngas through partial oxidation at high temperatures, typically above 700 °C. Unlike pyrolysis, gasification introduces a controlled amount of oxygen or steam to facilitate the conversion of solid biomass into a gas mixture composed primarily of carbon monoxide (CO), hydrogen (H_2), and carbon dioxide (CO_2). This syngas can be used for various applications, including electricity generation, heat production, and the synthesis of biofuels.

One of the key advantages of gasification is its high efficiency in converting biomass into a versatile energy carrier—syngas. Gasification can achieve energy

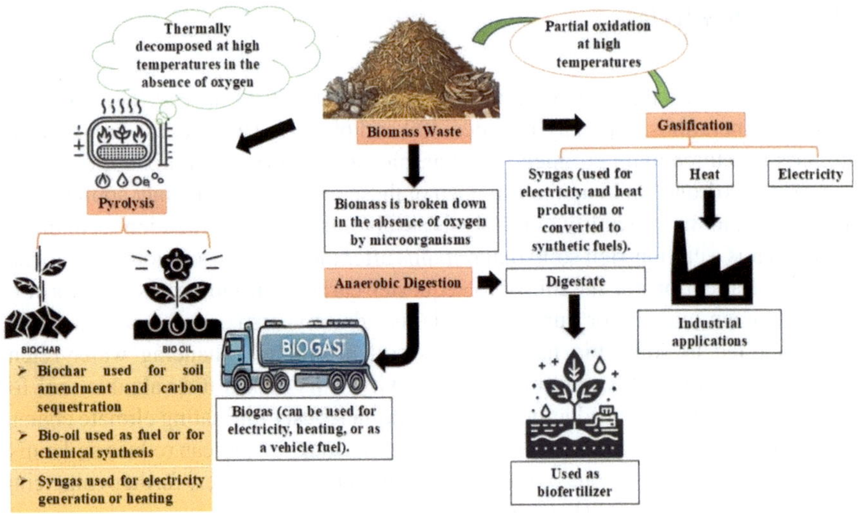

Fig. 15.1 Technologies for biomass conversion

conversion efficiencies of up to 80%, making it one of the most efficient technologies for bioenergy production (Ptasinski 2008). Moreover, gasification can handle a wide range of biomass feedstocks, including agricultural residues, wood chips, and even municipal solid waste, providing flexibility in waste management and energy production.

The syngas produced from gasification can be used in combined heat and power (CHP) systems to generate both electricity and heat, improving the overall energy efficiency of the process. Additionally, syngas can be further processed to produce liquid biofuels, such as biodiesel and bioethanol, through chemical synthesis routes. These biofuels can serve as renewable alternatives to fossil fuels in the transportation sector, helping reduce GHG emissions and enhance energy security (Demirbas 2009).

Recent advancements in gasification technology have focused on improving the process's environmental and economic viability (Fig. 15.1). For example, plasma gasification, which uses an electric arc to generate extremely high temperatures, can achieve nearly complete conversion of biomass into syngas, with minimal production of hazardous by-products (Dimitrakellis et al. 2022). Furthermore, integration with carbon capture and storage (CCS) technologies can make gasification a net-negative emission process, further contributing to climate change mitigation (Full et al. 2021).

15.3.4 Emerging Technologies and Innovations

In addition to the established technologies of anaerobic digestion, pyrolysis, and gasification, several emerging technologies are being developed to further enhance the utilization of biomass waste. One such innovation is hydrothermal carbonization (HTC), a process that mimics natural coal formation by converting wet biomass into hydrochar under high pressure and moderate temperatures. HTC can efficiently process high-moisture biomass, such as food waste and sewage sludge, without the need for extensive drying, making it an energy-efficient option for managing organic waste (Wang et al. 2019). The hydrochar produced through HTC can be used as a solid fuel or as a precursor for activated carbon, which has applications in water filtration and air purification.

Another promising area of innovation is microbial fuel cells (MFCs), which use microorganisms to convert organic matter into electricity. MFCs can generate bioelectricity from biomass waste, providing a sustainable and low-energy solution for decentralized power generation (Kurniawan et al. 2022). While still in the experimental phase, MFCs hold potential for scaling up in the future, particularly for small-scale, off-grid applications (Table 15.2).

Table 15.2 Comparison of biomass conversion technologies

Technology	Efficiency	Cost	Applications	References
Anaerobic Digestion (AD)	50–70% energy efficiency in biogas production	Moderate to high (depends on scale)	Biogas production for energy, biofertilizer from digestate	Sharma et al. (2023), Siddique and Wahid (2018), Bordelanne et al. (2011)
Pyrolysis	30–50% for bio-oil, 50–60% for biochar	High (due to advanced reactors)	Biochar for soil amendment, bio-oil for fuel, syngas for electricity	Lorenz and Lal (2014), Yang et al. (2015), Kammann et al. (2017)
Gasification	70–80% energy conversion efficiency	Moderate to high (depends on feedstock)	Syngas for electricity, heat, or fuel production	Ptasinski (2008), Dimitrakellis et al. (2022), Demirbas (2009)
Hydrothermal Carbonization (HTC)	50–65% for hydrochar and bioenergy	Moderate (cost of pressure and temperature control)	Hydrochar for solid fuel, soil amendment, and activated carbon	Wang et al. (2019), Kurniawan et al. (2022), Wang et al. (2022)

15.4 Economic and Environmental Benefits of Biomass Waste in the Circular Economy

The incorporation of biomass waste into the circular economy presents a multitude of economic and environmental benefits, positioning it as a key driver of sustainability. By converting biomass waste into renewable energy and valuable bioproducts, the circular economy model helps reduce greenhouse gas emissions, enhance resource efficiency, and create economic opportunities, particularly in rural areas. These benefits collectively contribute to achieving global sustainability goals, such as those outlined in the United Nations Sustainable Development Goals (SDGs).

15.4.1 Reducing Greenhouse Gas Emissions

One of the most significant environmental benefits of biomass waste utilization is its ability to reduce greenhouse gas (GHG) emissions. In conventional waste management systems, organic waste that is left to decompose in landfills generates methane, a potent GHG with a global warming potential approximately 28–36 times greater than that of carbon dioxide over a 100-year period (IPCC 2021). Biomass waste, when diverted from landfills and processed into bioenergy through technologies such as anaerobic digestion, not only mitigates methane emissions but also produces biogas, a renewable and cleaner alternative to fossil fuels.

Anaerobic digestion of organic waste, particularly agricultural residues and food waste, produces biogas that can be used for electricity, heating, or as a vehicle fuel, reducing the reliance on carbon-intensive fossil energy sources. For example, studies indicate that anaerobic digestion has the potential to reduce GHG emissions by up to 90% compared to traditional waste management practices, depending on the type of feedstock and the efficiency of the system (Shirzad et al. 2019). Furthermore, biochar produced from biomass waste through pyrolysis can be used as a soil amendment, sequestering carbon for long periods and improving soil health, further contributing to carbon capture and storage efforts (Elkhlifi et al. 2023).

In addition to biogas and biochar, biomass waste can also be converted into liquid biofuels such as bioethanol and biodiesel, which offer lower lifecycle GHG emissions compared to conventional gasoline and diesel. For instance, second-generation biofuels produced from lignocellulosic biomass have been shown to reduce GHG emissions by 60–80% compared to fossil fuels, depending on the production processes and feedstock types (Hirani et al. 2018).

15.4.2 Enhancing Resource Efficiency

The circular economy's emphasis on resource efficiency is well-aligned with the principles of biomass waste utilization. Rather than allowing biomass waste streams, such as agricultural residues, forestry by-products, and organic municipal waste, to decompose or be incinerated, they can be repurposed to create high-value products. This process not only reduces the demand for virgin resources but also extends the lifecycle of organic materials, promoting more sustainable consumption and production patterns.

Lignocellulosic biomass, in particular, holds significant potential for conversion into a wide range of bio-based products, including bio-based chemicals, materials, and advanced biofuels. Through processes like fermentation, hydrolysis, and pyrolysis, lignocellulosic waste can be broken down into its component sugars and cellulose fibers, which can then be used to produce bioplastics, bio-based chemicals, and other sustainable materials. These bio-based products serve as renewable alternatives to their petrochemical counterparts, helping to reduce reliance on finite fossil resources and minimize the carbon footprint of industrial production (Chowdhury et al. 2024).

For instance, the production of bioplastics from lignocellulosic biomass has emerged as a promising solution to the global plastic pollution crisis. Bioplastics derived from agricultural residues, such as corn stover or sugarcane bagasse, offer a more sustainable alternative to conventional plastics, as they are biodegradable and derived from renewable feedstocks (Coppola et al. 2021). Moreover, using lignocellulosic biomass for biofuel production enhances the overall efficiency of bioenergy systems by utilizing non-food feedstocks, thus avoiding competition with food production and contributing to a more sustainable energy landscape (Lin and Lu 2021).

The use of biomass waste in a closed-loop system also reduces the need for landfills and incineration, both of which contribute to environmental degradation. Incineration, while reducing waste volumes, generates toxic emissions, including dioxins and furans, which can harm air quality and public health (Kara et al. 2022). In contrast, the circular economy approach leverages the inherent value of biomass waste by turning it into useful products, minimizing environmental impacts and optimizing the use of natural resources.

15.4.3 Creating Economic Opportunities

Biomass waste utilization within the circular economy not only contributes to environmental sustainability but also holds significant economic potential, particularly for rural and agricultural regions. Biomass waste, such as crop residues, forestry by-products, and organic waste, is often abundant in rural areas. By establishing bioenergy and bio-based industries that convert these waste streams into valuable products, rural regions can stimulate local economies, create jobs, and enhance energy security.

Bioenergy production, for instance, can generate significant employment opportunities across the value chain, from feedstock collection and processing to the operation and maintenance of bioenergy plants. According to the International Renewable Energy Agency (IRENA), the global bioenergy sector employed approximately 3.58 million people in 2020, with a significant portion of these jobs concentrated in rural areas where biomass feedstocks are readily available (IRENA 2021). In developing regions, bioenergy industries can provide a much-needed source of income and employment, contributing to poverty alleviation and rural development.

In addition to job creation, the bio-based economy offers opportunities for local businesses and entrepreneurs to engage in the production of biofuels, biochemicals, and bioplastics. Small- and medium-sized enterprises (SMEs) can play a crucial role in developing decentralized bioenergy systems that cater to local energy needs while reducing dependence on imported fossil fuels. Such systems can enhance energy independence and resilience, particularly in regions that are vulnerable to energy supply disruptions (Muthukumaran et al. 2024).

Furthermore, the bio-based economy can attract investment and foster innovation, as governments and industries increasingly recognize the value of transitioning to a circular economy. For instance, the European Union's Bioeconomy Strategy, which aims to promote the sustainable use of biomass and bio-based products, has stimulated research, development, and investment in the bio-based sector, generating new economic opportunities across Europe (European Commission 2018). Similar policy initiatives and incentives in other regions could unlock the economic potential of biomass waste, particularly in areas where bioenergy and bio-based industries have yet to be fully developed.

15.4.4 Promoting Energy Security and Resilience

The use of biomass waste for bioenergy production contributes to energy security by diversifying the energy mix and reducing reliance on fossil fuels. Countries with abundant biomass resources, such as agricultural residues or forest by-products, can develop bioenergy systems that provide a stable and reliable source of renewable energy. This is particularly important for regions that are dependent on energy imports, as it reduces vulnerability to global energy price fluctuations and supply disruptions (Saleem 2022).

In addition, decentralized bioenergy systems that utilize locally sourced biomass waste can enhance energy resilience by providing communities with a reliable source of power. For example, small-scale anaerobic digestion plants that process agricultural or food waste can generate biogas for local electricity generation, reducing dependence on centralized power grids and enhancing community energy resilience in the face of natural disasters or infrastructure failures (Abniel 2024). By promoting the use of locally available biomass waste for energy production, the circular economy model strengthens both energy security and environmental sustainability.

15.5 Challenges and Future Prospects

Despite the considerable potential of biomass waste utilization within the circular economy, several challenges hinder its large-scale adoption. These challenges range from technological limitations to economic viability and regulatory bottlenecks, all of which require attention for biomass waste to fully realize its promise in driving sustainable development.

15.5.1 Technological Challenges

While technologies such as anaerobic digestion, pyrolysis, and gasification have made significant strides, several technical barriers continue to limit their widespread deployment. One of the primary issues is the pretreatment of lignocellulosic biomass. Lignocellulose, which forms the structural framework of plant cell walls, is a highly complex material, making its breakdown into simpler sugars for biofuel or biochemical production both energy-intensive and expensive. Enzymatic hydrolysis, for example, remains a costly and slow process despite advances in biotechnology (Taherzadeh and Karimi 2007). Improved pretreatment methods, such as steam explosion or advanced chemical treatments, are being explored to enhance efficiency but have yet to achieve scalability at competitive costs.

Another technological hurdle lies in improving the energy efficiency of biomass conversion processes. For instance, pyrolysis and gasification technologies still face challenges related to optimizing operating conditions to maximize energy yields while minimizing by-products and emissions (Dahlquist 2013). Continuous innovation is necessary to address these inefficiencies and to develop new technologies capable of handling a broader range of biomass waste streams. Moreover, there is a growing need for advancements in waste collection and sorting technologies to streamline the recycling of biomass waste and improve material recovery.

15.5.2 Economic Feasibility

The economic feasibility of biomass waste utilization is influenced by several key factors, including the cost of feedstock collection, transportation, processing, and the price competitiveness of the final bio-based products. One of the main challenges is that the logistics of collecting biomass from dispersed agricultural or forestry locations can be costly, particularly in regions lacking infrastructure. For instance, transportation costs can account for up to 40% of the total cost of bioenergy production (Hamelinck et al. 2005).

Furthermore, bioenergy and bio-based product industries often face high capital investment requirements. Setting up biorefineries or bioenergy plants requires significant upfront investments, making them less attractive compared to conventional fossil fuel-based industries, particularly in developing regions where financial resources are limited (Bento et al. 2024). While biomass-based energy and products can offer long-term cost savings and environmental benefits, the initial investment costs may deter both public and private investment without strong economic incentives.

Markets for bio-based products, such as biofuels and bioplastics, are still developing, and price volatility in traditional energy markets can negatively impact the competitiveness of biomass-derived products. In many cases, bio-based alternatives remain more expensive than their fossil-based counterparts, which hinders their large-scale adoption. To overcome these economic barriers, governments and financial institutions will need to introduce mechanisms such as subsidies, tax incentives, and green bonds to make bio-based industries more economically viable (Eckmann 2019).

15.5.3 Regulatory and Policy Support

Effective regulatory frameworks and policy support are crucial to accelerate the transition toward biomass waste utilization within a circular economy. In many regions, the lack of coherent policies that incentivize waste-to-energy initiatives and bio-based product development remains a significant barrier. Governments must implement policies that encourage waste diversion from landfills, including stricter regulations on landfill disposal, penalties for excessive waste generation, and incentives for recycling and composting.

Carbon pricing mechanisms, such as carbon taxes or cap-and-trade systems, could also enhance the economic viability of biomass waste utilization by making it more competitive with fossil-based alternatives. Countries that have adopted carbon pricing frameworks have seen a positive impact on the development of renewable energy sectors, including bioenergy (World Bank 2021). Additionally, introducing subsidies for renewable energy technologies, including those that process biomass waste, can stimulate market growth by reducing the financial risks associated with investing in bioenergy plants or biorefineries.

International cooperation and harmonization of standards for bio-based products will further drive the global market. As the demand for sustainable products rises, regulatory bodies must develop certifications and standards that verify the sustainability and environmental impact of bio-based products, ensuring transparency and boosting consumer confidence. For example, the European Union's Renewable Energy Directive (RED II) sets sustainability criteria for biofuels and biomass, providing a policy framework that supports market growth for bio-based industries (European Commission 2020).

Moreover, policies that promote research and development in biomass waste technologies are essential. Funding for innovation in biotechnologies, improved waste collection, and processing systems can play a pivotal role in overcoming the technological and economic barriers to scaling up biomass waste utilization. Collaborative initiatives between governments, industries, and academia can accelerate breakthroughs in efficient biomass conversion processes, ultimately making bio-based solutions more accessible and affordable.

15.5.4 Social and Public Perception

Another critical challenge is public acceptance of bio-based products and energy derived from biomass waste. In some regions, there is limited awareness of the environmental and economic benefits of bio-based alternatives. Public opposition to bioenergy projects, particularly in areas where land use changes are involved, can delay or derail project implementation. Education campaigns and public engagement are crucial to improving awareness and acceptance, highlighting how biomass waste can contribute to sustainability and energy security without competing with food resources (Lindgren 2021).

15.6 Conclusion

Biomass waste presents a significant opportunity to drive sustainable development within the framework of a circular economy. By transforming waste into renewable energy and bio-based products, societies can address pressing environmental challenges such as climate change, pollution, and resource depletion. This shift not only reduces waste and lowers greenhouse gas emissions but also creates economic opportunities, particularly in rural areas and developing regions where biomass resources are abundant. The circular economy model, supported by technologies like anaerobic digestion, pyrolysis, and gasification, enables the efficient use of biomass waste, fostering a closed-loop system that enhances resource efficiency and reduces dependence on finite fossil resources. Moreover, the production of bioenergy and bioproducts from waste can stimulate new industries, generate employment, and promote energy independence. However, achieving widespread adoption of biomass waste utilization requires overcoming challenges, particularly in terms of technological advancements, economic feasibility, and public acceptance. Continued innovation is necessary to improve the efficiency and cost-effectiveness of biomass processing technologies. Additionally, supportive policies, regulatory frameworks, and incentives will be essential to encourage investment and foster collaboration among governments, industries, and communities. With sustained efforts, biomass waste can play a pivotal role in the global transition toward a more sustainable, circular economy.

References

Abniel M (2024) Biomass utilization for energy generation in post-disaster scenarios: the case of Puerto Rico

Adami L, Schiavon M (2021) From circular economy to circular ecology: a review on the solution of environmental problems through circular waste management approaches. Sustainability 13(2):925

Agegnehu G, Srivastava AK, Bird MI (2017) The role of biochar and biochar-compost in improving soil quality and crop performance: a review. Appl Soil Ecol 119:156–170

Alrefai R, Benyounis KY, Stokes J (2017) Integration approach of anaerobic digestion and fermentation process towards producing biogas and bioethanol with zero waste: technical. J Fundam Renew Energy Appl 7(6):243

Bento HB, Tessaro Í, Theodoro JM, de Souza Matias Reis W, Policarpo G, Reis CER, de Carvalho AKF (2024) Economic, social, and organizational challenges in biorefineries. In: Clean energy transition-via-biomass resource utilization: a way to mitigate climate change. Springer Nature Singapore, Singapore, pp 205–235

Bordelanne O, Montero M, Bravin F, Prieur-Vernat A, Oliveti-Selmi O, Pierre H, Muller T (2011) Biomethane CNG hybrid: a reduction by more than 80% of the greenhouse gases emissions compared to gasoline. J Nat Gas Sci Eng 3(5):617–624

Chen L, Msigwa G, Yang M, Osman AI, Fawzy S, Rooney DW, Yap PS (2022) Strategies to achieve a carbon neutral society: a review. Environ Chem Lett 20(4):2277–2310

Chowdhury R, Hussain N, Mukherjee S, Barman S, Mandal H, Raul PK, Bhattacharya SS (2024) Bioethanol production from lignocellulosic waste without pre-treatment employing vermicompost and earthworm gut-isolated bacteria: insights on waste to wealth conversion efficiency towards cleaner lifestyle. Waste Biomass Valoriz 15(9):1–15. https://doi.org/10.1007/s12649-024-02548-7

Coppola G, Gaudio MT, Lopresto CG, Calabro V, Curcio S, Chakraborty S (2021) Bioplastic from renewable biomass: a facile solution for a greener environment. Earth Syst Environ 5:231–251

Dahlquist E (ed) (2013) Technologies for converting biomass to useful energy: combustion, gasification, pyrolysis, torrefaction and fermentation. CRC Press

Ddiba D, Andersson K, Rosemarin A, Schulte-Herbrüggen H, Dickin S (2022) The circular economy potential of urban organic waste streams in low-and middle-income countries. Environ Dev Sustain 24(1):1116–1144

Demirbas A (2009) Biofuels securing the planet's future energy needs. Energy Convers Manage 50(9):2239–2249

Dieterich V, Buttler A, Hanel A, Spliethoff H, Fendt S (2020) Power-to-liquid via synthesis of methanol, DME or Fischer–Tropsch-fuels: a review. Energy Environ Sci 13(10):3207–3252

Dimitrakellis P, Delikonstantis E, Stefanidis GD, Vlachos DG (2022) Plasma technology for lignocellulosic biomass conversion toward an electrified biorefinery. Green Chem 24(7):2680–2721

Eckmann C (2019) Strategies for the promotion of consumer acceptance within the bio-based economy. Master's thesis, University of Twente

Elkhlifi Z, Iftikhar J, Sarraf M, Ali B, Saleem MH, Ibranshahib I, Chen Z (2023) Potential role of biochar on capturing soil nutrients, carbon sequestration and managing environmental challenges: a review. Sustainability 15(3):2527

Ellacuriaga M, García-Cascallana J, Gómez X (2021) Biogas production from organic wastes: integrating concepts of circular economy. Fuels 2(2):144–167

Emenike EC, Iwuozor KO, Ighalo JO, Bamigbola JO, Omonayin EO, Ojo HT, Adeniyi AG (2024) Advancing the circular economy through the thermochemical conversion of waste to biochar: a review on sawdust waste-derived fuel. Biofuels 15(4):433–447

European Commission (2018) A clean planet for all: a European strategic long-term vision for a prosperous, modern, competitive and climate-neutral economy. European Union. https://eur-lex.europa.eu/legal-content/EN/TXT/?uri=CELEX:52018DC0773

European Commission (2020) The European green deal. European Union. https://eur-lex.europa.eu/legal-content/EN/TXT/?uri=CELEX%3A52019DC0640

Full J, Merseburg S, Miehe R, Sauer A (2021) A new perspective for climate change mitigation—introducing carbon-negative hydrogen production from biomass with carbon capture and storage (Hybeccs). Sustainability 13(7):4026

Gaur A, Gurjar SK, Chaudhary S (2022) Circular system of resource recovery and reverse logistics approach: key to zero waste and zero landfill. In: Advanced organic waste management. Elsevier, pp 365–381

Hamelinck CN, Suurs RA, Faaij AP (2005) International bioenergy transport costs and energy balance. Biomass Bioenerg 29(2):114–134

Hirani AH, Javed N, Asif M, Basu SK, Kumar A (2018) A review on first-and second-generation biofuel productions. Biofuels: Greenhouse Gas Mitig Glob Warm Next Gener Biofuels Role Biotechnol 141–154

Intergovernmental Panel on Climate Change (IPCC) (2021) Climate change 2021: the physical science basis. In: Masson-Delmotte V, Zhai P, Pirani A, Connors SL, Péan C, Berger S, Caud N, Chen Y, Goldfarb L, Gomis MI, Huang M, Leitzell K, Lonnoy E, Matthews JBR, Maycock TK, Waterfield T, Yelekçi O, Yu R, Zhou B (eds) Contribution of working group I to the sixth assessment report of the intergovernmental panel on climate change. Cambridge University Press. https://doi.org/10.1017/9781009157896

International Renewable Energy Agency (IRENA) (2021) World energy transitions outlook: 1.5 °C pathway. IRENA. https://www.irena.org/publications/2021/Jun/World-Energy-Transitions-Outlook

Kammann C, Ippolito J, Hagemann N, Borchard N, Cayuela ML, Estavillo JM, Wrage-Mönnig N (2017) Biochar as a tool to reduce the agricultural greenhouse-gas burden–knowns, unknowns and future research needs. J Environ Eng Landsc Manag 25(2):114–139

Kara S, Hauschild M, Sutherland J, McAloone T (2022) Closed-loop systems to circular economy: a pathway to environmental sustainability? CIRP Ann 71(2):505–528

Kurniawan TA, Othman MHD, Liang X, Ayub M, Goh HH, Kusworo TD, Chew KW (2022) Microbial fuel cells (MFC): a potential game-changer in renewable energy development. Sustainability 14(24):16847

Lee SH, Lum WC, Boon JG, Kristak L, Antov P, Pędzik M, Pizzi A (2022) Particleboard from agricultural biomass and recycled wood waste: a review. J Market Res 20:4630–4658

Lin CY, Lu C (2021) Development perspectives of promising lignocellulose feedstocks for production of advanced generation biofuels: a review. Renew Sustain Energy Rev 136:110445

Lindgren S (2021) Cookstove implementation and education for sustainable development: a review of the field and proposed research agenda. Renew Sustain Energy Rev 146:111184

Lorenz K, Lal R (2014) Biochar application to soil for climate change mitigation by soil organic carbon sequestration. J Plant Nutr Soil Sci 177(5):651–670

Manea EE, Bumbac C, Dinu LR, Bumbac M, Nicolescu CM (2024) Composting as a sustainable solution for organic solid waste management: current practices and potential improvements. Sustainability 16(15):6329

Millati R, Cahyono RB, Ariyanto T, Azzahrani IN, Putri RU, Taherzadeh MJ (2019) Agricultural, industrial, municipal, and forest wastes: an overview. Sustain Resour Recovery zero Waste Approaches 1–22

Mujtaba M, Fraceto LF, Fazeli M, Mukherjee S, Savassa SM, de Medeiros GA, Vilaplana F (2023) Lignocellulosic biomass from agricultural waste to the circular economy: a review with focus on biofuels, biocomposites and bioplastics. J Clean Prod 402:136815

Muthukumaran G, Passos MV, Gong J, Xylia M, Barquet K (2024) Decentralized solutions for island states: Enhancing energy resilience through renewable technologies. Energ Strat Rev 54:101439

Nagarajan D, Lee DJ, Chang JS (2021) Circular bioeconomy: an introduction. In: Biomass, biofuels, biochemicals. Elsevier, pp 3–23

Ptasinski KJ (2008) Thermodynamic efficiency of biomass gasification and biofuels conversion. Biofuels, Bioprod Biorefin 2(3):239–253

Saleem M (2022) Possibility of utilizing agriculture biomass as a renewable and sustainable future energy source. Heliyon 8(2)

Sarfraz R, Hussain A, Sabir A, Ben Fekih I, Ditta A, Xing S (2019) Role of biochar and plant growth promoting rhizobacteria to enhance soil carbon sequestration—a review. Environ Monit Assess 191:1–13

Sharma P, Bano A, Verma K, Yadav M, Varjani S, Singh SP, Tong YW (2023) Food waste digestate as biofertilizer and their direct applications in agriculture. Bioresour Technol Rep 23:101515

Shirzad M, Panahi HKS, Dashti BB, Rajaeifar MA, Aghbashlo M, Tabatabaei M (2019) A comprehensive review on electricity generation and GHG emission reduction potentials through anaerobic digestion of agricultural and livestock/slaughterhouse wastes in Iran. Renew Sustain Energy Rev 111:571–594

Siddique MNI, Wahid ZA (2018) Achievements and perspectives of anaerobic co-digestion: a review. J Clean Prod 194:359–371

Taherzadeh MJ, Karimi K (2007) Enzymatic-based hydrolysis processes for ethanol from lignocellulosic materials: a review. BioResources 2(4):707–738

Wang L, Chang Y, Li A (2019) Hydrothermal carbonization for energy-efficient processing of sewage sludge: a review. Renew Sustain Energy Rev 108:423–440

Wang Q, Wu S, Cui D, Zhou H, Wu D, Pan S, Wang Z (2022) Co-hydrothermal carbonization of organic solid wastes to hydrochar as potential fuel: a review. Sci Total Environ 850:158034

World Bank (2021) World development report 2021: data for better lives. World Bank

Yana S, Nizar M, Mulyati D (2022) Biomass waste as a renewable energy in developing bio-based economies in Indonesia: a review. Renew Sustain Energy Rev 160:112268

Yang L, Wang XC, Dai M, Chen B, Qiao Y, Deng H, Wang Y (2021) Shifting from fossil-based economy to bio-based economy: Status quo, challenges, and prospects. Energy 228:120533

Yang Z, Kumar A, Huhnke RL (2015) Review of recent developments to improve storage and transportation stability of bio-oil. Renew Sustain Energy Rev 50:859–870.eve

Correction to: Chemical Conversion Pathways for Enhanced Value of Biomass Waste: Catalytic Processes and Chemical Transformations

Hajar Rastegari

Correction to:
Chapter 6 in: A. Hind (ed.), Biomass Waste: *Processing, Valorization, Industrial Applications*, https://doi.org/10.1007/978-981-95-0991-1_6

The original version of this book was published with the wrong captions for Figures 1-4 and Tables 2 and 3 in Chapter 6 which have now been updated. The book and the chapter have been updated with the changes.

The updated version of this chapter can be found at https://doi.org/10.1007/978-981-95-0991-1_6

A. Hind (ed.), *Biomass Waste: Processing, Valorization, Industrial Applications*,
https://doi.org/10.1007/978-981-95-0991-1_16

MIX
Papier aus verantwortungsvollen Quellen
Paper from responsible sources
FSC® C105338

If you have any concerns about our products,
you can contact us on
ProductSafety@springernature.com

In case Publisher is established outside the EU,
the EU authorized representative is:
Springer Nature Customer Service Center GmbH
Europaplatz 3, 69115 Heidelberg, Germany

Printed by Libri Plureos GmbH
in Hamburg, Germany